Particles, Fields, Space-Time

Particles, Fields, Space-Time

From Thomson's Electron to Higgs' Boson

Martin Pohl

CRC Press
Taylor & Francis Group
Boca Raton London New York

CRC Press is an imprint of the
Taylor & Francis Group, an **informa** business

First edition published 2020
by CRC Press
6000 Broken Sound Parkway NW, Suite 300, Boca Raton, FL 33487-2742

and by CRC Press
2 Park Square, Milton Park, Abingdon, Oxon, OX14 4RN

© 2021 Taylor & Francis Group, LLC

CRC Press is an imprint of Taylor & Francis Group, LLC

Library of Congress Cataloging-in-Publication Data
Names: Pohl, Martin, author.
Title: Particles, fields, space-time : from Thomson's electron to Higgs' boson / Martin Pohl.
Description: Boca Raton : CRC Press, 2021. | Includes bibliographical references and index.
Identifiers: LCCN 2020017204 | ISBN 9780367347239 (paperback) | ISBN 9780367353810 (hardback) | ISBN 9780429331107 (ebook)
Subjects: LCSH: Particles (Nuclear physics)--History.
Classification: LCC QC793.16 .P63 2021 | DDC 539.7/2--dc23
LC record available at https://lccn.loc.gov/2020017204

ISBN: 978-0-367-35381-0 (hbk)
ISBN: 978-0-367-34723-9 (pbk)
ISBN: 978-0-429-33110-7 (ebk)

Typeset in CMR10
by Nova Techset Private Limited, Bengaluru & Chennai, India

Cover: « Strike » by Christian Gonzenbach, 2011, www.gonzenbach.net, reproduced by permission of the artist.

Contents

Preface

The idea for this book comes from my experience teaching physics and especially particle physics at the university level for more than 25 years. I noticed that historical background to physics livens up the subject, increases student interest by giving a human touch to an otherwise rather abstract matter. When I stumbled over Steven Weinberg's book, *The Discovery of Subatomic Particles* [548], which deals with the discovery of protons, neutrons and electrons, I realised that this experience is shared by others. When I taught the online courses "Particle Physics – An Introduction" (https://www.coursera.org/learn/particle-physics) and its French version with Mercedes Paniccia and other colleagues, I thought that a book using the history of particle physics as a guideline would be a good companion.

As an experimental particle physicist, I did not venture to make an original contribution to the history of science. That is why I did not really respect the chronology of discovery. Rather, historical facts, personalities and the often colourful stories of major discoveries serve as didactic tools to motivate more in-depth arguments about physics. In finding my narrative I have greatly profited from Helge Kragh's work *Quantum Generations* [514] and the chronological bibliography *Particle Physics: One Hundred Years of Discoveries* [480], carefully compiled by Vladimir V. Ezhela and many others in a U.S.-Russian collaboration.

The body of the text uses little mathematics. More technical discussions are concentrated in Focus Boxes, which – I hope – give the necessary condensed information to the reader, such that more ample treatment can be found elsewhere. Suggestions for this "elsewhere" are given at the end of each chapter. I generally refrained from explaining mathematical terms; adequate definitions can be found, e.g., on Wikipedia. I also rarely included biographical details for the many actors contributing to particle physics in the past hundred years. For most of them, concise entries exist, e.g., in the *Encyclopædia Britannica*; more elaborate biographies are referenced in the text or included in the Further Reading sections of each chapter.

The reader I had in mind when writing this book is a colleague of mine in spirit. Someone who has a background in physics, at least in classical physics, and an open mind for unusual concepts and facts. Maybe the reader has not followed particle physics for a while and wonders what happened since last in contact with academia. Maybe the reader is looking for input in his or her own teaching at whatever level, or just intellectual stimulus. In any case, the reader is someone who, like me, believes that physics should be based on observing nature, describing the results in mathematical language and ultimately condensing it into theory. That the order of these essential steps is rather often reversed, is an interesting fact all by itself.

I am grateful to Mercedes Paniccia from University of Geneva for carefully assessing the text from both the didactic and the physics points of view, as she had already done so for the online courses. She eliminated countless errors and flaws from the manuscript. The many remaining mistakes are, of course, all mine.

Martin Pohl
Hamburg

CHAPTER 1

Introduction

> The whole truth would be an infinite concatenation of mostly irrelevant facts. . . So we do not tell the whole truth; we tell carefully crafted stories, and we do this even when our moral purpose is to tell the truth.
>
> Justin E.H. Smith, *Irrationality*, 2019 [686][1]

I DO NOT KNOW how you feel, but I have often been bored by the history of sciences. Who exactly published what detail first does not really fascinate me. What I am interested in and what will consequently be the subject of this book is the great steps that brought us to where we are in particle physics today. The disruptive ideas, the unexpected experimental results that have determined the course of research, and what it took for the scientific community to accept and embrace them.

I am not very interested in the historian's viewpoint since I am an experimental particle physicist. I will not project myself (and you) back in time, but use today's knowledge to identify (and sometimes invent) a narrative for the progress that brought us to what is called the Standard Model of particle physics. I also consistently use terms in their modern definition, often not available to physicists at the time. You see, I take the Whig history point of view, often criticised by real historians.

So this is not a book about the history of my science; it is a book about particle physics. It uses historical landmarks as a vehicle to give an idea on how processes of innovation work in physics, instead of just explaining the end result. At any rate, the "end result" is never the end of the process, just a major step along the way. What is very important to me is the relation between theory and experiment, so you will find experimental details not only where you expect them. I have nonetheless tried to do justice to historical facts, mostly based on secondary literature, but also on the original publications I quote.

I owe you some clarification about my personal philosophy of sciences, even though I am far from pretending to be a philosopher. I have a reductionist

---

[1] Quoted by permission from Princeton University Press.

model of the progress in physics, which is well represented by the diagram in Figure 1. From the top down, more phenomena are explained by fewer concepts, laws and parameters, thus reducing their number, not necessary their complexity. In not much more than a century we have progressed impressively. As far as forces are concerned, electromagnetic, weak and strong interactions are now described by a common framework, although not completely unified. Gravity has been understood as a classical field theory, however with no quantum theory to describe its workings at small distances. Two layers of matter constituents have been identified, from the atom to electrons, protons and neutrons, and further on to six quarks and six leptons which may well be truly elementary. And the description of matter and forces has been unified in the quantum field theory of the Standard Model. Finally, mass has been understood to have a dynamical origin. I will sketch the fascinating concepts behind this progress in this book.

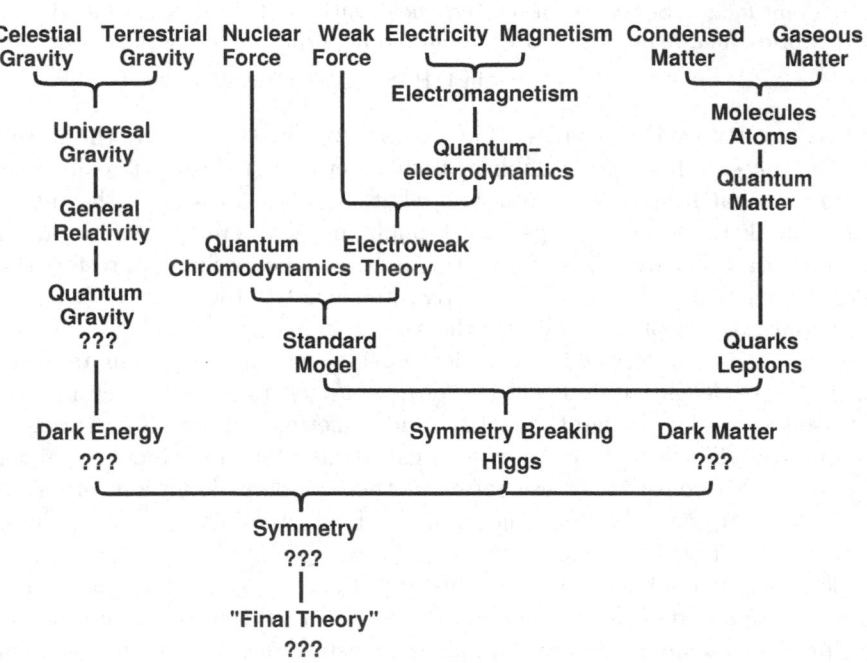

Figure 1.1 A reductionist view of the development of particle physics. Forces are represented to the left, matter to the right of the diagram. Missing elements known today are indicated by triple question marks.

Ideally—but definitely not in our lifetime, and probably never—this reduction process would culminate into a truly unified theory of the Universe, including elements dominating its evolution at cosmic scales. The impossible dream behind particle physics, to understand the Universe by its microscopic properties, would then have come true. As Einstein put it [289][2]:

> ...from the beginning there has always been the attempt to find a unifying theoretical basis for all these single sciences, consisting of a minimum of concepts and fundamental relationships, from which all concepts and relationships of the single disciplines might be derived by logical process. This is what we mean by the search for a foundation of the whole of physics. The confident belief that this ultimate goal may be reached is the chief source of the passionate devotion which has always animated the researcher.

At the very end, in Chapter 10, I will critically assess this approach as a guideline for further progress.

I will analyse events of Standard Model history from the standpoint of one of its victims. I learned much of physics from the "Feynman Lectures on Physics" and got my degree when weak neutral current interactions were discovered in the laboratory I was part of. As an experimental particle physicist in the late 20th and early 21st century, I have been running after Standard Model theoretical predictions for most of my professional life, trying in vain to find a breach in its fortifications. That causes some kind of fatalism in my view of its past and future, a bias that you should be aware of when critically assessing what I write here. I am a physicist by training so I cannot refrain from using the way physicists are thinking. As one of my teachers used to say: the truth is in the formulae, not in the blabla. I will nevertheless relegate most "technical" detail to Focus Boxes that should explain it to the necessary level and which you can skip if you are familiar with, say, classical electrodynamics, special relativity, quantum mechanics or quantum field theory. I suspect that you would find it most easy to follow my arguments if you had a basic science education, a major or a bachelor in a "hard" science subject.

Before we start to talk about the subjects of this book, I will remind you about elements of classical physics, which are necessary to understand what follows, like Newton's and Maxwell's laws and the concept of ether. If you are familiar with those, I invite you to skip section 2.1 and to follow me to what I took to be my starting point: the discovery of the electron by Walter Kaufmann, Emil Wiechert and most prominently Joseph John Thomson.

FURTHER READING

Richard P. Feynman, Robert B. Leighton and Matthew Sands, *The Feynman Lectures on Physics: The New Millennium Edition*, Basic Books, 2011.

---

[2]Quoted by permission from Science Magazine.

CHAPTER 2

The first particles

As in Mathematicks, so in Natural Philosophy, the Investigation of difficult Things by the Method of Analysis, ought ever to precede the Method of Composition. This Analysis consists in making Experiments and Observations, and in drawing general Conclusions from them by Induction, and admitting of no Objections against the Conclusions, but such as are taken from Experiments, or other certain Truths. For Hypotheses are not to be regarded in experimental Philosophy. And although the arguing from Experiments and Observations by Induction be no Demonstration of general Conclusions; yet it is the best way of arguing which the Nature of Things admits of, and may be looked upon as so much the stronger, by how much the Induction is more general.

Isaac Newton, *Opticks*, 1704 [2]

THE INDUSTRIAL REVOLUTION, set off by the development of practical steam engines in the 18th century, moved science –and especially physics– from a curiosity for aristocracy (used for amusement, warfare, or transmutation of elements to finance their follies) into the realm of practical applications pervading every day life. This transition was also fostered by the formation of scientific and physical societies in Europe and the USA, like DPG (Germany) in 1845, SFP (France) in 1873, IoP (Britain) in 1874 and APS (USA) in 1899. These were more open to the outside than the academies of science founded in the 18th century and served as platforms for discussion and more rapid dissemination of new physics. A good indicator for these dynamics is the wave of polytechnic schools that were created in the 19th century, starting with the Ecole polytechnique in Paris founded during the French Revolution in 1794, and followed by engineering schools all over Europe and in the USA, on public as well as private initiative. They quickly rose to high academic standards. Examples are ETH Zürich (1855), MIT Cambridge (1861), RWTH Aachen (1879) and Caltech Pasadena (1891). An exception is Britain where the training of engineers stayed with industry for a long period.

Physicists in the 19th century –and scientists in general– felt deeply imbedded in general culture. In particular to German physicists, music making and the appreciation of music was very important [562]. This stayed so well into the 20th century[1].

This first chapter sets the scene for the start of discoveries in particle physics by introducing necessary scientific ingredients: the understanding of motion and the mastering of electromagnetism. The former describes how forces change the movement of objects according to classical physics, with differential equations and their application in Newton's law. For the latter, Maxwell's laws completed classical electrodynamics with its enormous economic and societal impact. That impact had already set off with the revolution of communication by the electric telegraph in the 1830s, the introduction of commercial electric light in the 1870s and pioneering work on wireless communication by Guglielmo Marconi and Karl Ferdinand Braun in the 1890s. I will discuss experiments with cathode rays in the 1890s by Wiechert, Kaufmann and most prominently Joseph John Thomson, which led to the discovery of the electron, the first matter particle in modern terms. Millikan's experiment on charge quantisation, although later in time, will also be described in this context.

2.1 GIANTS' SHOULDERS

The aim of classical physics is to predict the evolution of a system of objects, when the initial conditions of all ingredients and the forces that act upon them are known. The idealised object of classical physics is the mass point, an object with no extension, but with properties like mass and electrical charge. Initial conditions are specified as the spatial location and velocities of all objects at some initial time, say $t = 0$. The positions and velocities of each object at any other time are then determined by Newton's law [1], as specified in Focus Box 2.1. The law is named after Isaac Newton, the father of classical physics. Newton's law is a differential equation linking the acceleration of a mass point, i.e. its infinitesimal change of velocity per infinitesimal time period, to the forces that act at the place and time where the mass point is. Acceleration and force are in fact proportional to each other, the proportionality factor is the mass. It thus takes twice as much force to give the same acceleration to an object which is twice as massive.

Position $\vec{r}(t)$, velocity $\vec{v}(t)$ and acceleration are vectors in the three-dimensional Euclidian space. The coordinate system has three axes that a Cartesian will chose as orthogonal. Origin and direction of the axes are arbitrary. Vectors specify a value as well as a direction in the chosen coordinate system. Consequently, also the force is a vector, with value and direction. If there are several forces, their vector sum determines the acceleration. You may remember the parallelogram of forces: it is the two dimensional representation of the vector sum.

---

[1] For a well-known example see J. Weidman, The Night I Met Einstein, Readers' Digest November 1955.

In classical physics, forces determine how objects move. If there are none, the acceleration is zero; the object moves at constant velocity, equal to its initial value, and in a straight line along its initial direction. If a force acts, the velocity changes in value and/or direction. If the force is always aligned with the object's velocity, only the speed will change. Otherwise speed and direction will change. A special case is a force always orthogonal to the object's velocity. Such a force will only change the direction of the velocity, but let it travel with constant speed. See Focus Box 2.1 for examples of these special cases.

Newton's law specifies how the momentum $\vec{p}(t)$ of an object changes under the influence of a force $\vec{F}$:

$$\frac{d\vec{p}(t)}{dt} = \vec{F}$$

The momentum of a mass point is the product of its mass m and velocity $\vec{v}(t)$ (see Focus Box 2.2). Velocity is the rate of change of position $\vec{r}(t)$ per time t, $\vec{v} = d\vec{r}/dt$. Thus if the mass is constant, the rate of change of momentum is $m\, d^2\vec{r}/dt^2 = \vec{F}$. The acceleration of a mass point, i.e. the rate of change of its velocity $d^2\vec{r}/dt^2$, is proportional to the applied force; the proportionality factor is its mass.

If there is no force at all, the momentum of the mass point remains constant. If there is a force $\vec{F}$, but it has no component along a given axis, the momentum component along this axis stays unchanged. Otherwise, the force component causes acceleration in that direction. Mind that depending on the sign, acceleration may also be reducing the speed.

A first example is the action of a force constant in space and time, such as the electric force on a point-like charge inside a parallel plate capacitor (see Focus Box 2.5). This will lead to a constant acceleration in the direction of the force.

As a second example let us consider a force $\vec{F}$ constant in value and always orthogonal to the velocity $\vec{v}$, such as the force on a charged mass point moving in a constant magnetic field (see Focus Box 2.5). In this case, since $\vec{F} \perp \vec{v}$ at all times, the speed $|\vec{v}|$ does not change, but its direction receives a constant shift. The force then constrains the mass point to move on a circular path with radius $R = mv^2/F$ at constant angular velocity $\omega = v/R$ in the plane defined by the force and the initial velocity vector.

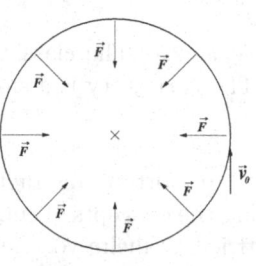

If the mass point had an initial velocity component orthogonal to that plane, it will stay constant, leading to a helical movement in three dimensions.

Focus Box 2.1: Newton's law

Since Newton's law is an ordinary differential equation –i.e. the position of a mass point does not depend on any variable other than time– it can simply be integrated over time provided we know an analytical expression for the force as a function of time and space. A first integration will deliver the velocity vector as a function of time, on the left side of the equation we thus get the momentum of the mass point (see Focus Boxes 2.1 and 2.2). A second integration will give us the position of the object as a function of time. Each time we integrate, a constant needs to be fixed. The initial conditions of the system specify them for us, i.e. fix the initial position and velocity of the object.

We thus have a law of motion where forces act continuously and inevitably on the objects, as sketched in Fig. 2.1. There is no option: if a force can act on an object, it will. And it will do so wherever it reigns. However, not all forces act on all objects alike. For example, the gravitational force acts if the object has mass, the electromagnetic force if it has electric charge.

Classical physics thus predicts what is called a trajectory for our mass points. A trajectory is a smooth curve in space as a function of time, since it is determined by a differential equation. And it is predicted accurately and reliably by Newton's mechanics, otherwise you could not drive your car responsibly. That remains so as long as all distances are macroscopic, i.e. large compared to the size of an atom, about 50 pm for hydrogen[2] as an example. And as long as its velocity stays far below the speed of light, about 300,000 km per second. So, in your car you are safe, at least as far as these conditions are concerned.

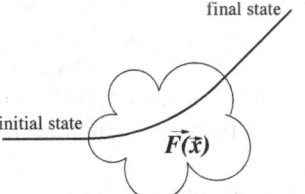

Figure 2.1 The classical action of a force on the movement of a mass point. The trajectory is a smooth curve as a function of time.

Important abstract characteristics of motion developed by classical physics are the concepts of energy and momentum, as explained in Focus Box 2.2. Gottfried Wilhelm von Leipniz was probably the first in the 1680s to use a mathematical formulation of energy close to its modern sense, calling it *vis viva*, a name which stuck for a long time. Thomas Young in the 19th century coined the term energy. Julius Robert von Mayer, James Prescott Joule, Hermann von Helmholtz and William Thomson (Lord Kelvin) established the conservation

---

[2]A picometer (pm) is 10^{-12} m, one millionth of a millionth of a meter.

of energy and its role in physics: Energy is a condition that describes the capacity of a system to do work. It cannot be created or destroyed, but it can be transformed from one form to another.

The formulation of momentum as a characteristic of motion took longer. In the 17th century René Descartes and Galileo Galilei introduced a quantity proportional to size and speed, which they called *quantitas motus* and *impeto*, respectively. The first term was retained by Isaac Newton in his Principia Mathematica [1], as a constant of motion which can only be changed by a force.

Apart from heat, the phenomena that classical physics is mostly concerned with are the gravitational and the electromagnetic interactions. Both can be described by an action-at-a-distance law as well as a local one. The latter involves the notion of the classical force field and thus deserves our special attention. Let us start with the gravitational force. Newton's action-at-a-distance law of the universal gravitational force between a mass M at location $\vec{0}$ and another mass m at $\vec{r}$ is:

$$\vec{F}(\vec{r}) = -G\frac{mM}{r^2}\hat{r}$$

The gravitational constant G determines the order of magnitude of the force, its strength. The masses m and M of the two objects are both the source and the object of the interaction; they are interchangeable. The force diminishes quadratically with the distance $r = |\vec{r}|$ between the two objects; when the distance is doubled, the force is four times less. The vector $\hat{r}$ of unit length points from the location of M (the origin of our coordinates) towards where m is. The minus sign thus indicates that we are dealing with an attractive force. Had we put m at the origin instead, the direction of $\hat{r}$ would reverse, but the strength of the force would be the same. The force thus only depends on the *relative* position of the two masses. We rightly talk about an *inter*-action: the force is completely symmetric, source and object of the force cannot be distinguished.

Thus the Moon attracts the Earth with the same force as the one by which the Earth attracts the Moon. But the consequences are quite different. Since the Earth's mass is about 80 times larger than that of the Moon, it is accelerated 80 times less by the same force. Thus we can usually assume that the Moon rotates around the Earth centre, while in reality both rotate around a common centre, displaced from the centre of the Earth by about 3/4 of the Earth radius.

With this law we know how large the gravitational force is and in which direction it acts. But how does an object know that the other is there? A first step towards the answer is the classical gravitational field. A field is a physical quantity attributed to a point in space, or in space and time as we will see later. This can be a scalar quantity such as temperature. When we map out the temperatures in a room, we will probably find that it is larger close to a heat source (because of radiation) and close to the ceiling compared

Kinetic energy E_{kin}, proportional to mass m and square of velocity v of an object, is the motor of motion:

$$E_{\text{kin}} \;=\; \frac{1}{2}mv^2$$

One injects energy into a system by doing work on it. Working against a force $\vec{F}$ by transporting an object along a trajectory $\vec{r}$, its potential energy, E_{pot}, is increased:

$$E_{\text{pot}} = \int \vec{F}\, d\vec{r}$$

It is "potential", because it can be stored by holding and realised by releasing the object. It is also called potential because it can be seen as an energy associated to *position inside a potential* Φ, if the force results from such a potential, i.e. $\vec{F} = -\vec{\nabla}\Phi$. The force is then called a conservative force. This term was coined by William Rankine in 1853 [20].

The importance of momentum in characterising classical motion took a long time to be realised. Several attempts to construct a "quantity of motion" finally resulted in Isaac Newton's conservation law for the product of mass and velocity, $\vec{p} = m\vec{v}$. It does not change if there is no force acting (see Focus Box 2.1). But a force can change its absolute value, its direction, or both, depending on the angle between force and momentum.

Indeed the importance of energy and momentum for kinematics is that both are strictly conserved quantities in closed systems. In case of energy, one must however include all its forms. When all intervening forces result from a potential, only kinetic and potential energies occur. The conserved quantity is then the sum of the two, $E = E_{kin} + E_{pot}$, but an exchange between the two remains possible. Potential energy can be transformed into kinetic one and vice versa. Likewise, the total momentum of a closed system is strictly conserved. Thus the vector sum of all momenta is a constant. How these conservation laws come about is clarified by Noether's theorem (see Focus Box 9.6).

Focus Box 2.2: Energy and momentum

to the floor (because of convection, unless the floor itself is the heat source). Interpolating the map, we can probably define an analytic function $T(\vec{r})$ that describes it, as a function of position $\vec{r}$. The large scale numerical equivalent is one of the ingredients that weather models use to make predictions. The field can also be a vector as in the case of gravitation. Let us treat M as the source of a gravitational field $\vec{g}$:

$$\vec{g}(\vec{r}) = -GM\frac{\hat{r}}{r^2}$$

You will notice that we have simply taken the field to be $\vec{g}(\vec{r}) = \vec{F}(\vec{r})/m$, but the approach changes more radically than that. We have now declared M to be the *source* of the field and m to be a *probe* that can be used to measure its strength. But the field $\vec{g}(\vec{r})$ is independent of the probe, it has become a property of each point in space. We can then express the gravitational force as a *local interaction* between field and probe:

$$\vec{F}(\hat{r}) = m\,\vec{g}(\hat{r})$$

You will rightly complain that we have still not specified *how* the source transforms the space around it to have the required property. And since time does not appear in the laws quoted so far, the action even appears to be instantaneous. Indeed classical physics is unable to answer these questions. Newton himself was well aware of this problem. In a letter to Bentley he wrote in 1692[3]:

> It is inconceivable that inanimate Matter should, without the Mediation of something else, which is not material, operate upon, and affect other matter without mutual Contact...That Gravity should be innate, inherent and essential to Matter, so that one body may act upon another at a distance thro' the Vacuum, without Mediation of any thing else, by and through which their Action and Force may be conveyed from one to another, is to me so great an Absurdity that I believe no Man who has in philosophical Matters a competent Faculty of thinking can ever fall into it. Gravity must be caused by an Agent acting constantly according to certain laws; but whether this Agent be material of immaterial, I have left to the Consideration of my readers.

Coulomb's law, named after Charles-Augustin de Coulomb, which describes the force between two static electric charges Q and q in the same spatial configuration as the two masses above, has a completely analogous form to Newton's gravitational law:

$$\vec{F}(\vec{r}) = \frac{1}{4\pi\epsilon_0}\frac{qQ}{r^2}\hat{r}$$

---

[3]See: Stanford Encyclopedia of Philosophy,
https://plato.stanford.edu/entries/newton-philosophy/

It thus shares the same properties: it describes action at a distance, in a symmetric way between source and probe. And even when reformulated using the local electric field $\vec{E}$ created by the electric charge Q, i.e. $\vec{F}(\vec{r}) = q\vec{E}(\vec{r})$, there is no information on what bridges the gap between the source Q and the probe q. Electromagnetic interactions take a leading role in what follows. The reason is that there is a complete classical field theory of electromagnetism, first formulated by James Clerk Maxwell in his "Treatise on Electricity and Magnetism" [38], published in 1873. The seminal importance of this work was immediately recognised by his contemporaries. A book review in Nature [39] in the same year celebrated the Treatise enthusiastically:

> We could easily point to whole treatises (some of them in many volumes) still accepted as standard works, in which there is not (throughout) a tithe of the originality and exhaustiveness to be found in any one of Maxwell's chapters.

The book discusses electrical and magnetic phenomena in a way that clearly separates the generation of the electric and magnetic fields from their action on electric charges. In his original publication, Maxwell presented the equations in less compact form than what we present in Focus Box 2.3. However, it was quickly realised that they can be reformulated using vector calculus, as we do here. This is natural, since both electric and magnetic fields are characterised by intensity and direction. Using the differential operators of divergence and curl, Oliver Heaviside, a largely self-taught engineer, mathematician and physicist, arrived at the four partial differential equations we quote. They show how electric fields converge towards or diverge from their source, positive and negative electric charges (Gauss' law). And how magnetic fields curl around electric currents, charges in motion (Ampere's law). See Figure 2.2 for a simple graphic representation of these ways to generate the two fields.

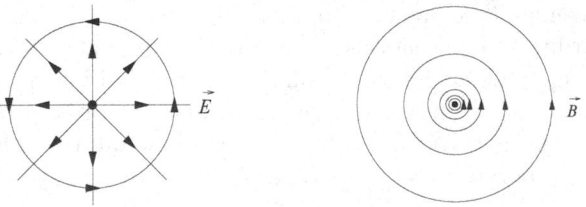

Figure 2.2 Left: A positive electric charge causes the electric field to diverge from its position (Gauss' law). A variable magnetic field causes an electric one that curls around it (Faraday's law). Right: An electric current causes a magnetic field that curls around the current (Ampere's law). A variable electric field also does that (Maxwell's completion of Ampere's law).

A dramatic feature of Maxwell's equations is that they also describe how electric and magnetic fields interact directly with each other. For that, they

Maxwell's equations *in vacuo* read as follows:

$$\vec{\nabla}\vec{E} = \frac{1}{\epsilon_0}\rho \quad \text{(Gauss' law)} \qquad\qquad \vec{\nabla}\vec{B} = 0$$

$$\vec{\nabla}\times\vec{E} = -\frac{\partial\vec{B}}{\partial t} \quad \text{(Faraday's Law)} \qquad \vec{\nabla}\times\vec{B} = \mu_0\left(\vec{j} + \epsilon_0\frac{\partial\vec{E}}{\partial t}\right)$$

$$\text{(Ampere/Maxwell law)}$$

Gauss' law specifies how the electric field vector $\vec{E}(\vec{r})$ diverges from its source, a local charge density $\rho(\vec{r})$. The magnetic field $\vec{B}(\vec{r})$ has no such sources, the magnetic field has no divergence. Faraday's law specifies how electric fields curl around a magnetic field, $\vec{B}(t)$, varying in time. Ampere's law, corresponding to the first term in the last equation identifies the current density $\vec{j}(\vec{r})$ as the prime source of magnetic fields. Currents being made of charges in motion, there is thus a single source of electromagnetic fields, electric charge. A second term was added by Maxwell to make the whole set consistent. It says that magnetic fields also curl around variable electric fields, $\vec{E}(t)$, and establishes symmetry between electric and magnetic fields as far as that is concerned. The constants permittivity ϵ_0 and permeability μ_0 are universal properties of empty space, fixed by measurement. Together they specify how fast electromagnetic phenomena propagate through space-time, i.e. the speed of light $c = 1/\sqrt{\epsilon_0\mu_0}$.

Since electromagnetic forces are conservative, we can express Maxwell's laws in terms of the electric and magnetic potentials V and $\vec{A}$, such that $\vec{E} = -\vec{\nabla}V - \frac{\partial\vec{A}}{\partial t}$ and $\vec{B} = \vec{\nabla}\times\vec{A}$. The potentials form a four-vector $A_\mu = (V/c, \vec{A})$. Four-vectors are explained in Focus Box 3.6. Using the four-vector of the electromagnetic current density $j^\mu = (c\rho, \vec{j})$, one can write Maxwell's equations in their most compact form: $\partial_\nu\partial^\nu A^\mu = (4\pi/c)j^\mu$, where $\partial_\nu\partial^\nu = \frac{1}{c^2}\frac{\partial^2}{\partial t^2} - \nabla^2$. They are invariant under gauge transformations:

$$\vec{A} \rightarrow \vec{A}' = \vec{A} + \vec{\nabla}\Lambda$$

$$V \rightarrow V' = V + \frac{\partial\Lambda}{\partial t}$$

with an arbitrary scalar function $\Lambda(x)$ of the space-time four-vector x. These transformations can be written in covariant form as $A^\mu \rightarrow A^{\mu\prime} = A^\mu + \partial^\mu\Lambda$. They leave the electric and magnetic fields untouched.

Focus Box 2.3: Maxwell's equations

do not need charges and currents. In the absence of both, variable magnetic fields cause curling electric fields to appear as sketched on the left of Figure 2.3. This fact had been shown earlier by Michael Faraday in a series of brilliant experiments on induction. Maxwell added a term to Ampere's law which removed a mathematical inconsistency in the set of equations known beforehand. His term claimed that also varying electric fields cause magnetic fields to curl around them, much as electric currents do (see Figure 2.3). The dramatic consequence of this addition is that electric and magnetic fields can now propagate in empty space without intervening charges and currents, as explained in Focus Box 2.4. If we set all charge and current densities to zero in Maxwell's equations, we arrive at electric and magnetic fields with no divergence, but a curl proportional to the rate of change in the *other* field, i.e. magnetic and electric, respectively. The solution to this set of equations is an electromagnetic wave as sketched on the right of Figure 2.3. Maxwell identified visible light with this electromagnetic phenomenon. When you are measuring the electric field at a given place, it oscillates with frequency $\nu = c/\lambda$ along a fixed axis, orthogonal to the direction of the wave propagation. At the same place, the magnetic field also oscillates with the same frequency, in phase with the electric one, but along an orthogonal transverse axis. When you take a snapshot of the wave at a fixed time, you see what the figure sketches: maxima and minima for both fields occur at the same place, separated by a wavelength λ. The maxima and minima of the wave propagate at a fixed speed in vacuum, $c = 1/\sqrt{\mu_0\epsilon_0}$; inside matter, this speed is reduced.

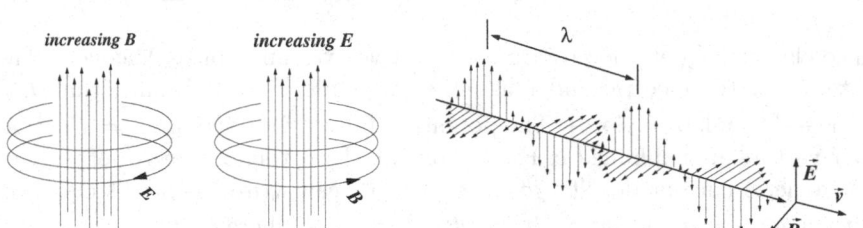

Figure 2.3 Left: A magnetic field $\vec{B}$ varying in time causes an electric field to curl around it. An electric field $\vec{E}$ varying in time causes a magnetic field to curl around it. Right: This leads to an electromagnetic wave propagating in space and time with the speed of light and a wave length λ. The oscillations of the electric and magnetic field are in phase, the electric and magnetic field vectors are orthogonal to each other and to the direction of propagation $\vec{v}$. The propagation velocity is equal to the speed of light, $|\vec{v}| = c$, and the amplitudes of the two fields are related by $E_0 = cB_0$.

After Maxwell's publication the Prussian Academy of Sciences offered a prize in 1879 to whoever could demonstrate the existence of electromagnetic waves. However, the prize expired uncollected in 1882. That electromagnetic waves indeed exist was shown by the physicist Heinrich Hertz, the son of a

Hamburg senator, in a series of experiments collectively described in a paper from 1887 [50]. Just two years earlier he had been nominated professor at the Karlsruhe polytechnic institute, which offered excellent experimental facilities. Figure 2.4 shows examples of the emitters and the receivers he used. Both were based on a spark gap. When the emitter was excited to show a spark, the receiver also showed one, with an intensity diminishing with increasing distance.

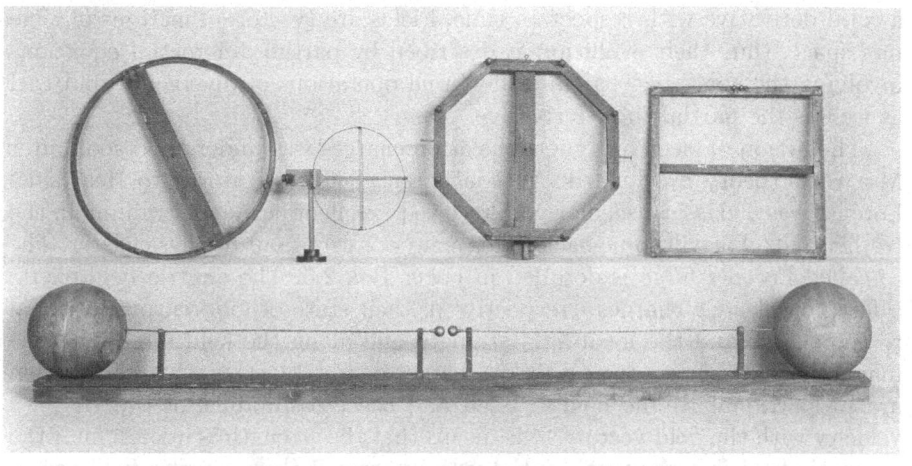

Figure 2.4   Foreground: An emitter oscillator used by Hertz for his transmission experiments. The two balls at the extremes regulate the capacitance and thus the resonance frequency of the open circuit. When the antenna is excited, the gap in the middle will spark and a short electromagnetic wave pulse will be emitted. Background: Receiver resonators used by Hertz, with different shapes and sizes. When the wave pulse arrives at the antenna, a spark will bridge the small gap. (Credit: Deutsches Museum, München, Archiv, BN02391, reprinted by permission)

The existence of electromagnetic fields propagating by themselves in vacuum has important implications for the interpretation of fields in general. Before Maxwell, one could have assumed that gravitational, electric and magnetic fields were a mathematical device, simply splitting in two the unexplained transmission of force from source to destination. Maxwell's electromagnetism and Hertz' confirmation of the transmission of electromagnetic waves fundamentally changed the role of fields. Einstein [239][4] summarises this change of paradigm in 1931:

> ... before Maxwell people conceived of physical reality –in as far as it is supposed to represent events in nature– as material points,

---

[4]Reproduced with permission of the Licensor through PLSclear.

whose changes consist exclusively of motions, which are subject to total differential equations. After Maxwell they conceived physical reality as represented by continuous fields, not mechanically explicable, which are subject to partial differential equations.

The reason for the shift from total to partial differential equations is simple. Mass points are exhaustively described by their position as a (vector-) function of time alone, $\vec{r}(t)$, thus of a single free variable. Changes are thus described by a total derivative with respect to time. Fields are (vector-) functions of time *and* space, thus their evolution is described by partial differential equations involving the space-like partial differential operations of divergence and curl as well as the partial time derivative.

The action of electromagnetic fields on charges was understood soon after Maxwell's theory by Hendrik Antoon Lorentz [60]. In contrast to Heaviside, Lorentz was a classical scholar who received excellent academic training in the Netherlands. We will come back to his many contributions to physics later. The so-called Lorentz force is detailed in Focus Box 2.5. The electric field exerts a force on electric charges irrespective of their state of motion, proportional to the charge and the local field strength and along the field direction. The magnetic field exerts a force only on charges in motion, strength and direction are proportional to the charge itself and the crossproduct of the charge's velocity with the field vector. This means that the strength is maximum if the magnetic field is orthogonal to the velocity, zero if the magnetic field vector points in the direction of the velocity. The magnetic force is also zero when the charge probing it is at rest. But at rest with respect to what?

It may seem natural even today to assume that there is a universal reference system for space and time, a fixed stage on which things happen. The celestial and terrestrial motion due to gravity, the generation and action of electromagnetic fields could then all be described with respect to that absolute space and time. In the simplest case, this universal coordinate system would be somehow fixed to a substance that would be detectable, the ether. There were many variants of ether theory. In general an ideal incompressible liquid was assumed, more or less frictionless with respect to the motion of matter, fixed to absolute space. Ether could be dragged along by the motion of matter or not, depending on the author's ideas. Ether would also be the ideal medium to transport electric and magnetic fields (and maybe also gravitational ones). Fields would be transmitted in vague analogy to sound in gasses or liquids. However, this analogy hurts limits rather early on: sound is a density wave, a scalar quantity which varies longitudinally with respect to the wave's evolution; electric and magnetic fields are vectors transverse to the direction of motion of an electromagnetic wave. It thus takes a more elaborate mechanical quantity, like shear stress, to transmit such phenomena. But theories of this kind were nonetheless accepted by prominent physicists, including Maxwell [38]:

> From these considerations Professor W. Thomson [Lord Kelvin] has argued, that the medium [ether] must have a density capable

In the absence of electric charges and currents, Maxwell's equations take on a particularly simple and symmetric form:

$$\vec{\nabla}\vec{E} = 0 \quad ; \quad \vec{\nabla}\vec{B} = 0$$

$$\vec{\nabla} \times \vec{E} = -\frac{\partial \vec{B}}{\partial t} \quad ; \quad \vec{\nabla} \times \vec{B} = \mu_0 \epsilon_0 \frac{\partial \vec{E}}{\partial t}$$

We take the curl of the second pair of equations:

$$\vec{\nabla} \times \left(\vec{\nabla} \times \vec{E} \right) = -\frac{\partial}{\partial t} \vec{\nabla} \times \vec{B} = -\mu_0 \epsilon_0 \frac{\partial^2 \vec{E}}{\partial t^2}$$

$$\vec{\nabla} \times \left(\vec{\nabla} \times \vec{B} \right) = \mu_0 \epsilon_0 \frac{\partial}{\partial t} \vec{\nabla} \times \vec{E} = -\mu_0 \epsilon_0 \frac{\partial^2 \vec{B}}{\partial t^2}$$

Using the vector identities $\vec{\nabla} \times (\vec{\nabla} \times \vec{V}) = \vec{\nabla}(\vec{\nabla}\vec{V}) - \vec{\nabla}^2\vec{V}$ and the fact that neither $\vec{E}$ nor $\vec{B}$ has a divergence, we find the wave equations:

$$\frac{1}{c^2}\frac{\partial^2 \vec{E}}{\partial t^2} - \vec{\nabla}^2\vec{E} = 0 \quad ; \quad \frac{1}{c^2}\frac{\partial^2 \vec{B}}{\partial t^2} - \vec{\nabla}^2\vec{B} = 0$$

with the speed of light $c = 1/\sqrt{\mu_0 \epsilon_0}$ in vacuum. Using the four-vector potential $A_\mu = (V/c, \vec{A})$ with the so-called Lorenz gauge condition, $\partial_\mu A_\mu = 0$, the same equations become $\partial_\nu \partial^\nu A_\mu = 0$. The solutions to the wave equations describe a monochromatic plane wave:

$$\vec{E} = \vec{E}_0 e^{-i(\vec{k}\vec{r} - \omega t)} \quad ; \quad \vec{B} = \vec{B}_0 e^{-i(\vec{k}\vec{r} - \omega t)}$$

with $\vec{E}_0 \perp \vec{B}_0 \perp \vec{k}$ and $E_0 = cB_0$, depicted in Figure 2.3. Its angular frequency is $\omega = \nu/(2\pi)$, its direction and wavelength are characterised by the wave vector $\vec{k}$ with $k = 2\pi/\lambda$. Equal phases of the wave are located in planes orthogonal to the direction of propagation, $\vec{k}/k$, thus the name *plane* wave. The observable values of the magnetic and electric field are of course given by the real values of these complex functions. The time averaged intensity of the wave is $I = c\epsilon_0 E_0^2/2$. In terms of the potentials, the plane wave solutions are:

$$A^\nu = A_0^\nu(\vec{k}) \, e^{-ik_\mu x^\mu}$$

with $k_\mu = (\omega, \vec{k})$ and $k_\mu k^\mu = 0$ so that the wave moves at the speed of light. The four-vector of the potential is thus orthogonal to the momentum four-vector, and it has only three degrees of freedom.

Focus Box 2.4: Electromagnetic waves

Lorentz' law describes how the local electric ($\vec{E}$) and magnetic fields ($\vec{B}$) act on a mass point with electric charge q moving at a velocity $\vec{v}$:

$$\vec{F} = q\left(\vec{E} + \vec{v} \times \vec{B}\right)$$

The electric field acts on charges at rest and in motion, in or against the direction of the electric field vector, depending on the sign of the charge. The magnetic force only acts on moving charges, its direction is orthogonal both to the velocity and to the magnetic field direction. Since the state of motion depends on the frame of reference in which it is described, one might suspect trouble with the distinction between these two fields. Indeed, the split into electric and magnetic forces is not invariant under Lorentz transformations (see Chapter 3).

Focus Box 2.5: Lorentz' law of the classical electromagnetic force

of comparison with that of gross matter, and has even assigned an inferior limit to that density. We may therefore receive, as a datum derived from a branch of science independent of that with which we have to deal, the existence of a pervading medium, of small but real density, capable of being set in motion, and of transmitting motion from one part to another with great, but not infinite, velocity.

However, at least Michael Faraday had doubts already in the 1850s. In an unpublished note [388] preserved at the Royal Society he wrote:

The ether –its requirements– should not mathematics prove or shew that a fluid might exist in which lateral vibrations are more facil than direct vibrations. Can that be the case in a homogeneous fluid? Yet must not the ether be homogeneous to transmit rays in every direction at all times. If a stretched spring represent by its lateral vibrations the ether and its vibrations –what is there in the ether that represents the strong cohesion in the line of the string particles on which however the lateral vibration(s) essentially depend. And if one tries to refer it to a sort of polarity how can that consist with the transmission of rays in every direction at once across a given ether.

If the classical theorem on the addition of velocities holds, i.e. if they simply add up vectorially, an experiment on Earth cannot always be at rest with respect to the ether. Thus if the speed of light is measured in two different directions, a difference ought to be observed if light propagates in this medium. Several experiments, most prominently by Fizeau, Michelson and Morley, tried to establish this dependence on moving sources or moving physical media. They had no success. The speed of light is constant under all circumstance, as Maxwell's laws predict. We will come back to the significance of these findings in Chapter 3.

2.2 RAYS

Particle physics started with the discover of rays: cathode rays, X-rays and uranium rays. Cathode rays came first, because the technology to produce them was already in place in the form of discharge tubes. These consist of glass tubes which contain a rarefied gas, equipped with electrodes. When high voltage is applied to the electrodes, a discharge is triggered and colourful light is produced by the interaction between fast electrons and the rest gas, creating a plasma. They were named after the German glassblower Heinrich Geissler, who beginning in 1857 constructed artistic cold cathode tubes with different gases in them. Geissler tubes were mostly novelty items, made to demonstrate the new science of electricity. The technology was commercialised by French engineer Georges Claude in 1910 and evolved into neon lighting.

In the middle of the 19th century, the nature of cathode rays was not known and attracted serious research interest. Pioneers were the Swiss Auguste de la Rive [40] and the German Julius Plücker. The former made the connection between the luminous effects of cathode rays and the *aurora borealis* [21]. The latter and his student Johann Wilhelm Hittorf worked with Geissler to develop more efficient vacuum pumps, since lowering the pressure at first enhances the luminous effect [32]. However, when a high enough vacuum was achieved, the light disappeared from most of the tube, instead a greenish glow appeared on the glass tube near the anode. Something was coming out of the cathode, hit the glass before being collected by the anode. The phenomenon was named cathode rays ("Cathodenstrahlen") by Eugen Goldstein [41]. Plücker also discovered that cathode rays could be deflected by magnetic fields. In parallel, similar phenomena were observed by the British physicist, chemist and spiritualist Sir William Crookes (see [85]). We will see many more examples of parallel concurrent studies in Britain and on the continent during this period.

Hertz, on the other hand, failed to deflect cathode rays by charged plates and wrongly concluded that they were more like electromagnetic radiation. He also observed that they could penetrate thin metal foils. Hertz' assistant Philipp Lenard used this to construct a discharge tube with a thin metal window allowing the cathode rays to penetrate into open air, and studied their absorption by various materials. He was awarded the 1905 Nobel prize for physics for his work on cathode rays [123]. A few years later he also discovered that cathode rays were emitted when the cathode is illuminated by ultraviolet light. This is now called the photoelectric effect and played an important role in the development of quantum physics (see Chapter 5). Lenard was a nationalist and later became a fervent supporter of Nazi ideology (see Section 6.2).

Following Lenard's findings, Wilhelm Conrad Röntgen in Würzburg started to be interested in cathode rays in 1895. Like many others, he used as detector a screen covered with barium platinocyanide, $Ba[Pt(CN)_4]$, a substance which fluoresces[5] when hit by cathode rays. When he wrapped his cathode rays tube in heavy paper in a dark room, he accidentally discovered that the screen started to fluoresce even when cathode rays could not possibly have hit it. He introduced the first presentation of his findings to the Physical and Medical Society in Würzburg on December 28, 1895 [62] with the words:

> If the discharge of a larger Ruhmkorff [inductor] is passed through a Hittorf vacuum tube, or a sufficiently evacuated Lenard-, Crooke- or similar apparatus, and the tube is covered with rather tightly fitting thin black cardboard, one will see, in a completely darkened room, a paper screen painted with barium platinocyanide light up brightly at every discharge, fluoresce regardless if the painted or the opposite side of the screen is directed towards the discharge tube. The fluorescence is still visible at 2m distance from the apparatus.

The new radiation was highly penetrating and attenuated only by dense material. It blackened photographic plates, was refracted by crystals, but not noticeably reflected. Röntgen concluded that the new rays presented a "kind of kinship to light." He named the new rays "X-rays". But already in the discussion after his presentation [62], which was received with roaring applause by the audience, a colleague proposed to call them "Röntgen'sche Strahlen", a name that has stuck in German, Russian and other northern European languages. The importance of his discovery for medical imaging was immediately realised and his second communication [71] included an X-ray photo of the hand of the Society's co-founder, an anatomist.

Uranium rays were discovered by Henri Becquerel in 1896, again using photographic plates as a detector. Apparently he realised immediately that he was onto something important, because he notified the weekly meeting of the Académie des Sciences in Paris with a short note, barely one page [68]. The description of his experiment reads:

> One wraps a gelotino-bromide photographic Lumière plate[6] with two sheets of very thick black paper, such that the plate does not grow hazy after an exposure to the Sun during one day. One puts on the outside of the paper a sheet of the phosphorescent[7] substance [in his case double potassium-uranium sulfate, $K(UO)SO_4$] and one exposes the whole to the Sun, during several hours. When one then develops the photographic plate, one recognises that the silhouette of the phosphorescent substance appears in black on the

---

[5]Fluorescence is the emission of visible or near-UV light by molecules which have been excited by X-rays or ionising radiation.

[6]A dry photographic plate commercialised by the Lumière brothers between 1890 and 1950.

[7]Phosphorescence is similar to fluorescence, but with a delay between irradiation and emission of light. The emitted glow slowly fades away, with a time scale ranging from minutes to hours.

picture. If one poses a coin between the phosphorescent substance
and the paper, ... one sees the image of this object appear on the
picture.

In a series of communications to the Académie [64, 65, 66, 67, 69, 70] he elaborated on his experiments. He found that the invisible radiation was able to discharge charged plates and penetrate various opaque substances. Also uranium salts which did not phosphoresce emitted the radiation, so unlike phosphorescence it was independent of an outside energy source. Using metallic uranium he confirmed that it was this component of the salts which emitted the radiation.

Marie and Pierre Curie, together with Gustave Bémont [81, 82, 83] discovered in 1898 other elements (like thorium and radium) contained in uraninite (called pitchblende at the time) and chalcolite, which emitted the "uranium" radiation much more intensely than uranium itself. Aware of the importance to publish new findings quickly in this very productive period, Marie Curie presented a short communication to the Académie. In there, she coined the term "radioactive" to characterise elements emitting the new radiation. She was unaware that a few months earlier, Gerhard Carl Schmidt had published similar findings in Berlin [80]. Becquerel and the Curies shared the 1903 Nobel prize for their discoveries of spontaneous radioactivity.

It soon became clear that radioactive substances emitted more than one kind of rays. Ernest Rutherford classified the charged ones according to their penetration capabilities [86]. He named the components alpha- and beta-rays in order of increasing penetration power. The former were stopped by paper. The beta-rays penetrated paper, but not metal, and turned out to be the ones first detected by Becquerel.

Villard [95, 96] and Becquerel [93] identified in 1900 a third component of radioactivity, which Rutherford [108] named gamma-rays. He and Edward Andrade [154] measured their spectrum in 1914 and found that they consisted of light with a much shorter wavelength than X-rays.

2.3 ELECTRONS

The nature of the newly discovered charged rays remained unclear and became a prominent subject of speculation and research. An important part of 19th century physicists imagined phenomena like the electric charge and current as consisting of charged particles. Prominently among them, Wilhelm Weber called the electric current a "liquid of electric molecules" [56]. On the basis of Michael Faraday's law of electrolysis, Hermann von Helmholtz argued that there existed "atoms of electricity", although he did not go so far as to identify these units of electric charge with a particle[8]. Moreover, in 1895 as a young graduate student, Jean Baptiste Perrin showed that when collected in

---

[8]The history of Faraday's electrochemical work and its interpretation is covered in [435].

a Faraday cup, cathode rays would charge up the cup negatively [61]. He also succeeded in deflecting them magnetically.

On the other hand, eminent physicists like Goldstein, Hertz, and Lenard thought that cathode rays were more light-like, i.e. associated with phenomena of the ether. This view seemed especially supported when Hertz, in one of the more famous wrong experiments, failed to deflect them electrically [47], because of the dielectric properties of the glass or conduction by the rest gas in his tube. It took progress in obtaining high vacuum by William Crookes [443] and others to prove him wrong.

There were thus two fundamentally different interpretations of cathode ray phenomena, which motivated Joseph John Thomson[9] at the Cavendish Laboratory to make further and careful investigations in 1896-97. He announced his findings to the Royal Institution [74] on April 30, 1897. In the introduction to his more elaborate account in the Philosophical Magazine [75] he wrote:

> The experiments discussed in this paper were undertaken in the hope of gaining, some information as to the nature of the Cathode Rays. The most diverse opinions are held as to these rays; according to the almost unanimous opinion of German physicists they are due to some process in the æther to which –inasmuch as in a uniform magnetic field their course is circular and not rectilinear– no phenomenon hitherto observed is analogous: another view of these rays is that, so far from being wholly ætherial, they are in fact wholly material, and that they mark the paths of particles of matter charged with negative electricity. It would seem at first sight that it ought not to be difficult to discriminate between views so different, yet experience shows that this is not the case, as amongst the physicists who have most deeply studied the subject can be found supporters of either theory.
>
> The electrified-particle theory has for purposes of research a great advantage over the ætherial theory, since it is definite and its consequences can be predicted; with the ætherial theory it is impossible to predict what will happen under any given circumstances, as on this theory we are dealing with hitherto unobserved phenomena in the æther, of whose laws we are ignorant.

In a series of experiments in the early 1890s, Thomson had already gathered information about the nature of cathode rays. He had e.g. determined that they were deflectable by a magnetic field not only close to the cathode but all along their trajectory. He had roughly measured their speed by a stroboscope. He found about $1.9 \times 10^7 \text{cm/s}$, large compared to the typical speed of ions in a discharge tube, but small compared to the speed of light [59]. In the 1897 series of experiments [74, 75], he was able to more accurately measure both the speed and the charge-to-mass-ratio e/m of the rays using a clever

---

[9]For a detailed history of this discovery, see [462].

combination of electric and magnetic deflection shown in Figure 2.5. Today we would call such a set-up an electromagnetic spectrometer for charged particles. The rays emitted by discharge between cathode and anode on the extreme left of the tube pass through the cylindrical anode and enter into the elongated straight section of the tube, with constant velocity. They enter into a deviation region where both a vertical electric field and a horizontal magnetic field act on them. Adjusting the voltage applied to the parallel plate capacitor, the electrical deviation can be varied. Regulating the current in the Helmholtz-type magnet coils, the magnetic deviation can be changed. Adjusting both deviations to be the same, and knowing the necessary field strengths, the velocity of the rays and the ratio of charge to mass can be simultaneously measured. How this is done is detailed in Focus Box 2.6. The results Thomson obtained for the speed, using various rest gases and three tubes, confirmed his earlier conclusion. The mass-to-charge results were all in the region of $-0.5 \times 10^{-11} \text{kg/C}$. Calculated by modern standards his result can be quoted as $m/e = -(0.51 \pm 0.16) \times 10^{-11} \text{kg/C}$ (or $e/m = -(1.96 \pm 0.31) \times 10^{11} \text{C/kg}$). As the number of electrons is very large, the error is entirely systematic in nature.

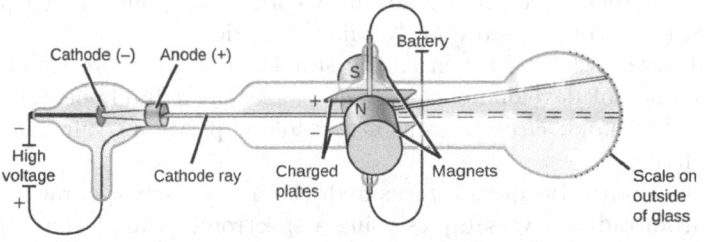

Figure 2.5 Schematic of a cathode ray tube equipped with deflector electrodes and magnets [670]. (Credit: openstax.org, access for free at https://openstax.org/books/chemistry-2e/pages/1-introduction)

The speed of the rays can also be determined by using the fact that the potential difference U between the first anode and cathode of the discharge tube confers a fixed kinetic energy $E_{kin} = E_{pot} = qU$ to them, independent of cathode material or rest gas. If on the other hand the rays consisted of cathode or rest gas ions, the speed would depend on these materials. This was tested by Walter Kaufmann [72, 73], simultaneously to Thomson's experiments. Kaufmann found a more accurate value of $e/m = -1.77 \times 10^{11} \text{C/kg}$, close to modern values (see Focus Box 2.6) and with smaller errors. Similar results were obtained by Emil Wiechert in Königsberg [76]. Kaufmann, however, shied away from firmly concluding that cathode rays were light charged particles emitted by the cathode, the same for all cathode materials. He merely called this conclusion "satisfactory". Steven Weinberg suspects that this timidity is due to the strong influence of the Viennese physicist and philosopher of science Ernst Mach [548], which was widespread at the time among German and Austrian physicists. Mach's philosophy was that hypothetical entities like particles had no role in physics and only directly observable entities did. We will find similar attitudes when we discuss the interpretation of quantum mechanics in Chapter 5.

That these results can be obtained in a consistent way does in fact show that cathode rays consist of the "electrified particles" mentioned in Thomson's introduction, emitted by any cathode material. Thomson concluded boldly [75]:

> ...we have in the cathode rays matter in a new state, a state in which the subdivision of matter is carried much further than in the ordinary gaseous state: a state in which all matter –that is, matter derived from different source such as hydrogen, oxygen, etc.– is of one and the same kind; this matter being the substance from which the chemical elements are built up.

He thus correctly speculated that the new particles were a component of all matter, yet existed also outside matter [99]. Thomson called them "corpuscles". We call them electrons, following George Johnstone Stoney [58] who introduced the term to denote an "atom of electricity".

This discovery can be taken as the starting point for two tremendously important lines of development: *particle physics* as a fundamental research subject on one hand; *electronics* as a technology pervading our daily life on the other hand.

In 1900, Henri Becquerel measured the mass-to-charge ratio of beta rays [92] from radioactive samples using a spectrometer analogous to Thomson's. He found that it was the same as the m/e of cathode rays. Thus beta particles were also identified as electrons. Kaufmann, like essentially all other physicists then changed sides [102], the electron as a real particle was accepted.

2.4   QUANTUM CHARGE

Let us jump ahead by a few years to see how the measurement of the electron e/m was converted to separate values for the two quantities, by a direct measurement of the elementary electric charge e. Important as the electron charge and mass are all by themselves, the impact of this measurement goes beyond. As Weinberg points out [548, chapter 3], the measurement of the electron charge opens the door to the atomic scale. Ratios of the molecular masses were known for many chemical elements by the work of John Dalton and the followers of his molecular theory (see Chapter 4). Likewise, ratios of charge and mass were known for many ions, mostly through electrolytic determination. Considering crystals and metals to be densely packed arrangements of atoms, the ratio of their mass to their volume could also be estimated. The single isolated measurement of e gave access to a host of atomic and molecular characteristics.

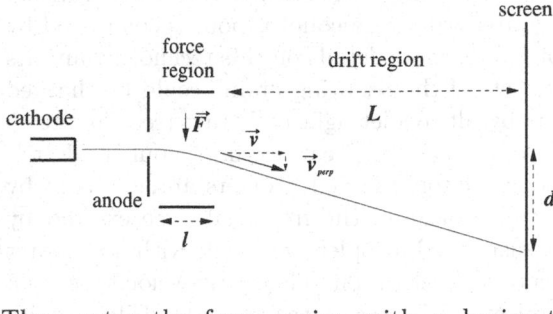

This sketch shows the principle of Thomson's measurement of the electron charge-to-mass ratio e/m. The electrons exit the cathode on the left and are accelerated by the electric field between cathode and anode.

They enter the force region with a horizontal velocity v. With a constant force F orthogonal to v applied in this region during a time $t = l/v$, they exit with a perpendicular velocity component:

$$v_{perp} = t\frac{F}{m} = \frac{l}{mv}F$$

Since $v_{perp} \ll v$, they take a time $T \simeq L/v$ to traverse the field-free drift region. The displacement d of the cathode ray beam (largely exaggerated in the sketch) is thus:

$$d = Tv_{perp} = \frac{Ll}{mv^2}F$$

Thomson had a parallel plate capacitor inside the evacuated glass tube, with a roughly constant electric field E pointing vertically upward, exerting a constant force eE pointing downward. Solenoidal magnets outside the tube created a roughly constant magnetic field B in the horizontal direction pointing towards you. The magnetic force was thus evB. Both resulted in displacements:

$$d_{el} = \frac{e}{m}\frac{Ll}{v^2}E \quad ; \quad d_{mag} = \frac{e}{m}\frac{Ll}{v}B$$

Measuring the two displacements, we can solve for the unknowns v and e/m:

$$v = \left(\frac{E}{B}\right)\left(\frac{d_{mag}}{d_{el}}\right) \quad ; \quad \frac{m}{e} = \frac{B^2Ll}{E}\left(\frac{d_{el}}{d_{mag}^2}\right)$$

With Thomson's set-up and the electric and magnetic fields he used, the displacement was a few centimetres. He obtained roughly correct values of e/m around -2×10^{11} C/kg. Todays value recommended by the CODATA group is $e/m = -1.75882001076(53) \times 10^{11}$ C/kg (see www.codata.org). A gram of electrons thus has a negative charge of almost 200 million Coulomb.

Focus Box 2.6: Thomson's measurement of the electron e/m

It is thus not astonishing that eminent researchers involved themselves in trying to measure the charge separately from mass. In the early years of the 20th century, Charles Thomson Rees Wilson invented the expansion cloud chamber, in which a saturated water or alcohol vapour is condensed by adiabatic expansion (see Chapter 8 for more details on this technology and its use). It was noted that the droplets of the resulting "rain" could be charged by picking up gas ions produced by ultraviolet light or X-rays (see Chapter 5 for this mechanism, the photoelectric effect). Creating the vapour inside the electric field of a parallel plate capacitor, charged droplets are attracted by the plate of the opposite charge. Because of the frictional force exerted by the gas in which the vapour is suspended, droplets will drift with a constant velocity after a very short period of acceleration. Their drift velocity is then simply proportional to the total exerted force. Measuring the drift velocity under gravity alone and under the sum of gravity and electric field, one can deduce the electric charge. For a whole cloud, this will then correspond to the average charge of the drops inside. J.J. Thomson [84, 91, 110] and Harold A. Wilson [113] were pioneers of this technique.

The Chicago physicist Robert A. Millikan had learned about this methodology during a visit to Cavendish lab. Upon return he started his own investigation by narrowing down the velocity measurement to single drops in a similar experimental set-up. He used water droplets condensed in an expansion cloud chamber moving with and without electric field [134], and ions produced by X-rays from radium exposure. He confirmed that all droplet charges were multiples of an elementary charge e with good accuracy. However, he also realised that there were important systematic limitations to the accuracy of this measurement, most prominently by air turbulence and convection following the expansion and mass change of the droplets by evaporation during the velocity measurement.

He thus improved the method [139], using mineral oil as a less volatile substance, charged by friction in an atomiser, as shown in Figure 2.6. His set-up let single drops fall into the capacitor through a hole, thus separating the turbulent inlet volume from the calm air between the plates. Ultraviolet or X-rays ionised the air between the plates, giving the droplets a chance to pick up further charged ions[10]. He proudly states in the end of his paper that "so far as I am aware, there is no determination of e ... by any other method which does not involve an uncertainty at least 15 times as great as that represented in the above measurements." Of course he also realised the importance of his precision measurement in determining other fundamental constants, which he evaluated in a separate section. A comparison with modern determinations of e makes little sense, since in today's SI system e serves as the definition of the

---

[10]Electrons freed by ionisation are very mobile and unlikely to be picked up by the droplets.

The balance of forces in Millikan's oil drop experiment is shown in the adjacent figure for two cases. On the left, the drop is falling freely inside the shorted capacitor. After a short acceleration by gravity, an equilibrium with air friction is reached.

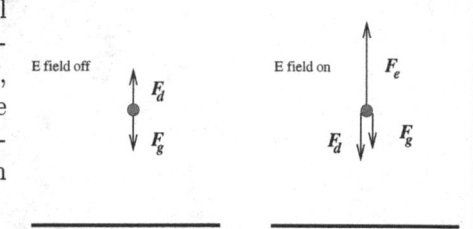

The downward gravitational force is diminished by buoyancy, such that $F_g = (4\pi/3)R^3(\rho_{oil} - \rho_{air})g$, with the drop radius R and the mass densities ρ. The gravitational acceleration, $g = GM_\oplus/r_\oplus^2 \simeq 9.81\,\mathrm{m/s}$, is given by the mass and radius of the Earth and the gravitational constant G. Gravity is compensated by the upward frictional force F_d. For a perfectly laminar flow around a spherical drop, Stoke's law gives $F_d = 6\pi\mu_{air}Rv$, proportional to its velocity v and pointing in the opposite direction. The viscosity of the medium is μ. In equilibrium, the two forces are equal and opposite. Solving for v we get the drift velocity of the falling drop:

$$v_1 = \frac{2}{9}\frac{\rho_{oil} - \rho_{air}}{\mu_{air}}gR^2$$

When the capacitor is charged such that there is an upward force on the drop, the equilibrium sketched on the right is reached. The upward electrical force on a singly charged drop, $F_e = eE$, is balanced by the sum of the gravitational force and the frictional force, which now points downward. The equilibrium is reached for $F_e = F_g + F_d$ and corresponds to the rising drift velocity v_2. Inserting the known forces and solving for e we get the relation used (and corrected) by Millikan [139]:

$$e = \frac{4}{3}\pi\left(\frac{9\mu_{air}}{2}\right)^{\frac{3}{2}}\left(\frac{1}{g(\rho_{oil} - \rho_{air})}\right)^{\frac{1}{2}}\frac{(v_1 + v_2)v_1^{\frac{1}{2}}}{E}$$

In his systematic approach to the problem, Millikan found a deviation from Stokes' law used above. It occurs for drop radii which are smaller than the mean free path of molecules in the gas. He took this linear deviation into account in his final results [152].

Focus Box 2.7: Millikan's oil drop experiment to measure the electron charge

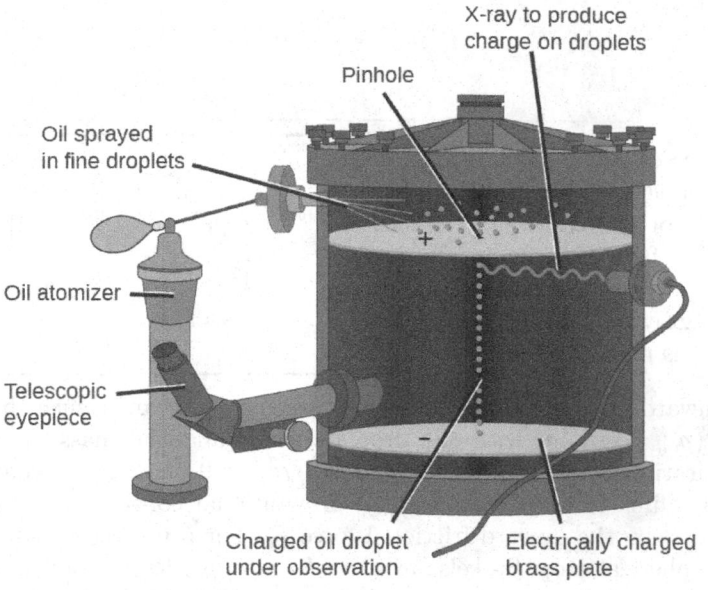

Figure 2.6 Schematic of the Millikan oil drop experiment [670]. (Credit: openstax.org, access for free at `https://openstax.org/books/chemistry-2e/pages/1-introduction`)

charge unit Coulomb, and the translation into the CGS units of 1911 involves the speed of light (also fixed in SI units)[11].

Needless to say that such an important measurement was up for scrutiny by concurrent methods. The Austrian Felix Ehrenhaft indeed claimed to have found charges smaller than e and not in multiples of an elementary charge [131]. Millikan kept refining his method, especially measuring and parametrising deviations from Stokes' law for small drop radii. In his final publication on the subject [152], he discussed the systematics of his measurement in depth. Among all his notebooks, he kept the one with these results; it is conserved at the Caltech library and can be consulted online[12]. Millikan's result stood up to all challenges. He was awarded the 1923 Nobel prize for physics for this work and his research on the photoelectric effect (see Chapter 5).

Strictly speaking, what was measured is the electric charge of ions, i.e. of molecules stripped by a few of their electrons or with extra electrons attached. The greatest common denominator of these charge measurements is thus equal

---

[11]For a lively and interesting discussion of units for physical properties see the blog `blog.wolfram.com` of Nov. 16, 2018.

[12]`caltechln.library.caltech.edu/7/1/Millikan_1.pdf`

to the electron charge if and only if atoms are strictly neutral. Millikan of course realised that this assumption needed experimental verification [174, p. 80-83]. A few of his drops first had a negative charge, he then succeeded to charge them positively by X-ray exposure (photoelectric effect, see Section 4.3). By direct comparison he concluded that common denominators of positive and negative charges were the same to one part in 2000. This is indeed confirmed today with staggering accuracy. Modern measurements of the neutrality of atoms find that $|q_e - q_p|/e < 10 \times 10^{-21}$, by showing that an AC excitation of a gas does not cause an acoustic resonance [584].

FURTHER READING

Edward A. Davis and Isobel J. Falconer, *J.J. Thomson and the Discovery of the Electron*, Taylor and Francis, 1997.

Jed Z. Buchwald and Andrew Warwick (Edts.), *Histories of the Electron: The Birth of Microphysics*, MIT Press, 2001.

Steven Weinberg, *The Discovery of Subatomic Particles*, Cambridge University Press, 2003.

CHAPTER 3

Relativity

...not only in mechanics, but also in electrodynamics, no properties of the phenomena correspond to the concept of absolute rest, but rather that for all coordinate systems for which the mechanical equations hold, the equivalent electrodynamical and optical equations hold also ...In the following we will elevate this guess to a presupposition (whose content we shall subsequently call the "Principle of Relativity") and introduce the further assumption, –an assumption which is only apparently irreconcilable with the former one– that light in empty space always propagates with a velocity V which is independent of the state of motion of the emitting body.

Albert Einstein, *Zur Elektrodynamik bewegter Körper*, 1905 [118]

I N THIS CHAPTER I discuss the consequences of the fact that no absolute space-time exists, but that all physics laws should be the same in all reference frames. Newton had indeed shown this to be the case for mechanics in 1687 [1, Corollary 5], for reference systems moving relative to one another at constant velocity. These are called inertial systems since the inertial resistance of a body to an accelerating force will be the same, i.e. Newton's law will hold the same way. The fact that only relative velocity matters is often called Galilean relativity since it is based on the homonymous transformations of coordinates (see Focus Box 3.4). Newton assumed that there was a "natural" coordinate system, absolute space and time, with respect to which all others would be defined, the ether.

It was thus logical to assume that when Newton's law holds in any inertial coordinate system, so should Maxwell's laws. This is what Einstein calls the "principle of relativity" in the above quote. This cannot be achieved without modification, since Newton's law requires that velocities simply add up vectorially, in obvious conflict with the speed of light being the same in all reference frames. Thus a light ray emitted from a fast train evolves with the same speed when seen from the train and from the train station. We will see how Einstein modified classical mechanics to admit this, while preserving the principle of relativity. But beforehand I will review experimental evidence that both facts hold. As examples I describe three experiments: the one by Fizeau, the well-known Michelson and Morley experiment and the one by Trouton and Noble. I then describe special relativity, which refers to inertial systems. General relativity, covering all coordinate systems, is also briefly covered, since it is necessary to understand cosmological facts detailed in Chapter 10.

3.1 ETHER

In 1851, Hippolyte Fizeau set up an experiment to measure the speed of light c in flowing water [19]. He used an interferometer to measure the difference in speed between light propagating in and against the direction of water flowing through a glass pipe with velocity v, as sketched in Figure 3.1. Taking into account that light propagates slower in a medium with refractive index n, he expected a modification according to the Galilean addition of velocities, $c/n \pm v$ in the two arms. Fizeau indeed observed a small shift of the interference

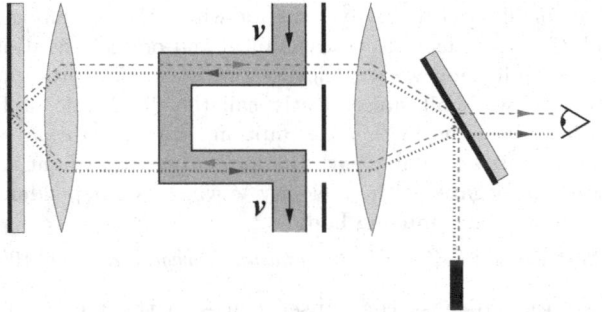

Figure 3.1 Sketch of the Fizeau experiment to measure the speed of light in flowing water. The dotted light path always goes with the water flow, the dashed one always against it. Their interference pattern is observed on the right. If the speed of light c were influenced by the movement of the medium, the interference pattern would change when the water speed v is varied. Indeed it does, but only by a very small amount.

pattern, but the velocity of the water only entered reduced by a large factor, $c/n \pm v(1 - 1/n^2)$. This had been anticipated by Fresnel for a medium at rest with respect to the ether [422][1]. In 1886, Albert A. Michelson and Edward W. Morley confirmed the experiment with an improved apparatus [48]. The results clearly showed that the velocity of a medium does not simply add to the speed of light.

Thirty years after Fizeau's experiment, the American physicist Albert A. Michelson set up an interferometer in Berlin, shown in Figure 3.2, to compare the speed of light in two orthogonal directions [45]. The idea behind this experiment was the following: the Earth moves around the Sun with a velocity

---

[1]The correct explanation for moving dielectrics was finally found by the Dutch Pieter Zeeman after the formulation of special relativity [162, 167].

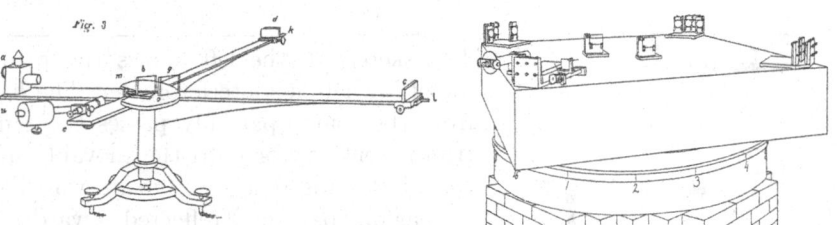

Figure 3.2 Left: Interferometer of the 1881 Michelson experiment in Berlin and Potsdam [45]. Right: Sturdy interferometer mounted on a stone slab floating on a mercury bed, used by Michelson and Morley in 1887 in Cleveland [51]. (Credit: American Journal of Science, reproduced by permission)

of about 30 km/s. If a luminiferous ether were at rest with respect to anything else than the Earth itself, comparing the speed of light in the direction of rotation and orthogonal to it would reveal a difference. Tiny differences in speed are sensitively detected (or not!) by interferometry, as Fizeau had demonstrated. However, vibrations of the ground in busy Berlin disturbed the sensitive Michelson apparatus. Even moving it outside the city to Potsdam did not solve the problem. A much more sturdy set-up, also shown in Figure 3.2, was thus mounted on a stone slab by Michelson and Edward W. Morley in Cleveland/Ohio in 1887 [51]. Focus Box 3.1 explains how it works. The experiment convincingly showed that there is no dependence of the speed of light on the direction in which it is emitted. Since the Earth constantly moves with respect to the Sun (which is not at rest with respect to other stars either), it is not conceivable that an ether transmits electromagnetic waves. Electromagnetic waves do not require a medium, they propagate in empty space. In 1930, Georg Joos repeated the experiment [238] with unprecedented precision and confirmed the result. It is thus clear that neither linear nor quadratic influences of motion on c exist. It is simply the same in any system of reference.

3.2 MOVING FRAMES

We had noted in Chapter 2 that the split into electric and magnetic fields cannot be independent of the reference frame. Charges must be moving to make a current which causes a magnetic field. The current is proportional to both their spatial density and their velocity, so is the magnetic field. To feel a force from this magnetic field, the probe must also be moving. Again, the force is proportional to the probe velocity. When we fix our reference frame to the moving probe, magnetic forces magically disappear. What happens to the principle of relativity?

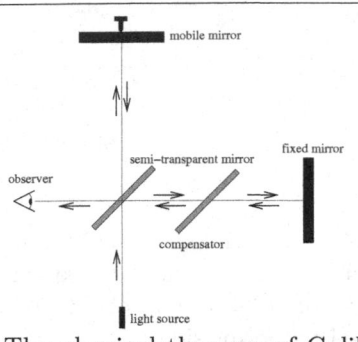

The sketch on the left shows the principle of Michelson's interferometer. The light from the source partially passes the semi-transparent mirror onto the movable mirror where it is reflected. On its way back it is again (partially) reflected towards the observer. The other part of the source light is reflected into the right arm of the interferometer, where it meets the fixed mirror and passes back towards the observer.

The classical theorem of Galilean relativity predicts that two velocities add up vectorially. If the Earth moves parallel to the right arm, through an ether (assumed to be at rest) with velocity v, the speed of light c will be enhanced on the way towards the fixed mirror and reduced on the way back to the observer. For the upper arm, the velocity of the interferometer will be orthogonal to the light path, such that the speed of the light ray will be c, but the path slightly increased by the relative motion. The times of arrival of the two beams with respect to an arbitrary time of emission will thus be:

$$T_\| = \frac{L}{c+v} + \frac{L}{c-v} = \frac{2L}{c}\frac{1}{1-v^2/c^2} \quad ; \quad T_\perp = \frac{2L}{\sqrt{c^2-v^2}} = \frac{2L}{c}\frac{1}{\sqrt{1-v^2/c^2}}$$

L is the length of both arms, adjusted to be the same. The time difference between the two arms for this direction of motion will thus be:

$$\Delta_1 = \frac{2L}{c}\left(\frac{1}{1-v^2/c^2} - \frac{1}{\sqrt{1-v^2/c^2}}\right)$$

Now we rotate the interferometer by 90°; the upper arm will now be in the direction of motion of the Earth, the right arm will be orthogonal to it. Consequently, the two terms in the bracket will be interchanged. The rotation will thus change the phase between the two light rays by a time difference:

$$\Delta_1 - \Delta_2 - \frac{4L}{c}\left(\frac{1}{1-v^2/c^2} - \frac{1}{\sqrt{1-v^2/c^2}}\right) \simeq \frac{2L}{c}\frac{v^2}{c^2}$$

The approximation is valid in the limit of $v \ll c$, which is certainly the case since Earth moves inside the solar system with a velocity of about $v \simeq 30$ km/s, compared to the speed of light $c \simeq 300,000$ km/s. The corresponding phase shift in units of wavelength will thus be $n = (c/\lambda)\,(\Delta_1 - \Delta_2) \simeq (2L/\lambda)(v^2/c^2)$. This is the fraction of a line spacing that the shift should amount to. None is observed.

Focus Box 3.1: The experiments of Michelson and Morley

A demonstration of the principle of relativity in electrodynamics is already included in Faraday's experiments on induction [511]. Faraday analysed three cases:

- A conductor is moved through the inhomogeneous field of a magnet at rest.

- The magnet and its field are moved through the conductor at rest.

- With both magnet and conductor at rest, the magnetic field is varied in time.

In each case, the voltage or current induced in the conductor is measured, thus the induced electric field. What happens is explained in modern terms in Focus Box 3.2. The three cases are found to be qualitatively and quantitatively equivalent, what matters is the relative motion, leading to a variation of the magnetic flux in time as in the third case. In other words: Faraday's law holds (see Focus Box 2.3).

When the conductor is at rest, the charges in it do not move. How can they be sensitive to a magnetic field? The answer is that the changing magnetic field now looks to them like an electric one, as shown in Focus Box 3.2. The relative role of electric and magnetic field depends on the reference frame, but the force that acts remains the same.

Trouton and Noble searched for effects of this transformation from electric to magnetic fields in 1903 [112, 111], as explained in Focus Box 3.3. Classically, a dipole should feel a torque when moving at a constant velocity and turn until it is orthogonal to the direction of motion. No such torque is observed, in agreement with Einstein's first axiom, the principle of relativity.

The negative results of these two experiments, among others, led Hendrik A. Lorentz in 1904 [114] to derive a set of transformations from one inertial system to another that would accommodate these facts. He came up with the spectacular result that moving length scales would be contracted when seen from an observer at rest. Likewise, time would depend on the relative motion of two systems. We detail the mathematics of these Lorentz transformations in Focus Box 3.5. They deform the electric field of a moving electron. The effects are of second order in v/c, the ratio of the relative speed of the two systems to the speed of light, and explain the absence of effects in Trouton and Noble's experiment. Lorentz still believed, however, that there was an absolute space-time, which defines the "right" lengths and times.

Lorentz' work was generally taken as a theory of the electron, with his results on the deformation of the field considered as due to a change of its shape. It was generalised to gravitational properties of the electron by Henri Poincaré in 1905 [121], who concluded that also the gravitational field would propagate at the speed of light, in the form of gravitational waves.

We consider the first two cases of Faraday's induction experiments, once fixing our coordinate frame to the magnet, $(\vec{r}, t)$, and then to the conductor, $(\vec{r}\,', t')$. **Magnet frame:** In the rest frame of the magnet, the magnetic field is inhomogeneous but fixed, $\vec{B}(\vec{r})$. The electric field is zero. According to Lorentz' law (see Focus Box 2.5), the force on a particle of charge q and velocity $\vec{v}$ (as given by the motion of the conductor) is $\vec{F} = q\vec{v} \times \vec{B}$.

Conductor frame: In the conductor frame, there is a time-varying magnetic field $\vec{B}'$ related to the (fixed) magnetic field $\vec{B}$ in the magnet frame: $\vec{B}'(\vec{r}\,', t') = \vec{B}(\vec{r}\,' + \vec{v}t')$. In this frame, there is an electric field: $\vec{\nabla} \times \vec{E}' = -\frac{\partial \vec{B}'}{\partial t'}$. We thus find that $\vec{E}' = \vec{v} \times \vec{B}$. A charge q in the conductor is at rest in the conductor frame, thus $\vec{F}' = q\vec{E}' = q\vec{v} \times \vec{B} = \vec{F}$.

To first order, the force is thus the same in both frames, so are induced voltage and current. However the force is seen to be electric in the conductor frame, magnetic in the magnet frame.

When using Lorentz transformations between the two reference frames, effects of second order in v/c are revealed. For the conductor frame we find $\vec{E}' = \gamma \vec{v} \times \vec{B}$, $\vec{F}' = q\vec{E}' = q\gamma \vec{v} \times \vec{B}$, with $\gamma = 1/\sqrt{1 - v^2/c^2}$ and the speed of light c. The relativity principle is thus preserved in that Maxwell's equations have the same form in both frames.

Focus Box 3.2: Relativity in induction

3.3 SPECIAL RELATIVITY

The concept of absolute space and time was radically abandoned in Albert Einstein's June 1905 paper "On the electrodynamics of moving bodies" [118], as seen in the introductory quote to this Section. Einstein motivates his study by the facts we mention above. But his paper does not have a single reference, since "it is known that Maxwell's electrodynamics –as usually understood at the present time– when applied to moving bodies, leads to asymmetries which do not appear to be inherent in the phenomena." Instead "the unsuccessful attempts to discover any motion of the earth relatively to the 'light medium' suggest that the phenomena of electrodynamics as well as of mechanics possess no properties corresponding to the idea of absolute rest." Einstein traces back the problems to a lack of understanding for simultaneous processes. What is to be understood by *at the same time* when one deals with two different coordinate systems in relative motion? How does one attribute a time to an event elsewhere when absolute time and space do not exist?

The answer is that one needs a prescription for the synchronisation of two clocks. Let us assume, like Einstein does in Section 2 of his paper, that there are two identical clocks in two places A and B. We now emit a light signal from point A at time t_A (measured by the A clock). It is reflected from point B at t_B (measured by the B clock) and arrives back at A at time t'_A (measured by the A clock). The two clocks are synchronous if $t_B - t_A = t'_A - t_B$. In addition, the speed of light defines the relative units of space and time, because twice

To understand the idea of the experiment conducted by Trouton and Noble, we consider an electric dipole, which moves with a constant velocity v in the horizontal direction with respect to the laboratory system. The charge $+q$ at its upper end causes a current I in the laboratory frame, which does not exist in the rest frame of the dipole. The magnetic field caused by the current results in a force F pointing vertically downward, and felt by the lower charge.

Likewise, the lower charge causes a current flowing in the opposite direction, the corresponding magnetic field pulls the upper charge upward. In the laboratory system, the dipole should thus feels a torque aligning it orthogonal to its direction of motion. No such torque is observed.

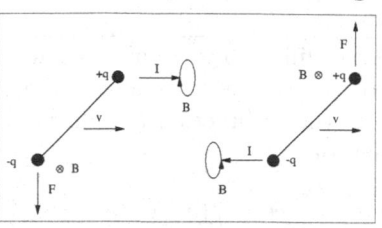

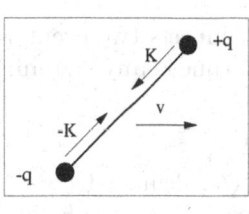

A magnetic torque also exists when treating the problem relativistically. However, a second force is to be taken into account, which has no effect in the non-relativistic limit. As the sketch on the left shows, the Coulomb force K in the rest frame of the dipole acts in the direction of its axis and has no effect on its orientation.

Seen from the laboratory system, however, the vertical coordinates do not change, but the horizontal ones are contracted. The electrostatic force is thus no longer parallel to the dipole axis and compensates the magnetic forces. There is thus no resulting torque, in agreement with observation.

Focus Box 3.3: The Trouton and Noble experiment

According to Newtonian mechanics, the transformation from a system of coordinates at rest to a moving system is given by the Galilean transformation:

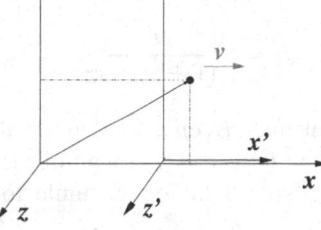

When the system (x', y', z') moves at constant velocity v in the $+x$ direction of system (x, y, z) and the origins of the two coïncide for $t = t' = 0$ we have:

$$t = t' \; ; \; x = (x' + \beta ct) \; ; \; y = y' \; ; \; z = z'$$

with $\beta = v/c$.

This leaves lengths $\Delta \vec{r} = \vec{r}_1 - \vec{r}_2$ and elapsed times $\Delta t = t_1 - t_2$ invariant, i.e. $\Delta \vec{r} = \Delta \vec{r}'$, $\Delta t = \Delta t'$. Velocities add up vectorially, such that the velocity $\vec{u}'$ of a mass point in the moving system will be measured in the system at rest as $\vec{u}$:

$$u_x = u'_x + v \; ; \quad u_y = u'_y \; ; \quad u_z = u'_z$$

Momentum $\vec{p} = m\vec{u}$ and kinetic energy $E = mu^2/2$ measured in the two systems behave accordingly.

Focus Box 3.4: Galilean transformations

According to Einstein's special relativity, the transformation from a system of coordinates (x, y, z) at rest to a system (x', y', z') moving at constant velocity v in the x direction (like in Focus Box 3.4) is given by the Lorentz transformation:

$$ct = \gamma\left(ct' + \beta x\right) \quad ; \quad x' = \gamma\left(x + \beta ct\right) \quad ; \quad y = y' \quad ; \quad z = z'$$

with $\beta = v/c$ and $\gamma = 1/\sqrt{1 - v^2/c^2}$. When a light ray connects two events, $x_1 = (ct_1, \vec{x}_1)$ ans $x_2 = (ct_2, \vec{x}_2)$, it takes the same elapsed time in any system:

$$\left((x_1 - x_2)^2 + (y_1 - y_2)^2 + (z_1 - z_2)^2\right)^{\frac{1}{2}} = c\left(t_1 - t_2\right)$$

This transformation thus leaves the space-time length $\sqrt{s}$, defined by $s = c^2 t^2 - \vec{r}^2$, invariant. Time and space thus form a four-vector (see Focus Box 3.6). Lorentz transformations are the rotations and translations in space-time, which leave s invariant. Since s is to be positive definite, all velocities have to be smaller than the speed of light. Lengths and elapsed times, on the contrary, are not invariant separately. Instead we have time dilatation and length contraction in the direction of relative motion:

$$\Delta t = \gamma \Delta t' \quad ; \quad \Delta x = \gamma \Delta x' \quad ; \quad \Delta y = \Delta y' \quad ; \quad \Delta z = \Delta z'$$

Accordingly, velocities do no longer add up simply. Instead we have:

$$u_x = \frac{u'_x + v}{1 + u'_x v/c^2} \quad ; \quad u_y = \frac{u'_y}{\gamma(1 + u'_x v/c^2)} \quad ; \quad u_z = \frac{u'_z}{\gamma(1 + u'_x v/c^2)}$$

Another important four-vector is energy-momentum, $(E, c\vec{p})$. Its length m gives the mass of the corresponding object: $m^2 c^4 = E^2 - c^2 \vec{p}^2$. For an object at rest in a given system ($p = 0$), we find back Einstein's famous formula for the equivalence of mass and energy, $E = mc^2$.

Focus Box 3.5: Lorentz transformations

the distance between A and B, $2r_{AB}$, is traversed by the light ray in the period $t'_A - t_A$, such that $2r_{AB}/(t'_A - t_A) = c$, a universal constant. When the second clock is moving, this leads to a loss of synchronism, unless redefined by light rays.

With this definition of synchronism, Einstein derives the Lorentz transformations between two inertial systems in Section 3, without reference to Lorentz' work. Lengths are defined by rulers at rest in any given systems. When one needs to known them in a moving system, they must be defined by synchronous clocks at both ends. In summary, lengths in a moving system appear contracted in a system at rest. Times measured in a moving system appear longer than measured by a clock at rest[2]. Of course the role of "the moving system" and the "system at rest" can be readily reversed, since there is no absolute rest. Thus symmetry is restored, only relative motion matters.

In the second part of his paper, Einstein applies these findings to electromagnetism and derives the transformation properties of electric and magnetic fields. He concludes his paper by looking into the consequences of his findings for the energy of a moving body and its resistance to acceleration, its inertia. This is treated more in-depth in a short paper submitted in September 1905 [117]. There he considers the energy balance of a moving body which emits light of energy E. His result is that indeed its inertial mass reduces by $\Delta m = E/c^2$. This is the first example of the famous equivalence of mass and energy, $E = mc^2$.

The theory of special relativity found an almost immediate support by leading authorities like Max Planck. Especially Hermann Minkowski [128, 165] strongly emphasised its disruptive character and introduced the four-vector notation of relativistic kinematics that we use throughout this book and explain in Focus Box 3.6. It allows for a particular elegant and compact representation of Maxwell's equations, which is quoted in Focus Box 3.7. Today, special relativity is used and confirmed in every experiment using particles travelling at a velocity close enough to the speed of light. A popular example is the fact that unstable particles created by cosmic rays high up in the atmosphere reach Earth despite their short lifetimes [295], due to time dilatation.

Special relativity connects inertial systems, those on which no forces act. But do such systems actually exist? Earth is constantly accelerated by the Sun's gravitational force such that it stays on its elliptical path, so no coordinate system fixed on Earth is really inertial. The Sun is also rotating and attracted by other stars in the Milky Way, so it cannot hold an inertial system either. Since the range of the gravitational force is infinite, there is indeed no object in the whole Universe on which it does not act. So Einstein asks a valid question [280, p. 223]: "Can we formulate physical laws so that they are valid for all coordinate systems, not only those moving uniformly, but also those

---

[2]Please keep in mind that we measure elapsed time by counting "ticks" of a clock. When the second is shorter, we measure a longer time.

Covariant four-vectors transform like the space-time four-vector x_μ under Lorentz transformations:

$$x_\mu = (x_0, x_1, x_2, x_3) = (ct, x, y, z)$$

An important example is the energy-momentum four-vector p^μ :

$$p_\mu = (p_0, p_1, p_2, p_3) = (E, cp_x, cp_y, cp_z)$$

Contravariant four-vectors x_μ are related to covariant ones by the metric tensor $g^{\mu\nu}$:

$$x^\mu = g^{\mu\nu} x_\nu \quad ; \quad g^{\mu\nu} = \begin{pmatrix} 1 & 0 & 0 & 0 \\ 0 & -1 & 0 & 0 \\ 0 & 0 & -1 & 0 \\ 0 & 0 & 0 & -1 \end{pmatrix}$$

$$\begin{aligned} x^\mu = (x^0, x^1, x^2, x^3) &= (ct, -x, -y, -z) \\ p^\mu = (p^0, p^1, p^2, p^3) &= (E, -cp_x, -cp_y, -cp_z) \end{aligned}$$

In Minkowski space, the scalar product between two four-vectors is:

$$\begin{aligned} x_\mu x^\mu &\equiv \sum_\mu x_\mu x^\mu = c^2 t^2 - x^2 - y^2 - z^2 = c^2 t^2 - \vec{x}^2 = s \\ p_\mu p^\mu &= E^2 - c^2 p_x^2 - c^2 p_y^2 - c^2 p_z^2 = E^2 - c^2 \vec{p}^2 = m^2 c^4 \\ p_\mu x^\mu &= Ect - cp_x x - cp_y y - cp_z z = cEt - c\vec{p}\vec{x} \end{aligned}$$

The first equation defines squared length, the second squared mass (see Focus Box 3.5). The third product appears e.g. in equations describing the evolution of particle systems. We use the convention that when upper and lower greek indices are the same, summation is implicit. Scalars are invariant under Lorentz transformations, they stay the same when changing from one inertial frame to another.

Focus Box 3.6: Minkowski's four-vectors

The electromagnetic charge and current density as well as the electromagnetic potentials form four-vectors, transforming like space-time under Lorentz transformations:

$$j_\mu = (c\rho, \vec{j}) \quad ; \quad A_\mu = (V/c, \vec{A})$$

The time-like element of j_μ is the charge density ρ, i.e. the electric charge per unit volume. Its space-like components are the current density $\vec{j}$, i.e. the electric charge per unit time flowing through a unit surface. Maxwell's laws (see Focus Box 2.3) connect it to the electromagnetic four-vector potential A^μ. Its time-like component is the scalar electric potential V, with $\vec{E} = -\vec{\nabla}V - \frac{\partial \vec{A}}{\partial t}$, its space-like component the magnetic vector potential $\vec{A}$, with $\vec{B} = \vec{\nabla} \times \vec{A}$. Introducing the electromagnetic field tensor $F_{\mu\nu}$ and using the four-gradient $\partial_\mu = (\partial t/c, \vec{\nabla})$:

$$F_{\mu\nu} \equiv \partial_\mu A_\nu - \partial_\nu A_\mu = \begin{pmatrix} 0 & E_x/c & E_y/c & E_z/c \\ -E_x/c & 0 & -B_z & B_y \\ -E_y/c & B_z & 0 & -B_x \\ -E_z/c & -B_y & B_x & 0 \end{pmatrix}$$

one can condense the four Maxwell equations into one:

$$\partial^\mu F_{\mu\nu} = \mu_0 j_\nu$$

In words: the local current density causes the field tensor to diverge.

Focus Box 3.7: Maxwell's equations in Minkowski space

moving quite arbitrarily, relative to each other?" Einstein himself gave the answer: general relativity.

3.4 EQUIVALENCE

In Chapter 2 we have tacitly introduced an important symmetry between Newton's law linking force and acceleration and Newton's law of universal gravity: the mass m, denoting inertial mass in the first and gravitational mass in the second, is the same. It means in particular that the same weight of different materials shows the same inertia. This fact is far from trivial and Newton checked on it using a pendulum. He measured the period of its motion with identical weights of materials, like metals, glass, sand, salt, water and even wood or grain. He found no difference, thus establishing that the inertial mass of all these materials is the same. In fact the period of the pendulum, $T = 2\pi\sqrt{L/g}$, depends only on its length if and only if inertial and gravitational mass are the same. You are familiar with this result from your physics course at school, where a stone and a feather take the same time to fall through an evacuated glass tube. The masses of the two objects are vastly different, but so are the gravitational forces. Mass cancels, the acceleration g is the same, so is then the velocity $v(t)$ and the total time the fall takes (see Focus Box 2.1).

Friedrich Bessel (better known for the periodic functions named after him) later improved on these experiments using his own version of the Kater pendulum [12]. Towards the end of the 19th century, the Hungarian physicist Roland Baron Eötvös found that modern technology warranted a measurement many orders of magnitude more precise [54]. He used the fact that in the northern hemisphere the centrifugal force due to the Earth's rotation slightly deviates the direction of the gravitational force towards the south (see the left sketch in Figure 3.3). The centrifugal force is of dynamical origin, $F = m\omega^2 r$ with the inertial mass m, the angular velocity ω of the rotation and the distance r to the Earth's rotational axis. In Budapest, where Eötvös conducted his experiments, it deviates about 5'56" from the vertical direction. If one orients the horizontal bar of a torsion balance along the east-west direction and uses different materials of the same weight on both ends, the balance feels a torque if a difference in inertial mass exists. Using a mirror to project a light ray over a large distance, a very sensitive angular measurement can be achieved. The torsion balance Eötvös used is shown on the right in Figure 3.3. He estimated its angular accuracy to be better than 1.6×10^{-6} arc seconds. Eötvös found no difference when comparing the inertial mass of materials as different in density as brass and cork. He concluded that inertial and gravitational mass are the same to one part in 20 million.

That inertial and gravitational mass are the same to this stunning accuracy must be more than a coïncidence. Following this line of thought Einstein was led to a completely novel theory of the gravitational force.

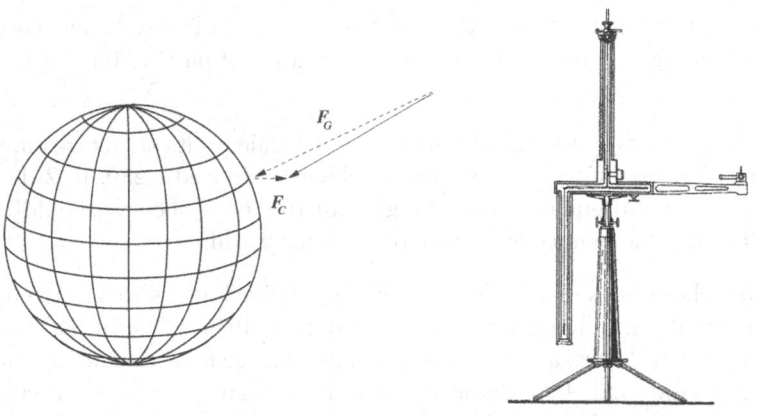

Figure 3.3 Left: The vertical gravitational force $\vec{F}_G$ is slightly deviated towards the South by the centrifugal force $\vec{F}_C$ caused by the Earth rotation. Right: Cut through a torsion balance used by Eötvös in his experiment in 1898 (Credit: Springer Nature [228], reprinted by permission). The probe is located in the vertical tube. The torsion angle is read via a mirror on the axis.

3.5 GRAVITY

It is thus impossible to distinguish a system with no gravity from a system in free fall inside a homogeneous gravitational field. When you are in free fall, you do not feel your own weight. This is what we call "weightlessness" in space, like on board the International Space Station: it is in free fall around the Earth, its rotation is caused by a (nearly constant) initial horizontal velocity. Its inhabitants thus float freely inside the ISS as a reference system[3]. Einstein [280, p. 228] called such a system a "pocket edition" of an inertial system, since it is limited in space and time. The inability to tell the difference between the two cases, no gravity or free fall, is called the equivalence principle.

With this in mind, Einstein developed a generalisation of relativity from the special case of inertial systems to any system of reference, i.e. general relativity as the theory of space-time and gravity. Three conditions had to be fulfilled by the new theory:

1. The gravitational theory should be applicable to any reference system. In the absence of gravity, special relativity had to be included.

2. The gravitational laws should be structural ones. Just as Maxwell's electromagnetic theory surpassed Coulomb's and Ampère's laws by describing the spatial and temporal change of the electromagnetic field, so

---

[3]Its inhabitants do feel a very small residual of the gravitational force, called microgravity. It is caused by horizontal deceleration due to the very much diluted atmosphere at 400 km altitude.

should the new theory surpass Newton's law of gravity by describing the change of the gravitational field by a set of partial differential equations.

3. The geometry of space should not be Euclidean, its structure should be determined by masses and their velocities. Indeed [280, p. 251]: "The gravitational equations of the general relativity theory [should] try to disclose the geometrical properties of our world."

Following these lines, Einstein painstakingly arrived at his revolutionary theory of general relativity during the years 1907 to 1915. General relativity is a metric theory at heart, a theory that describes the generation and propagation of gravitational fields by a deformation of Euclidean space to a curved space. So after replacing absolute time by a prescription for the synchronisation of clocks, Einstein found that space was not what we naively think. Instead of moving along straight lines, bodies in free fall follow geodesics, the shortest paths between two points in a curved geometry.

In Newtonian gravity, mass is the source of the gravitational field (see Section 2.1). In general relativity this role is taken by the energy-momentum tensor. A tensor is a general concept in vector algebra, originally invented to describe the deformation of solids. Here it includes the densities of three quantities –mass, energy and momentum– and expresses surprising features of theirs, like stress, pressure and shear. It does not need reference to a given coordinate system and is readily generalised to a curved space-time. An example is the Ricci tensor, which describes the change in volume of a cloud of mass points, which are initially at rest and then fall freely. Einstein's theory relates these two tensors, the energy-momentum tensor and the Ricci tensor, to the metric tensor of space-time. The results are called the Einstein field equations. In a nutshell, space-time tells matter along which path to fall and matter tells space-time how to curve[4].

There are numerous experimental proofs for the validity of this concept. Most prominently, the frequency of photons is shifted towards the blue when they fall towards a massive body, towards the red when they are emitted away from it. This effect is known as gravitational time dilatation and means that processes in general run more slowly when close to a massive body than away from it. Precision measurements of this frequency shift have been made by Robert V. Pound and collaborators [345, 347, 381] using as frequency standard the nuclear magnetic resonance, which he had helped to discover. Gravitational time dilatation in the Earth field has been verified by atomic clocks. It must be corrected for to achieve the impressive accuracy of the Global Positioning System (GPS) in your car and your smartphone.

Light follows a geodesic instead of a straight line. This fact was verified by observing that light from distant astronomical objects is deviated by the Sun, first by Arthur Eddington and collaborators [184] during the total eclipse of

---

[4]This slogan and variants thereof are often attributed to John Archibald Wheeler [409].

May 29, 1919. Through this bending of light rays, massive astronomical objects form gravitational lenses. Depending on their mass distribution, they can produce multiple or characteristically deformed images of objects that lie behind them in the line of sight. Gravitational lensing is used in modern astronomy to look into the distribution of mass in galaxies and clusters, including mass that does not radiate, absorb or reflect light, which is called dark matter. We will come back to this mysterious form of matter in Chapter 10.

At the end of a massive star's life, when its fuel runs out and the ratio of its mass to its radius becomes sufficiently large, gravitational collapse creates an environment where gravity rules. General relativity then predicts the formation of what is called a black hole, a region where gravitational forces are so large that neither matter nor radiation can escape. This appears to be a ubiquitous phenomenon, all usual galaxies including the Milky Way have a massive black hole in their centre, weighing a few million to a few billion solar masses. In current models of structure formation, black holes play a prominent role when galaxies and their clusters form.

An early spectacular prediction of general relativity [169, 175] is the emission of gravitational waves. These are ripples in the metric of space-time which are emitted by dynamic gravitationally bound systems. An early indication of their existence was the slowing of the emission frequency of pulsars, rotating neutron stars where the axis of rotation is misaligned with their magnetic axis. They emit electromagnetic radiation like the beacon of a light house, the period of light pulses arriving on Earth indicates the rotation frequency. Their spin down is due to a loss of rotational energy, in the form of both electromagnetic and gravitational waves. Observations agree quantitatively with the predictions of general relativity [604]. In 2016, the giant interferometers Advanced LIGO and Virgo reported the first observation of a gravitational wave, from the merger of two black holes [630]. A triumph of general relativity 100 years after it was first published as a theory.

The fiftieth anniversary of the Apollo 11 moon landing in 2019 reminded me of other spectacular tests of general relativity by the Lunar Laser Ranging experiments. They use mirrors deposited on the lunar surface by the manned Apollo and robotic Lunokhod missions [513]. The timing of very strong laser light, about 2.5s delay between emission on Earth and arrival of the reflected signal, allows to determine the distance of the Moon to millimetre accuracy [578], even though only one in 10^{17} photons arrives back. It is found that the Moon orbit agrees with general relativity within the accuracy of the measurement [549]. Earth and Moon fall towards the Sun with equal acceleration, confirming the equivalence principle [467]. And Newton's gravitational constant is indeed constant within $(2 \pm 7) \times 10^{-13}$ per year [571].

In practical terms, effects of gravity are negligible in the context of particle physics; this is why we do not include technical detail here. The gravitational force is many orders of magnitude weaker than concurrent forces at microscopic distances. Indeed, we still do not know how gravity works at short distances, since general relativity resists quantisation efforts for more than a century. However, gravity is the dominating force at cosmological distances. We will come back to important consequences of general relativity for the evolution of our Universe in Chapter 10.

FURTHER READING

Albert Einstein, *Relativity: The Special and the General Theory: Popular Exposition*, Methuen & Co. Ltd., 1920.

Steven Weinberg, *Gravitation and cosmology: principles and applications of the general theory of relativity*, Wiley, 1972.

John A. Wheeler, *A Journey Into Gravity and Spacetime*, Scientific American Library, 1990.

Hans C. Ohanian and Remo Ruffini, *Gravitation and Spacetime*, W. W. Norton & Company, 1994.

S. James Gates Jr. and Cathie Pelletier, *Proving Einstein Right: The Daring Expeditions that Changed How We Look at the Universe*, Public Affairs, 2019.

CHAPTER 4

Atoms and nuclei

> Shortly after electrons were discovered it was thought that atoms
> were like little solar systems, made up of a ... nucleus and electrons,
> which went around in "orbits," much like the planets ... around the
> sun. If you think that's the way atoms are, then you're back in 1910.
>
> Richard P. Feynman, *The Strange Theory of Light and Matter*,
> 1985 [457]

YOU HAVE NOTED by now that I do not care too much about chronology. I rather care about the continuity or discontinuity of ideas as we follow our reductionist narrative. In this chapter we look at how classical physics arrived at the notion of atoms in the 18th and 19th centuries. And how Rutherford, Geiger and Marsden discovered their substructure.

The uncertainty about the ultimate nature of matter is of course much older than physics. Indian and Greek philosophers already disputed whether matter was granular or continuous, made of atoms or made of symmetries. But their theories were based on pure thinking, without much input from Nature itself. So we jump ahead to the beginning of the scientific study of matter, in physics and chemistry, in the 18th century.

4.1 ATOMISM

For many, scientific atomism starts with Daniel Bernoulli's book "Hydrodynamica" of 1738 [3]. The bulk of the book deals with the dynamics of fluids as the title says, based on the conservation of energy. This conservation law was then still far from established. In Chapter 10 of his book, Bernoulli introduced the kinetic theory of gases, describing them as made of microscopic rigid bodies in motion, which scatter elastically from one another and from the boundaries of the gas volume. He demonstrated that pressure is caused by their impacts on the boundaries, as we show in Focus Box 4.1. Thus heat was identified as the kinetic energy of the gas atoms or molecules.

Important steps along the way to scientific atomism are due to chemists. Prominent among them, Antoine Lavoisier introduced quantitative measures to chemistry in the late 18th century [6] and came up with a novel nomenclature for chemical elements [4]. Carefully weighing reactants and products of chemical reactions in sealed glassware, he established the conservation of mass in chemical reactions. He coined the term "elements" for substances which cannot be broken down further by chemical reactions. Lavoisier was also a political reformer and philanthropist. On false accusations, he was beheaded during the French revolution.

John Dalton, of modest origins and educated by Quakers, was appointed teacher of mathematics and natural philosophy in 1793 at the "New College" in Manchester, a dissenting academy not conforming to the Church of England. At the turn of the century he resigned and began a new career as a private tutor in Manchester, then a thriving city thanks to textile industry. Dalton was a versatile researcher interested in subjects from metrology to colour blindness. He was, however, often content with rough measurements, from which he drew bold conclusions. His important contribution to atomism is the law of multiple proportions in chemistry [8, 9]. Consider two substances which form more than one compound. If the same mass M of the first element combines with masses m_1 and m_2 of the second, then the law says that the ratio m_1/m_2 is a small integer. A good example are the oxides of carbon, CO and CO_2. If you know that atoms form molecules, it is clear that oxygen masses needed for the two molecules have a ratio of 1:2. If you do not know that, the observation will lead you, like Dalton, to become an atomist. He estimated relative atomic weights based on this law, but got almost all numbers wrong because of his lack of experimental rigour. Nevertheless his conclusions survived more rigorous measurements.

In 1808, at the same time when the first volume of Dalton's magnum opus "A New System of Chemical Philosophy" appeared, Joseph Louis Gay-Lussac found a novel feature of the chemistry of gases. Not only did definite proportions of weight combine to form compounds, but also definite proportions of volume did [10]. The explanation of this fact came with another revolution, initiated in 1811 by Conte Amedeo Avogadro [11], professor of physics at the University of Torino. His hypothesis that equal volumes of any gas at a given temperature and pressure always contain the same number of gas particles exactly fit both Dalton's law and Gay-Lussac's findings. To distinguish between atoms and molecules, Avogadro adopted terms including "molécule intégrante" (today molecule of a compound, like CO_2), "molécule constituante" (molecule of an element, like O_2), and "molécule élémentaire" (atom, like O)[1]. Avogadro's number, the number of molecules per mol of substance, is fixed by the international system of units[2] to exactly $N_A = 6.02214076 \times 10^{23}$ mol^{-1}. The role of this constant in gas pressure is explained in Focus Box 4.1.

The discovery of Brownian motion by the botanist Robert Brown in 1827 could have lent strong support to the atomist view, but passed almost

---

[1]See: https://www.brittanica.com/biography/Amedeo-Avogadro
[2]https://physics.nist.gov/cgi-bin/cuu/Value?na

Gases where only elastic scattering among molecules and with the confining boundaries occurs are called *ideal gases*. Noble gases are a particularly good example. We calculate their pressure as a function of temperature according to the kinetic theory. Imagine an average molecule bouncing off the container wall with an orthogonal component $\bar{v}_x$ of its velocity. When the gas is in equilibrium, half of the molecules will move with an average velocity component $+\bar{v}_x$, the other half with $-\bar{v}_x$. Only the former will impact the right wall.

The number of molecules which can impact on the boundary surface A within a total time t is the number moving right in a volume $A\bar{v}_x t$. It amounts to $\rho(A\bar{v}_x t)/2$, where the factor $1/2$ comes from the fact that only half of the total number density ρ moves right. Elastic scattering will reverse the orthogonal velocity component, as sketched in the figure. The change in velocity by the scatter is thus $2\bar{v}_x$. Consequently, the force that the average molecule with mass m exerts is $\delta F = m2\bar{v}_x/\delta t$, where δt is the average contact time during which the force acts.

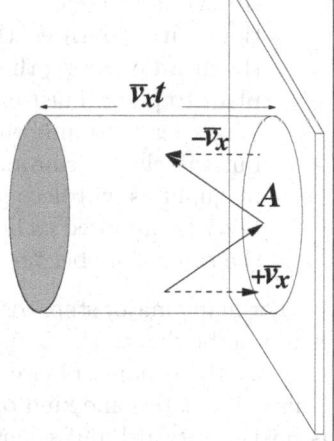

During that short time period, a fraction of $\delta t/t$ of the molecules moving right hits the wall. The total force per unit surface, the pressure, is thus $p = \rho m \bar{v}_x^2$. The average velocity in each direction is given by the law of equipartition, which states that every degree of freedom of the gas molecule on average has the same energy, $\bar{E}_x = kT/2$, where T is the temperature and k is the Boltzmann constant, with its modern value of 1.380649×10^{-23} J K^{-1}. The average orthogonal speed is thus $\bar{v}_x^2 = kT/m$, and the pressure is $p = \rho kT$. The number density ρ of the gas molecules is the number of moles, n, times Avogadro's number of molecules per mole, N_A, divided by the container volume V. With this we obtain the more familiar form of the ideal gas equation $pV = nN_A kT$. This equation contains no particular property of the gas except that a mole consists of N_A molecules, it is valid for any ideal gas. The combination $R = N_A k$ is thus rightly called the universal gas constant.

Focus Box 4.1: Pressure in ideal gases

unnoticed by physicists. Brown had observed random motion of pollen suspended in water [13, 14]. Finally in 1863 Christian Wiener [29] provided a kinetic theory of the phenomenon[3]. The pollen on the liquid surface are pushed around by water molecules. For that to happen, they must be small enough so that impacts in different directions do not average out, and light enough for the small forces to cause visible displacements. Jean Perrin [135] quotes work by Father Carbonelle SJ from the late 1870s to summarise this approach:

> In the case of a surface having a certain area, the molecular collisions of the liquid which cause the pressure, would not produce any perturbation of the suspended particles, because these, as a whole, urge the particles equally in all directions. But if the surface is of area less than is necessary to ensure the compensation of irregularities, there is no longer any ground for considering the mean pressure; the inequal pressures, continually varying from place to place, must be recognised, as the law of large numbers no longer leads to uniformity; and the resultant will not now be zero but will change continually in intensity and direction. Further, the inequalities will become more and more apparent the smaller the body is supposed to be, and in consequence the oscillations will at the same time become more and more brisk ...

For many major steps forward in science the parenthood is disputable. This may well be due to the fact that novel concepts do not just appear overnight, they are the product of circulating experimental results and theoretical ideas. This can create some kind of *l'air du temps*, an atmosphere in which progress is pushed forward and scientific consensus is formed [404]. In particular that seems to be the case for the discovery of periodicity in the sequence of chemical elements [403]. In the early 19th century, William Prout and Johann Wolfgang Döbereiner had noted that atomic weights appeared in rough multiples of the hydrogen mass. They formulated the "triad law": elements with similar chemical characteristics formed groups of three, the triads, in which the atomic weight of the middle element is the arithmetic mean of the lighter and heavier one. Examples of triads are the halogens chlorine, bromine, and iodine or the alkali metals lithium, sodium and potassium. These sequences were found to extend beyond triads in the first half of the 19th century.

In 1863, the English chemist John Newlands published a classification of the 56 then established elements in 11 different groups in a series of letters to the journal Chemical News [28, 30, 31, 33]. As the chief chemist of a sugar factory, he had little academic standing. When he finally read a paper to the Chemical Society in 1866, it was not taken seriously. One of the attendants even asked "whether he had ever examined the elements according to the order of their initial letters" [34]. Newlands certainly contributed to the

---

[3] For the history and theory of Brownian motion see: Edward Nelson, Dynamical Theories of Brownian Motion, https://web.math.princeton.edu/~nelson/books/bmotion.pdf, Princeton University Press (2001).

idea of the periodic system, if he counts among the fathers of the system itself is debatable. The break-through came in 1869 with the work of the Russian chemist Dmitri Ivanovich Mendeleev [35], which was clearer and more consistent than what had been published through the 1860s. Almost simultaneously, the German Lothar Meyer published similar findings [36]. Both had come across the periodicity in chemical properties as a function of atomic weight while preparing textbooks on chemistry. On the basis of empty places in their tables, Mendeleev and Meyer predicted further elements that had yet to be discovered. Mendeleev also discussed their properties. In 1882, both Meyer and Mendeleev received the Davy Medal from the Royal Society in recognition of their work on the periodic law. Figure 4.1 shows the oldest surviving example of a classroom periodic table from 1885, kept at the University of St. Andrews in Scotland and discovered in 2014 in a storage room of its School of Chemistry[4]. Today's table counts 118 elements, with all up to element 101, mendelevium, identified by chemical analysis. The discovery of plutonium Pu during research for nuclear weapons (see Chapter 6) started a race for the synthesis of man-made elements, which is still ongoing at major accelerator centres around the world, like LBNL in Berkeley, USA, JINR in Dubna, Russia, GSI in Darmstadt, Germany, and RIKEN in Wakō, Japan.

We do not want to jump ahead too far, but keep in mind that the chemical properties of elements are determined by the configuration of their electrons, thus by their nuclear charge, not their atomic mass. You will notice that an entire group is missing from the early table, the noble gasses. Helium was discovered by the astronomers Pierre Janssen and Joseph Norman Lockyer in 1868, identified by its characteristic spectral lines in the light from the Sun's chromosphere, only visible during eclipses. Four others, argon, krypton, xenon and neon, were discovered by the Scottish physical chemist Sir William Ramsay in collaboration with John William Strutt, Lord Rayleigh, who also established that they formed an entirely new group of elements, because of their chemical inertness. Both received a Nobel prize in 1904 to honour their discovery, Ramsay for chemistry, Rayleigh for physics.

So by the end of the 19th century, the atomic composition of matter was largely accepted. At that time, however, atoms were considered by many physicists to be purely hypothetical constructs, rather than real objects. An important turning point was Albert Einstein's (1905) [119] and Marian von Smoluchowski's (1906) [124] papers on Brownian motion, which succeeded in making accurate quantitative predictions based on the kinetic theory. We sketch the argument and Jean Perrin's verification of the result in Focus Box 4.2.

Why was Brownian motion more effective in fostering atomism than other arguments? I think that the cause may well be the direct visual observation. When observing the jitter of small suspended particles through a microscope one has a direct impression of the random impacts of molecules. This is a valid

---

[4]https://www.thesaint-online.com/2019/01/worlds-oldest-surviving-periodic-table-found-in-st-andrews/

Einstein treated the problem of Brownian motion as a diffusion process. He showed that the mean quadratic displacement $\overline{x^2}$ (in one dimension) of a small suspended particle of radius r, observed over a time period τ, is:

$$\overline{x^2} = \frac{RT}{3\pi\eta r N_A}\tau$$

R is the universal gas constant (see Focus Box 4.1), T the temperature, η the viscosity of the liquid and N_A Avogadro's number of molecules per mole. The relation can be used to determine Avogadro's number in a way independent of the ideal gas equation. A simple derivation in the Newtonian spirit of Focus Box 2.7 was given by Paul Langevin [125]. A suspended particle with mass m has the following one-dimensional equation of motion:

$$m\frac{d^2x}{dt^2} = F_x - f\frac{dx}{dt}$$

The thermal force F_x in the x direction works against the frictional force fv_x. We multiply this equation by x and obtain:

$$\frac{1}{2}m\frac{d}{dt}\left(\frac{dx^2}{dt}\right) - m\left(\frac{dx}{dt}\right)^2 = xF_x - \frac{1}{2}f\frac{dx^2}{dt}$$

When one takes the average of this equation over multiple observations of the same particle during time τ, the first term on the right averages to zero. The average of dx^2/dt is the mean quadratic deviation $\overline{x^2}$ per unit time, the second term on the left is twice the mean kinetic energy $kT/2$:

$$\frac{1}{2}m\frac{d}{dt}\left(\frac{\overline{x^2}}{\tau}\right) - kT = -\frac{1}{2}f\left(\frac{\overline{x^2}}{\tau}\right)$$

This is a differential equation for $\overline{x^2}/\tau$, a quantity which grows over time to reach an asymptotic value of:

$$\frac{1}{2}f\frac{\overline{x^2}}{\tau} = kT$$

Replacing $f = 6\pi\eta r$ according to Stokes Law (see Focus Box 2.7) and $k = R/N_A$, we obtain Einstein's formula quoted above. The linear relation between mean displacement and observation time was verified experimentally by Jean Perrin in 1909 [135], using a *camera lucida* to project and measure the motion of suspended particles. His experiment has been repeated in 2006 by Ronald Newburgh and collaborators [564] in a modern version using a CCD camera. Some systematic dependence on the particle size is observed for the obtained value of N_A, but the linear relationship is verified.

Focus Box 4.2: Brownian motion

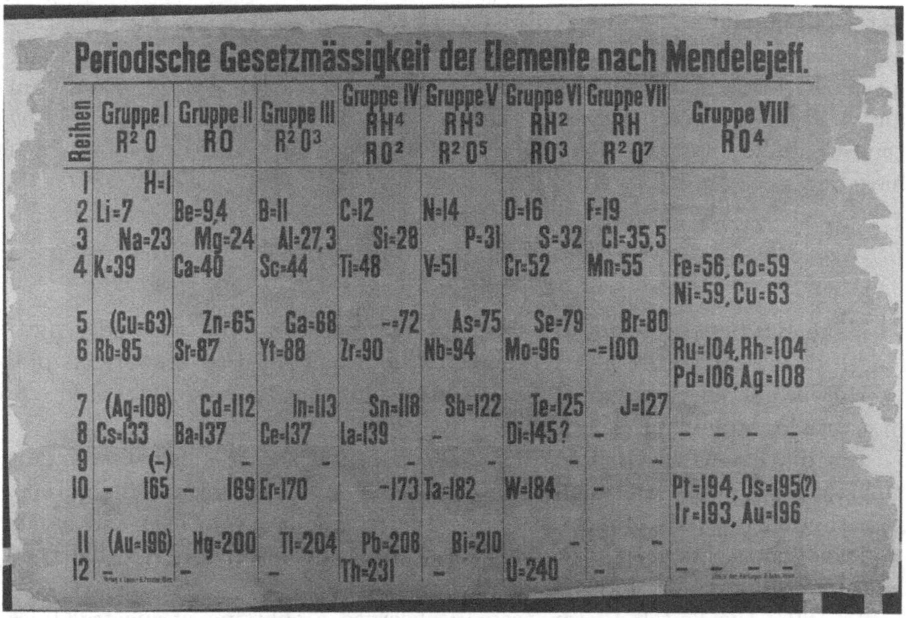

Figure 4.1 The oldest surviving classroom poster (according to the *Guinness Book of Records*) of Mendeleev's Periodic Table of Elements dated 1885, found at St Andrews University in Scotland in 2014. (Credit: University of St. Andrews Library Ms39012)

answer to Ernst Mach's famous (but undocumented) dictum to question the reality of atoms: "Have you seen one?" Or at least as close as one could get until the scanning tunnelling microscope (STM) was invented by Gerd Binnig and Heinrich Rohrer in 1986 [459].

4.2 ATOMIC SPECTRA

Another fact known about atoms in the second half of the 19th century was that they radiate light when heated. The emitted light can be observed through a prism as a series of coloured lines with dark spaces in between. Each element produces a unique set of spectral lines. Likewise, atoms absorb ultraviolet, visible or infrared light only for the wavelengths they emit. This fact was established by the Swedish physicist Anders Jonas Ångström in the 1850s [22]. In the 1860s, Ångström studied the emission spectrum of the Sun and identified more than 1000 spectral lines. Gustav Kirchhoff and Robert Bunsen [25], a physicist and a chemist, associated the wavelengths of lines to the elements that emitted them. It was thus found that four of the lines in the visible sunlight belong to the hydrogen spectrum [22], with wavelengths 410 nm, 434 nm, 486 nm, and 656 nm. These data were used by the Swiss mathematical physicist Johann Jakob Balmer to establish an empirical formula [46] for the hydrogen emission wavelengths: $\lambda = Bn^2/(n^2 - 4)$ with $B = 364.5$ nm and n an integer number greater than 2. The formula was later generalised by the Swede Johannes Rydberg [53] to all hydrogen-like ions, i.e. those which have one electron like to He^+, Li^{2+}, Be^{3+} and B^{4+}:

$$\frac{1}{\lambda} = R_H \left(\frac{1}{2^2} - \frac{1}{n^2} \right)$$

with the Rydberg constant $R_H = 4/B$. The typical atomic size and typical length of crystal bonds, $1\text{Å}=100\text{pm}=10^{-10}\text{m}$, was named to honour this pioneer of atomic spectroscopy. It is not an SI unit, but still regularly used e.g. in crystallography and chemistry.

In addition to the distinct lines, all elemental spectra also show bands at low wavelengths. And metals have especially rich spectra, with hundreds, sometimes thousands of narrowly spaced lines. The origin of lines and bands was not known, but atomists and especially J.J. Thomson believed that they had somethings to do with the electrons in matter.

In 1897, the Dutch Pieter Zeeman observed a splitting of spectral lines when a static external magnetic field is applied to the emitting substance [78, 79]. Sometimes lines split into two, sometimes into three separate weaker lines [77]. There were many attempts to explain this phenomenon, none really succeeded before the advent of quantum mechanics and the discovery of spin (see Chapter 5). The same is true for the analogous line splitting in electric fields, established by Johannes Stark [159] after initial observations by Woldemar Voigt [105].

4.3 ELECTRONS IN MATTER

We noted in Chapter 2 that soon after Thomson discovered the electron in 1897, he suggested that it was an ingredient of all matter. He conjectured this, since cathode rays were emitted regardless of the cathode material or the nature of the rest gas in his tubes. The convincing proof of this fact, however, had to wait for the study of direct interactions between light and atomic electrons, the photoelectric effect[5]. During his experiments on electromagnetic waves Heinrich Hertz had noticed that the intensity of sparks in his spark gap emitter (see Figure 2.4) was weaker when it was in the dark and stronger when illuminated [49]. His assistant Wilhelm Hallwachs observed that focussing ultraviolet light on zinc plate connected to a battery caused a current to flow [52]. Using differently pre-charged plates, he concluded that this was due to negative charges being emitted from the plate. J.J. Thomson showed in 1899 that these were the same as the "corpuscles" he had discovered [91]. Philipp Lenard followed up on these experiments using a powerful arc

---

[5]The history of photoelectric emission is discussed in detail in [476].

lamp to generate light of variable intensity. Using a set-up sketched in Focus Box 4.3, he measured the energy and intensity of the emitted electrons [106], even though he believed that electrons were immaterial. He found the astonishing fact that the electron energy had a fixed minimum, independent of the light intensity. He wrongly concluded that the electron energy was already present when they were bound in matter and that light only triggered their emission [432]. This stayed the generally preferred assumption until Millikan, using an arc lamp strong enough to allow spectral separation, established in 1914 that the electron kinetic energy was proportional to the wavelength of the light and used this to measure Planck's constant [157, 171] (see Chapter 5).

It was generally assumed at the time that if granular matter contained electrons, there would be a great many per atom. This guess was based on the small ratio of electron mass to atomic mass on one hand, and the rich atomic spectra on the other hand. The emission of heated substances displayed many spectral lines, which were assumed to come from the eigenfrequencies of electrons in atoms. For this to work, there would have to be hundreds per atom. A first atomic model inspired by astronomy was published in 1903 by Hantaro Nagaoka [115], a pioneer of modern Japanese physics in the Meiji era. His model was inspired by Maxwell's essay on the stability of Saturnian Rings [23], which won the Adams prize of the University of Cambridge for the year 1856. Maxwell had examined the possibility that the rings were solid, liquid or fragmented in nature and had concluded that mechanical stability of their motion required them to consist of many small fragments circulating Saturn independently. There was thus a tempting analogy to atomic models with a large positive charge in the centre surrounded by a densely populated electron ring, generically called Saturnian models. Except that electrons of course repel each other, creating a mechanical instability in the system, given the $1/r^2$ distance law for the Coulomb force. One can in principle balance the attractive central force by the repelling force among the ring particles. But the slightest deviation of one electron in the radial direction would make the ring collapse or explode. The latter event was conjectured to be the cause of radioactivity.

Clearly, electrons in the ring would have to be constantly accelerated by the central force to stay on their orbit. They would thus also lose energy due to electromagnetic radiation like all accelerated charges. Eventually they would fall into the positive central charge. However, for a very large number of electrons, this radiative instability is not very important. While individual circulating charges radiate, a continuous circular DC current does not. It was thus the mechanical instability, not the radiative one that was the main argument against Saturnian models.

The same year 1904, J.J. Thomson proposed a different model, in which positive charge and electrons would not be separated, but inhabit the atomic volume together. That would reduce the distance law to $1/r$, since only the positive charge within the electron ring radius would act according to Gauss' law. The title of his paper [116] resembles a modern restaurant menu, where

The experimental set-up used by Hertz, Hallwachs, Thomson, and Lenard to look into the photoelectric effect is sketched here. It consists of a discharge tube with a quartz window, which lets UV light from an arc lamp pass and hit the metallic cathode. This causes a current to flow if the energy of electrons liberated by the light is sufficient to reach the anode. The voltage between cathode and anode can be regulated to be positive as well as negative.

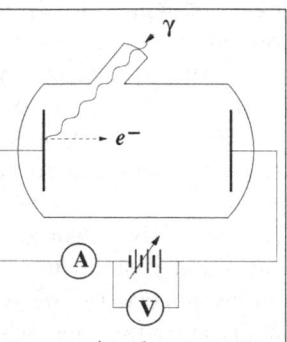

If the anode charge is negative, it will repel the electrons. At the negative stopping potential V_0, the current A will stop. In that case, the potential difference between cathode and anode times the electron charge, eV_0, is just equal to the kinetic energy E_{kin} of the electrons liberated from the cathode. For positive voltages, the photocurrent quickly saturates.

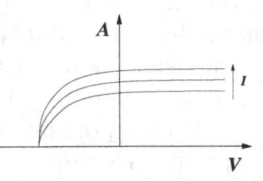

To the surprise of many, when the light intensity I is increased, the saturation current increases, but the stopping voltage does not change, as schematically shown by the VA-graph on the left. The electron kinetic energy is thus independent of the light intensity.

The stopping voltage only changes when varying the frequency ν of the light. It is in fact proportional to that frequency, or inversely proportional to the wavelength. The shorter the wavelength, the higher the kinetic energy of the liberated electrons. There is also a maximum wavelength above which no current flows irrespective of the voltage.

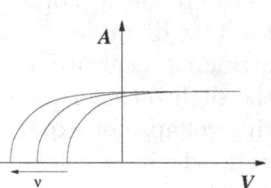

This cannot be understood by classical electrodynamics. There the energy of an electromagnetic wave is proportional to its intensity, thus the opposite should happen: the kinetic energy of kicked-out electrons should be proportional to the light intensity, the wavelength should have no influence. To understand the phenomenon, one needs to treat the beam of light as a collection of particles, photons, each with its own fixed energy (see Section 5.2).

Focus Box 4.3: Photoelectric emission

the name of the dish contains the whole recipe: "On the structure of the atom: an investigation of the stability and periods of oscillation of a number of corpuscles arranged at equal intervals around the circumference of a circle; with applications of the result to the theory of atomic structure." Thomson's model does not at all deserve the nick name "plum pudding model", which suggests a static scatter of electrons in a sticky, positively charged dough. Instead, his model consisted of an orderly distribution of corpuscles on rings, dynamically moving without friction inside a sphere of homogeneously distributed positive charge. The electron distribution was inspired by experiments of the American physicists Alfred M. Mayer in the later 1870s [43]. Mayer had studied the distribution of magnetic needles, made to float upright in water, under the influence of a strong magnet held above. He found that the needles arranged themselves in rings keeping regular distances, and suggested that this finding could have a bearing on atomic models. Thomson systematically studied the stability of electrons arranged in a ring as a function of their number[6]. He found that large numbers of electrons can only be stably accommodated in several concentric rings [116]:

> We have thus in the first place a sphere of uniform positive electrification, and inside this sphere a number of corpuscles arranged in a series of parallel rings, the number of corpuscles in a ring varying from ring to ring: each corpuscle is travelling at a high speed round the circumference of the ring in which it is situated and the rings are so arranged that those which contain a large number of corpuscles are near the surface of the sphere while those in which there are a smaller number of corpuscles are more in the inside.

Thomson's model, even though only calculable in a two-dimensional approximation, was the reference atomic model in the first decade of the 20th century. Max Born chose it as the subject of his 1909 habilitation lecture in Göttingen. The mechanical instability of Saturnian models was overcome, only the radiative one remained; it was taken as the cause of radioactivity. But the nature of the positively charged atomic substance remained mysterious. In any case, despite many efforts the Thomson model failed to explain atomic spectra.

4.4 NUCLEI

The nature of the positive charge in atoms was brought to light by Ernest Rutherford and his colleagues Hans Geiger and Ernest Marsden at the Physical Laboratories of the University of Manchester. After undergraduate studies in his native New Zealand, Rutherford joined Thomson's Cavendish laboratory in Cambridge and soon became its rising star. In 1898, he was nominated professor at McGill University in Montreal, Canada, where he started research

---

[6]A detailed translation of Thomson's calculation into modern terms is found in [595].

Rutherford and Soddy understood in 1903 that radioactive decay is a random process governed by the laws of probability [109]. The probability, that a given atom (or rather nucleus, as we know today) decays in the laps of time between t and $t+dt$, is a constant independent of t. That means that the nucleus has no memory, it does not "age". This fact is also valid e.g. for lotteries: the fact that a certain number has not been drawn for a number of weeks does *not* mean that it will be more likely to be drawn next week. In a well-implemented lottery, the probability for all numbers to be drawn is constant, and (hopefully!) the same.

All nuclei decay independently of each other. Thus out of a number $N(t)$ that exists at time t, a constant fraction dN/N will decay in the laps of time dt:

$$\frac{dN}{N} = \lambda\, dt \quad ; \quad N(t) = N(0)e^{-\lambda t}$$

The number of intact nuclei decreases exponentially with time, the sample ages, the atoms in it do not. The constant λ is called the decay constant or decay rate. It is a characteristic of each radioactive nucleus, its inverse is called the lifetime $\tau = 1/\lambda$ of the nucleus. Another often used measure is the half-life $t_{1/2} = \tau \ln 2$, the time at which half the nuclei have decayed.

If nuclei were not independent of each other, terms inversely proportional to $N(N-1)$, the number of pairs, or $N(N-1)(N-2)$, the number of triplets, etc. would have to be taken into account. If radioactivity were not a random process, the decaying fraction would not be constant. In a recent long term test, Dobrivoje Novković and collaborators have measured the decay rate of ^{198}Au, a radioactive isotope of gold, over more than two months [560], close to twenty times its lifetime. They found no deviation from the exponential decay law and λ to be constant within the statistical errors of their measurement.

Focus Box 4.4: Radioactive decay law

on radioactivity together with the radiochemist Frederick Soddy. They discovered several radioactive elements, including radium, and established the law that describes the timing properties of radioactive decay [109] (see Focus Box 4.4).

In 1907, Rutherford moved to the Victoria University of Manchester. He and Thomas Royds proved in 1909 that alpha particles are doubly ionised helium [129]. In parallel, Hans Geiger and Ernest Marsden started investigating the scattering of these particles by matter. They found that while the scattering angles were small in general, a tiny fraction of the alpha particles, of the order of one in several thousand, was scattered by large angles, even reflected backwards [127]. This is inconceivable, if the positive charge in an atom were smeared out over the atomic volume. The qualitative observation thus triggered Ernest Rutherford to propose a configuration of the atom where the positive charge is concentrated in a small nucleus at the centre of the atom [140], surrounded by only as many electrons as necessary to balance the central charge. The negative charge density of the electrons, small since they are few and take up the whole atomic volume, does not contribute to large angle scattering. He thus calculated the rate of scatters off the compact nucleus alone using classical electrodynamics. The results predicted the following features (see Focus Box 4.5):

- The number of deflected α particles per unit time, the scattering rate, is proportional to $1/\sin^4(\theta/2)$, where θ is the scattering angle. The rate of scatters thus decreases very rapidly with angle, but is still appreciable for angles larger than 90°.

- The rate is proportional to the thickness of scattering material for small thicknesses.

- It is proportional to the square of the central positive charge of the target atom.

- It is proportional to the inverse fourth power of the velocity of the incident α particles, i.e. inversely proportional to the square of their kinetic energy.

Guided by Rutherford's predictions, Geiger and Marsden perfected their experimental set-up (see Focus Box 4.6) and systematically tested these predictions. They measured the scattering rate as a function of angle for various target materials and kinetic energies of the α projectiles. Their quantitative experimental results [150] confirmed all of the above predictions concerning relative scattering rates. They were even able to obtain a rough result on the absolute scattering rate and established that the charge of the nucleus is about half the atomic number of the target. Their results clearly demonstrated that the positive nucleus is much smaller than the atom. The 5 MeV α particles approached the atomic centre to about 45fm in a head-on collision, a gold atom is about 3000 times larger, 135pm. The size of the gold nucleus is only

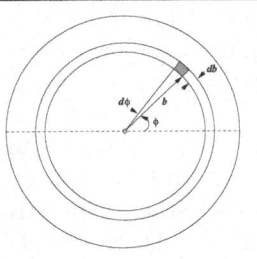

We calculate the fraction dI/I of the initial intensity I of projectiles with charge ze, scattered off a target which consists of nuclei with charge Ze. Classically, all particles which pass by with an impact parameter between b and $b+db$ are scattered by the same angle. The scattering angle θ depends only on this impact parameter, the transverse distance between projectile and target.

The scattering material is characterised by its density of atoms, ρ, and the foil thickness, Δx, which has to be small enough so that multiple scatters can be neglected. The fraction of the beam intensity which comes by a target particle is then $\rho\Delta x$. The scattered fraction is thus $dI/I = \rho\Delta x(2\pi b\, db)$.

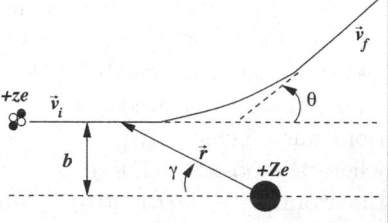

How does the scattering angle depend on b? We consider elastic scattering off a static target, the kinetic energy of the projectile is thus conserved and $|\vec{v}_i| = |\vec{v}_f| = v_0$. Newton's equation yields:

$$\vec{F} = m\frac{d\vec{v}}{dt} = m\frac{d\vec{v}}{d\gamma}\frac{d\gamma}{dt} = \frac{(ze)(Ze)}{4\pi\epsilon_0}\frac{\hat{r}}{r^2} \quad ; \quad \frac{d\vec{v}}{d\gamma} = \frac{zZe^2\hat{r}}{4\pi\epsilon_0 mr^2(d\gamma/dt)}$$

The angular velocity $d\gamma/dt$ can be deduced from the equally conserved angular momentum L:

$$L = mr^2\frac{d\gamma}{dt} = mv_0 b \quad ; \quad \frac{d\vec{v}}{d\gamma} = \frac{zZe^2\hat{r}}{4\pi\epsilon_0 L}$$

Integrating from initial to final state we have:

$$\int_{v_i}^{v_f} d\vec{v} = \frac{zZe^2}{4\pi\epsilon_0 L}\int_0^{\pi-\theta} \hat{r}d\gamma = \vec{v}_f - \vec{v}_i$$

The length of the vector on the left is $|\vec{v}_f - \vec{v}_i| = 2v_0\sin(\theta/2)$. The integral over $d\gamma$ gives a vector of length $2\cos(\theta/2)$, also in the direction $(\vec{v}_f - \vec{v}_i)$. Substituting this result we obtain the relation between the impact parameter b and the scattering angle θ:

$$b = \frac{zZe^2}{4\pi\epsilon_0 mv_0^2}\cot\frac{\theta}{2}$$

The scattered fraction of the beam intensity is thus:

$$\frac{dI}{d\Omega} = I\rho\Delta x\left(\frac{zZe^2}{4\pi\epsilon_0 2E_{kin}}\right)^2\frac{1}{\sin^4\frac{\theta}{2}}$$

where $d\Omega = \sin\theta\, d\theta\, d\phi$ is the solid angle subtended by the observer.

Focus Box 4.5: Classical electromagnetic scattering off a heavy nucleus

about 7fm, so the projectiles did not touch it, and the Coulomb force is the only one acting.

So far so good. Matter consists of atoms. Electrons somehow surround their compact nucleus without mechanical instability nor radiative collapse. But how do they manage to do that? And how do the positive charges in the nucleus clump together in a minute volume, vanquishing Coulomb repulsion? Classical physics has no answer to these questions. But quantum mechanics does.

FURTHER READING

Joseph John Thomson, *Recollections and Reflections*, G. Bell & Sons Ltd., London 1937; republished by Cambridge University Press, 2011.

Bernard Pullman, *The Atom in the History of Human Thought*, Oxford University Press, 1998.

Eric Scerri, *The Periodic Table, its Story and its Significance*, Oxford University Press, 2007.

Richard Reeves, *A Force of Nature: The Frontier Genius of Ernest Rutherford*, W. W. Norton, 2008.

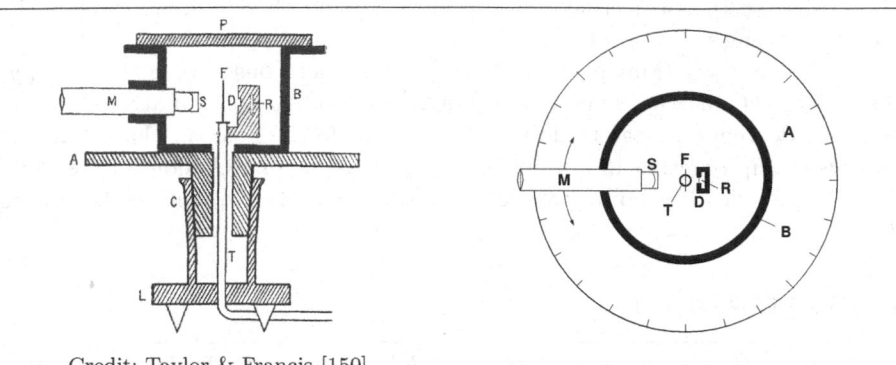

Credit: Taylor & Francis [150]

The Geiger and Marsden set-up shown above in a cut [150] and top view was enclosed in a cylindrical metal box B, sealed by the top cover P. It contained the source of α particles R, the scattering foil F, and a microscope M to which the zincsulphide screen S was rigidly attached. The box was fastened down to a graduated circular platform A, which could be rotated by means of the airtight joint C. By rotating the platform, the box and microscope moved with it, while the scattering foil and radiating source, attached to the tube T and fastened to the stand L, remained in position. The box could be evacuated through the tube. The radiation from source R could be attenuated by the diaphragm D, both in kinetic energy and in intensity.

Geiger and Marsden used several radioactive substances made available by Rutherford to measure the scattering rate off materials ranging from carbon to gold. Most of their measurements were made with "radium emanation" as a source, i.e. radon, ^{222}Rn, and its decay product polonium, ^{218}Po. Both emit α particles with a kinetic energy of about 5 MeV.

The number of impacts per unit time was counted on the phosphorescent screen which emitted a flash of visible light when hit by an alpha particle. This could be reliably done by eye for impact rates between 5 and 90 per minute. For small scattering angles, the microscope was thus pulled further out to decrease the solid angle $d\Omega$, for large angles it was pushed closer to the target to increase it. Also observations at small angles were done when the sources had weakened according to the decay law. The measurements obviously were corrected for these variations in solid angle and source strength. The rates were also corrected for stray β rays from the source, measured in absence of a foil.

The scattering rates were measured for angles between 5° and 150°, and target materials varying in thickness by a factor of 9 and in material from carbon ($Z = 6$) to gold ($Z = 79$). The kinetic energy of the α projectiles was degraded by inserting mica foils between source and target. All of Rutherford's predictions for the relative variation of the scattering rate as a function of these parameters were verified.

Focus Box 4.6: The experiment of Geiger and Marsden

CHAPTER 5

Quanta

> The beauty and clearness of the dynamical theory, which asserts heat and light to be modes of motion, is at present obscured by two clouds. I. The first came into existence with the undulatory theory of light ...; it involved the question, how could the earth move through an elastic solid, such as essentially is the luminiferous ether? II. The second is the Maxwell-Boltzmann doctrine regarding the partition of energy.
>
> Lord Kelvin, *Nineteenth Century Clouds over the Dynamical Theory of Heat and Light*, 1901 [103]

I N THIS CHAPTER we see how quantum mechanics kicks in when matter and forces are described at small distances. How perspective is changed by introducing quantisation of energy. How quanta move through space-time in the form of granular waves. What the difficulties were and still are in interpreting quantum mechanics.

There is no other way to introduce this subject than by considerations about heat and radiation. Thermodynamics is not exactly my favorite subject. The reason may be the proliferation of macroscopic notions, like entropy, enthalpy and other 'natural' state variables, which seem to defy reductionism. Where others see the emergence of novel features, when a system goes from few to many ingredients (like e.g. Ilya Prigogine [498] and his followers), I only see statistics at work. But for the same reason, thermodynamics had a crucial role in the development of quantum mechanics. In the beginning of the 20th century, the Right. Hon. Lord Kelvin (William Thomson, not to be confused with J.J.) identified two "clouds" hanging over the kinetic theory of heat, named in the initial quote of this chapter. The first "cloud" has been dealt with by Einstein's relativity theory in 1905 (see Chapter 3). His second "cloud" has been dealt with by Max Planck and Einstein in the same period. It refers to the Maxwell-Boltzmann distribution of the energy of molecules in an ideal gas in thermal equilibrium (see Focus Box 5.1). It turns out that their distribution law is one of a class of laws describing the energy distribution in thermal equilibrium. And what Lord Kelvin really meant is the challenge to understand why and how that equilibrium is reached.

5.1   HEAT

The main findings of 19th century thermodynamics can be summarised in three fundamental laws, which describe thermal systems in equilibrium:

The *first law* is an application of the conservation of energy, well known from the laws of motion (see Focus Box 2.2). It is based on the idea that as heat produces work (like in a steam engine), so should work be able to produce heat. The idea that heat and work are both forms of energy was proposed by Julius Robert von Mayer [15] in 1842. Independently James Prescott Joule [16] published his findings about heat produced by mechanical and electromagnetic work in 1843. In 1850 both published a numerical value for the mechanical heat equivalent, with the more reliable number coming from Joule [18]. There was a lengthy fight for precedence of these findings, but that does not need to concern us here. Joule's name was chosen for the energy unit of the SI, $J = Nm = kg\,m^2/s^2$.

What does concern us is the fact that conservation of energy requires that one can increase the internal energy U of a thermal system in two ways: by putting an amount Q of heat into it, or by applying an amount W of work. Conversely, the internal energy decreases, when heat or work is extracted from the system. In summary: $U = Q - W$, where the sign of W indicates that work is done by the system, which is normally what one wants.

The *second law* is concerned with the way that thermal equilibrium is reached. Imagine two thermal systems both in equilibrium, but not at the same temperature. When they are brought into contact, heat will naturally be transferred from the hot one to the colder one, not vice versa. Heat transfer will continue until both reach the same temperature. The temperatures of the two system do not drift apart, they drift towards a common one. The contact of the two systems may involve mixing, i.e. exchange of material, or just thermal contact. The law thus formalises the fact that thermal systems tend towards spatial homogeneity of matter, energy and temperature. The pilot of this irreversible action was called entropy by the German physicist Rudolf Clausius.

In the 1850s and 1860s, he argued that a change occurs not only in the system that is worked on, but also in the working engine. There is an inherent loss of "usable" heat when work is done [17]. Clausius described entropy as the dissipative energy use of a thermodynamic system or working body during a change of state. This is summarised in the fundamental thermodynamic relation:

$$dU = T\,dS - p\,dV$$

It relates the change of internal energy U to temperature T and change of entropy S, in addition to the mechanical work, $-p\,dV$, done by expanding the volume V at constant pressure p, e.g. by moving a piston. The minus sign indicates that the system is providing work, not receiving it. What entropy really is was finally found by Ludwig Boltzmann in the 1870s [63]. He interpreted entropy in the framework of statistical thermodynamics. If there are N different ways to form a system's state from microscopic ingredients, then $S = k \log N$ is the entropy of the state, where k is the Boltzmann constant we met earlier (see Focus Box 4.1). The number of realisations N can be seen as a probability of the macroscopic state: the more ways there are to realise it, the more likely it becomes. That entropy rises in systems which are left to evolve freely is thus simply due to the fact that they evolve from less probable to more probable states. In isolated systems the change of entropy between initial and final state is thus always positive and the evolution is irreversible. If the system is not isolated, then the sum of the entropies, system plus surroundings, increases.

The *third law* of thermodynamics finally takes care of the temperature scale. If the temperature of the macroscopic system is proportional to the average kinetic energy of its microscopic components, then there must be a minimum to temperature, $T = 0$ in temperature units of Kelvin. However, this also means $\lim_{T \to 0} S = 0$, such that this temperature can be approached, but not actually reached. The third law was formulated by the physical chemist Walter Nernst in 1905.

In a classical ideal gas with massive particles, the energy of the particles is distributed according to a Maxwell-Boltzmann distribution. Equilibrium is established by the particles themselves: they collide with each other, exchanging energy and momentum in the process. Maxwell [26, 27] had based his derivation of the energy distribution on a Newtonian approach to the kinetic theory, as we did in Focus Box 4.1. Ludwig Boltzmann [37, 42] gave it firm grounds in statistical thermodynamics. The Maxwell-Boltzmann distribution is an approximation for more general ones in the case of high temperatures and low densities (see Focus Box 5.1).

Another member of the family of equilibrium distributions is Planck's law of black body radiation, a special case for massless bosons which do not interact with each other at all. Its derivation by Planck, and its connection by Einstein to the properties of atoms, triggered the quantum revolution of the first decade of the 20th century.

A black body is an ideal substance which emits and absorbs all light, i.e. electromagnetic radiation in all frequencies. The corresponding radiation is indeed a perfect example of a thermodynamic system. First experimental approaches to measuring the energy distribution of this radiation, in the infrared for realistic temperatures, used roughened or blackened metal surfaces to realise a black body at least approximately. It was then realised that any enclosure of fixed temperature, which was impenetrable to electromagnetic radiation, contains a black body spectrum of light. Gustav Kirchhoff wrote in 1860 [24]: "When a volume is enclosed in bodies of equal temperature and when radiation cannot penetrate these bodies, then every bundle of rays in this volume has the same quality and intensity, as if it were emitted by a black body of the same temperature, it is thus independent of the nature and shape of the bodies and only depends on the temperature." In modern language, this statement means that black body radiation is a gas of photons. Since photons do not interact with each other for lack of electric charge, equilibrium is established by interaction with the walls of the cavity, the interaction between photons and matter. The temperature of the walls thus determines the temperature of the photon gas.

Measurements of the spectrum of black body radiation [520] require three ingredients: a black body constructed according to this recipe; a detector able to measure the intensity of radiation in a wide range of frequencies; and an analysing device capable of separating frequencies of light in the infrared. All three were developed in the 1890s at the Physikalisch-Technische Reichsanstalt (PTR) in Berlin[1], a research institute without teaching mission founded on the initiative of Werner von Siemens and Hermann von Helmholtz, which combined free fundamental research with services for industry. The industry tycoon von Siemens ceded private land in Berlin-Charlottenburg to the Reichsanstalt, Helmholtz became its first president. The motivation for research on black body radiation was the search for a reliable device to gauge light sources, a standard candle, driven by the increasing competition between gas light and electric light.

A black body cavity was successfully constructed by Otto Lummer, Ferdinand Kurlbaum and Ernst Pringsheim at PTR in the late 1890s. It consisted of an electrically heated platinum cylinder, blackened on the inside[2], enclosed in a second insulating cylinder. Radiation could leave the black body through diaphragms at the end of the cylinder.

A detector for a wide spectrum of light, the bolometer, was invented in 1878 by the American astronomer Samuel P. Langley. It is essentially a very sensitive calorimeter (see Section 8.3). The sensitive part consists of a thin strip of material with low heat capacity like platinum, such that small radiation intensities warm it up as much as possible. Its temperature is measured through its thermal resistance, compensated with a Wheatstone bridge. At PTR, Lummer and Kurlbaum [55] refined the bolometer such that it had an impressive resolution and stability, capable of measuring temperature differences of 10^{-7} degrees in a short exposure.

Spectroscopic separation of wavelengths was achieved with specialised prisms and lenses made of halite, fluorite and sylvine, which transmit light efficiently up to far infrared wavelengths. Combined with a bolometer with a narrow sensitive strip, one thus obtains a so-called spectrobolometer, apt to make precision measurements of the black body spectrum up to high temperatures.

In 1893, Wilhelm Wien derived a first law for the spectral radiance of a black body [57], i.e. the power per emitting area, solid angle and frequency

---

[1] See https://www.ptb.de/cms/en/about-us-careers/about-us/history-of-ptr-and-ptb.html

[2] Black surfaces were at the time obtained by using carbon pigments, typically lamp black produced by collecting soot from oil lamps. Modern technology based on carbon nanotubes obtains a record absorptivity of 99.995% from ultraviolet to terahertz frequencies [665].

range, it is quoted in Focus Box 5.1. From thermodynamical considerations he concluded that the wavelength at the maximum of the spectrum depended linearly on temperature. This relation is called Wien's displacement law.

5.2 QUANTUM LIGHT

Lummer and Pringsheim [87] at PTR found that Wien's law fitted their observed spectrum rather well between $T \simeq 800K$ and 1400K and wavelengths between 1μm and 6μm, but also observed systematic deviations. Enlarging the temperature and wavelength range to 1650K and 8.3μm, respectively, the deviations did not go away [88]. Figure 5.1 shows these results. The observed spectrum lies systematically above Wien's law for long wavelengths and high temperatures. However, data obtained by Friedrich Paschen and Hans Wanner supported Wien's law [90, 89, 100] in the visible and near infrared range, for low temperatures.

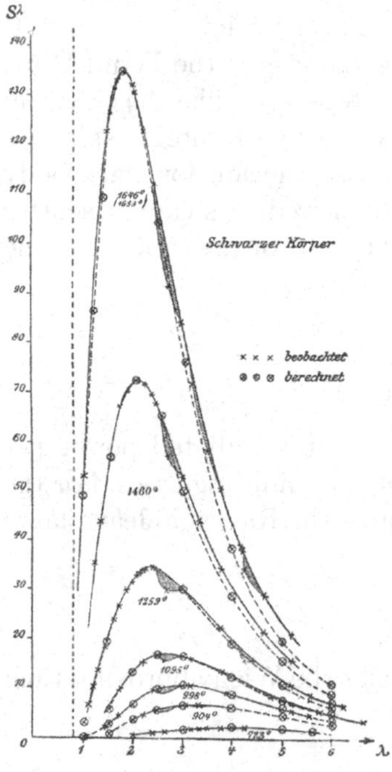

Figure 5.1 The spectral radiance of a black body as a function of wavelength for different temperatures [88]. The values measured by Lummer and Pringsheim (solid line) are compared to Wien's law (dashed line). Systematic deviations are observed for long wavelength and high temperatures. (Credit: Verhandlungen der DPG, 1899)

A breakthrough came in 1900 [188], when Lummer and Pringsheim obtained a sylvine prism (potassium chloride KCl) and extended the wavelength range of their experiments to long wavelengths, 12 to 18μm, and even higher temperatures, exceeding 1770K [94]. The wavelength range was further extended by Heinrich Rubens and Kurlbaum [101] using the so-called reststrahlen (residual ray) effect. Their method profited from the fact that certain materials totally reflect a certain narrow band of wavelengths. It allowed

A distribution law describes the way that a measurable quantity, velocity, energy, wavelength or frequency, is distributed among the microscopic ingredients of a thermodynamic system. The archetype is the Maxwell-Boltzmann distribution of the energy of rigid particles, scattering elastically off each other. The probability f of finding a particle with energy between ϵ and $\epsilon + d\epsilon$ in a given degree of freedom of the particles is:

$$f(\epsilon, T)d\epsilon = \sqrt{\frac{1}{\pi \epsilon k T}} e^{-\frac{\epsilon}{kT}} d\epsilon$$

Here T is the temperature and k denotes Boltzmann's constant. Note that this is a marginal probability, applicable to this particular degree of freedom when any energy is admitted for the others. All other thermal energy distributions have the Maxwell-Boltzmann distribution as their asymptote if the temperature is sufficiently high and the density sufficiently low.

If these conditions are not met, one finds two general cases: the Fermi-Dirac distribution for fermions, i.e. particles with half-integer spin like the electron; the Bose-Einstein distribution for bosons, i.e. particles with integer spin like the photon. The latter distribution has a special case relevant for black body radiation. Since that consists of a photon gas, the ingredients cannot scatter off each other. If there is interaction only with the walls of the container, one obtains Planck's radiation law:

$$B(\lambda, T) = \frac{2hc^2}{\lambda^5} \frac{1}{e^{\frac{hc}{\lambda k T}} - 1}$$

It describes the spectral radiance B of the gas, i.e. the radiated power per unit area, solid angle and wavelength. The law has two limiting cases. For the limit of low frequencies, Planck's law tends towards the Rayleigh-Jeans law:

$$B_{RJ}(\lambda, T) = \frac{2ckT}{\lambda^4}$$

In the limit of high frequencies Planck's law contains the Wien approximation:

$$B_W(\lambda, T) = \frac{2hc^2}{\lambda^5} e^{-\frac{hc}{\lambda k T}}$$

The maximum of this spectrum is inversely proportional to temperature, $\lambda_{max} = \frac{0.288}{T} \text{cm}$, for T in Kelvin, a fact which is also called Wien's displacement law.

Focus Box 5.1: Dictionary of radiation laws

a further extension of the measurements up to more than $50\mu m$ wavelengths. In the meantime Lord Rayleigh had "ventured to suggest a modification" [98] to Wien's law, suggesting that the spectral radiance be proportional to temperature at long wavelengths. This was indeed in agreement with Rubens' and Kurlbaum's results.

Meanwhile, Max Planck followed a completely different scientific program. In an article describing the winding road to his radiation law [298], he states:."What has always interested me in physics were the great universal laws, which are important for all natural processes, regardless of the properties of intervening bodies, and of the ideas one forms about their structure." He therefore took a special interest in the laws of thermodynamics. Again according to his own judgement, he was particularly inspired by the writings of Clausius, "due to their excellent clarity and persuasiveness of language." Entropy became the subject of his doctoral thesis [44] in Munich in 1879. After a first appointment in Kiel, in 1889 he was named the successor to Kirchhoff at the Friedrich-Wilhelms-Universität in Berlin. Starting in 1895, he published a series of papers in the reports of the Berlin Academy, documenting the progress of his work. The aim was to prove the second law of thermodynamics, based on the kinetic theory of heat. His working model was a black body cavity, with harmonic oscillators inside. He managed to prove Wien's approximation this way, but further progress was limited. When he learned about Lummer's and Pringsheim's findings of deviations from Wien's law of radiation and Rubens told him about the long wavelength behaviour of black body radiation, he approached the problem again starting with entropy (see Focus Box 5.2). That led him to an empirical radiation law interpolating between Wien's approximation and Rayleigh's conjecture. But how could the constants appearing in the law be calculated based on first principles?

This problem made him surmount his aversion to statistical thermodynamics and use Boltzmann's approach, which formulates entropy as a measure of the probability of a macroscopic state in terms of microscopic ingredients. Following this line, as sketched in Focus Box 5.2 he found that there was indeed a finite number of small portions of energy, quanta, which make up the internal energy of the system. The energy of a light beam is thus not continuous, but comes in quanta of $h\nu$, proportional to the frequency ν of the light with Planck's constant h intervening. He presented his findings to the German physical society in December of 1900 [97], as shown in the protocol of their session, Figure 5.2. It was published in 1901 [104].

This marks the birthdate of quantum physics. Arnold Sommerfeld [194, p. 36] wrote in hindsight: "The quantum theory is a product of the twentieth century. It came to life on 14th December, 1900." But according to Helge Kragh [521], "if a revolution occurred in physics in December 1900, nobody seemed to notice." The real importance of energy quantisation only came to light with Einstein's interpretation of the photoelectric effect [120], which linked black body radiation to matter properties.

The second law of thermodynamics, in the absence of external work, states how entropy varies with internal energy, $dS/dU = 1/T$. Proportionality of the energy with temperature, as suggested by Rayleigh, thus meant $d^2S/dU^2 = \text{const}/U^2$. Wien's approximation for short wavelengths, on the other hand, means that $d^2S/dU^2 = \text{const}/U$. A radiation law interpolating the two would have:

$$\frac{d^2S}{dU^2} = -\frac{\text{const}}{U(U+\text{const})} \quad \rightarrow \quad B(\lambda, T) = \frac{c_1}{\lambda^3}\frac{1}{e^{\frac{c_2}{\lambda T}} - 1}$$

Planck communicated this result to Rubens the same day on a post card [188]. It described the experimental data very well, especially improved later measurements [189], but it was not based on first principles, as Planck intended. He had so far ignored Boltzmann's statistical interpretation of entropy, but was now forced to use his approach, "since no other way out opened" [298]. This required counting the microstates that made the distribution. Instead of only one oscillator inside the cavity, he now based his calculation on a large number n of oscillators. The total internal energy U is also decomposed into p equal portions ϵ. Boltzmann's number N of microstates forming the macroscopic state can then be counted by combinatorics:

$$N = \frac{(p+n)!}{p!\,n!}$$

The interpolating radiation law comes out for $\epsilon = h\nu$, such that $c_1 = 2hc^2$ and $c_2 = hc/2\pi k$ with Boltzmann's constant k and the speed of light c. Planck's constant h has the dimension of a product of energy and time, he thus named it a quantum of action. The energy of the radiation in a black body is thus not continuous, it is quantised in tiny portions proportional to the frequency of the radiation.

Focus Box 5.2: Planck's radiation law and entropy

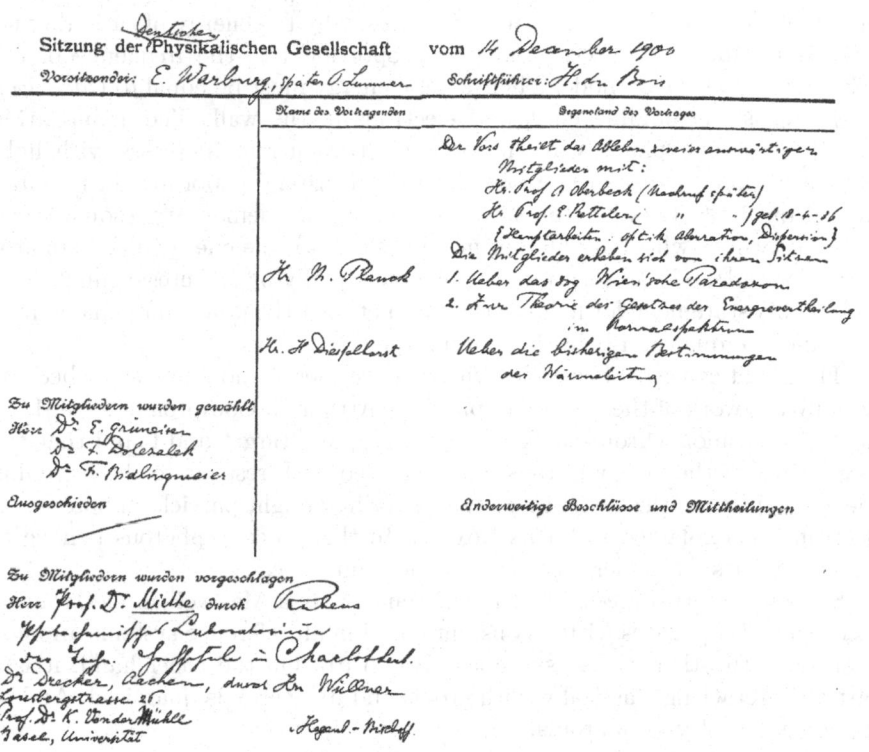

Figure 5.2 A copy of the protocol for the session of the German Physical Society DPG on December 14, 1900 [517]. It records two contributions by Max Planck, "About the so-called Wien paradox", and "On the theory of the energy distribution law in the normal spectrum." (Credit: Archiv der Deutschen Physikalischen Gesellschaft, Signatur 10008)

Einstein's heuristic argument goes as follows. If the light inside the black body cavity cannot interact with itself, it must be the reflection at the walls which establishes the thermal equilibrium. According to Planck's law, the energy stored in the light comes in quanta, tiny portions which can be counted. In Einstein's words [120]: "the energy of a light beam sent out from a point is not distributed to larger and larger volumes, but it consists of a finite number of energy quanta, localised in spatial points, which move without splitting and can only be absorbed or created as a whole." This means that the atoms in the walls can emit or absorb quanta of a definite energy. In his seminal paper Einstein first demonstrates Planck's radiation law based on this evidence. He then makes the connection to the photoelectric effect (see Focus Box 4.3). There, the same principle applies: "Quanta penetrate into the surface layer of the [cathode] body, and their energy is transmitted at least in part into kinetic energy of electrons." This obviously explains, why the energy of cathode rays emitted by the photoelectric effect is proportional to the frequency of the light shining on the cathode. Their kinetic energy must be equal to $(h\nu - E_b)$, where E_b is the binding energy of electrons in the wall. The quantisation hypothesis also explains why the photoelectric current increases with light intensity: more quanta eject more electrons. Einstein elaborates on this in a paper from 1906 [122]. He states[3]: "The energy of an elementary resonator can only take values which are integer multiples of $[h\nu]$; the energy of a resonator changes in a leap by absorption or emission, equally by an integer multiple of $[h\nu]$." This statement formulates the link between the quantum behaviour of light and the quantum behaviour of matter.

That light consists of particles with definite energy and momentum became clear by the work of the American physicist Arthur H. Compton in 1923 [192]. He scattered monochromatic X-rays off a graphite target and found that the wavelength of the outgoing rays was increased with respect to the incoming one. He explained this as an inelastic scattering of light particles, photons, off electrons, as explained in Focus Box 5.3. In this process, photons behave as single particles with energy $h\nu$ and momentum $h\nu/c$.

So light has two faces. On the one hand, it is a Maxwellian electromagnetic wave, propagates with a constant speed in all reference systems and can interfere with other waves (see Focus Box 5.10). On the other hand, it is a particle interacting elastically with atoms and its energy is quantised. And so are the energy levels of atoms.

5.3 QUANTUM ATOMS

Rutherford had refrained from formulating a theory of how electrons fill the atomic volume around the compact nucleus, only making vague reference to saturnian models à la Nagaoka. Early endeavours to add small numbers of electrons to a central nucleus using Planck's quanta, e.g. by Arthur Erich Haas [133] and John William Nicholson [143], failed since they were stuck with the idea that frequencies of revolution or axial vibration had to be proportional to the frequencies of the emitted light.

In came Niels Bohr. He received his doctorate at the University of Copenhagen in 1911 with a thesis on the electron theory of metals, which followed and improved on ideas of J.J. Thomson, treating metals essentially as a gas of quasi-free electrons. Endowed with a fellowship from the Carlsberg foundation, he went to Cambridge in September of 1911 and joined Cavendish laboratory, with the idea of translating his thesis into English and profit from advice by Thomson. His state of mind is rather well documented in numerous letters, mostly to his fiancée, future wife and lifetime love Margarethe Nørlund [594].

---

[3]In the spirit of this book I have converted Einstein's notation into the modern one for Planck's quantum of action h.

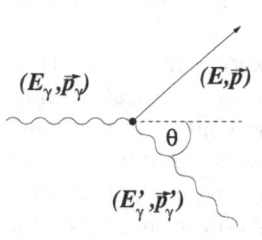

In the early 1920s, Arthur Compton scattered monochromatic X-rays off electrons. The kinematics of the process is shown on the left. The incoming photon has an energy of $E_\gamma = h\nu$ and a momentum $p_\gamma = h\nu/c$. The outgoing one leaves under an angle θ with $E'_\gamma = h\nu'$ and $p'_\gamma = h\nu'/c$. The outgoing energy is not the same as the incoming because of the electron recoil.

The electron (mass m) is initially at rest, its momentum is negligible compared to that of the photon. It leaves with energy E and momentum $\vec{p}$. Energy conservation in the process means:

$$E_\gamma + mc^2 = E + E'_\gamma \quad ; \quad h\nu + mc^2 = E + h\nu'$$
$$E^2 = m^2c^4 + \vec{p}^2c^2 \quad ; \quad p^2c^2 = (h\nu - h\nu' + m)^2 - m^2c^4$$

Momentum conservation means:

$$\vec{p} = \vec{p}_\gamma - \vec{p}'_\gamma \quad ; \quad p^2 = \left(\frac{h\nu}{c}\right)^2 + \left(\frac{h\nu'}{c}\right)^2 - 2\frac{h^2}{c^2}\nu\nu' \cos\theta$$

Using $\nu\lambda = \nu'\lambda' = c$, we find for the loss in frequency and the increase in wavelength:

$$\frac{\nu - \nu'}{\nu\nu'} = \frac{h}{mc^2}(1 - \cos\theta) \quad ; \quad \lambda' - \lambda = \frac{h}{mc}(1 - \cos\theta)$$

Focus Box 5.3: Compton scattering

Thomson listened to him politely but was uninterested to revisit his former subject of study. John Heilbron [430] identifies their difference in working style as one of the reasons. Bohr needed lengthy discussions to elaborate his ideas, often ending in monologue, while Thomson hardly involved even his close collaborators in the formation of his thinking. Thus Bohr accepted an invitation to Manchester in March of 1912 to join Rutherford's laboratory. There, radioactivity was the subject of the period. Bohr took courses and was assigned laboratory work, which he diligently executed. According to Michael Eckert [598], his attention was drawn to atomic theory by a study of multiple Coulomb scattering and energy loss of α particles when traversing matter [149].

In the summer of that year, Bohr drew up what is known as the "Rutherford Memorandum". It mainly addressed the question of mechanical stability à la Thomson, ignoring radiative instability. But it also contained first hints at his atomic model, introducing the "special hypothesis" that the kinetic energy of an electron be proportional to its revolution frequency. Bohr also thought he had a solution for the problem of periodicity in atoms, as exhibited by Mendeleev's table [430]. The rest of the memorandum addressed simple molecules and chemical binding. He did not argue in terms of quanta, but the work of Nicholson put him on the track. In March of 1913 he sent the first part of his famous 1913 trilogy of papers [146, 147, 148] to Rutherford, in which he set out what we call today Bohr's atomic model. It boldly left behind the classical treatment of the problem by what one can call heuristic axioms:

- Not all classical orbits are available to the electron, but only those called stationary, where it does not radiate electromagnetic energy. The stationary states can be numbered by a principle quantum number n, $n = 1$ denotes the ground state of the atom. A specific energy E_n belongs to each such state.

- The electron can jump from one stationary state i to another j. This "quantum leap" is outside the realm of classical physics. When the electron undergoes this process, energy is conserved by emitting or absorbing an electromagnetic wave, with a frequency ν such that its energy, $h\nu = E_i - E_j$, corresponds to the energy difference between the two stationary states.

It is thus not the frequency or energy of the orbiting electron, but the energy difference between the initial and the final stationary states which determines the frequency of the light. These hypotheses lead to the Rydberg formula for the Balmer series when one makes one of two additional assumptions. One hypothesis, sometimes called the third Bohr postulate, requires that in addition to the energy, also the angular momentum of the electron's orbit comes in quanta (see Focus Box 5.4). The other and more far reaching hypothesis is Bohr's correspondence principle [605]. Roughly speaking it requires that for large principle quantum numbers, the orbit tends towards a classical one [183].

What Bohr exactly meant by this is debatable, but the general requirement that a new theory must have its successful predecessors as a limiting case is not.

Bohr's model of the hydrogen atom put a heavy point-like nucleus with charge $+e$ in the centre, surrounded by orbiting electrons of charge $-e$ and mass m. Classically, the Coulomb force keeps them on an orbit of radius r:

$$\frac{e^2}{4\pi\epsilon_0}\frac{1}{r^2} = \frac{mv^2}{r}$$

Multiplying this equation with mr^3, the right side of the equation becomes the modulus of the electron's angular momentum $\vec{L} = \vec{r} \times \vec{p}$:

$$\frac{e^2}{4\pi\epsilon_0}mr = (mvr)^2 = L^2$$

Applying the quantisation requirement that L can only take values which are a multiple of $\hbar = h/(2\pi)$, one obtains a value for the radii of stationary orbits:

$$r_n = n^2\frac{4\pi\epsilon_0\hbar^2}{me^2}$$

The total energy of the stationary electron can be calculated using the virial theorem:

$$E_n = E_{pot} + E_{kin} = \frac{1}{2}E_{pot} = -\frac{1}{2}\frac{e^2}{4\pi\epsilon_0}\frac{1}{r} = \left(\frac{e^2}{4\pi\epsilon_0}\right)^2\frac{m}{2\hbar^2}\frac{1}{n^2} = -\frac{1}{n^2}E_R$$

The energy thus also comes out quantised. The coefficient of the n^2 term is called the Rydberg energy, it is proportional to the Rydberg constant, $E_R = hcR_H$. The energy difference between two energy levels n and m is thus:

$$\Delta E = \frac{hc}{\lambda} = E_m - E_n = E_R\left(\frac{1}{n^2} - \frac{1}{m^2}\right)$$

with $n = 1, 2, 3, \ldots$ and $m = n+1, n+2, n+3, \ldots$ For $n = 2$ we thus recover the Balmer series of hydrogen spectral lines (see Section 4.2), for $n = 1$ and $n > 2$ the other spectral series (named after Lyman, Paschen, Brackett, Pfund and Humphreys).

Focus Box 5.4: Bohr atom and spectral lines

Bohr's approach to atomic theory obviously breaks with classical physics in a bold way. None of his axiomatic hypotheses is compatible with classical electrodynamics and mechanics, nor rooted in an underlying new theory. Heilbron [163] traces back Bohr's boldness to the philosophy of Harald Høffding and William James. Høffding breaks down truth to an analogy between actual reality and its image in the human mind. There are multiple partial truths, which justify themselves by the fact that they work. In James' words quoted by Heilbron: "The true is the name of whatever proves itself to be good [that is, useful and thus valuable][4] in the way of belief." Thus two practical arguments justify abandoning the classical approach: the results explain the observed line spectra and the Rydberg constant comes out correctly in terms of known natural constants.

Figure 5.3 Physicists and chemists at a gathering in Berlin-Dahlem in 1920 in honour of James Franck's appointment to a professorship at the University of Göttingen [672]. Back row, left to right: Walter Grotrian, Wilhelm Westphal, Otto von Baeyer, Peter Pringsheim, Gustav Hertz. Front row, left to right: Hertha Sponer, Albert Einstein, Ingrid Franck, James Franck, Lise Meitner, Fritz Haber, Otto Hahn. (Credit: Archiv der Max-Planck-Gesellschaft, Berlin-Dahlem)

Strong support to Bohr's theory of distinct energy levels for electrons was given by the 1914 experiments [156, 155] of James Franck and Gustav Hertz, the nephew of Heinrich Hertz (see Figure 5.3). They demonstrated energy levels in mercury, using inelastic electron scattering, as explained in Focus Box 5.5[5]. The experiment demonstrates the existence of discrete energy levels

---

[4]Explanatory comment by Heilbron.

[5]The experiment and its results are thoroughly analysed in modern language by Robert E. Robson, Malte Hildebrandt and Ronald D. White [613].

in mercury atoms. Inelastic collisions of electrons transfer just the energy to atomic electrons which is required to pass from one energy level to another. When they fall back to the ground state, light is emitted with the same energy that it took to excite them.

The period of World War I, from 1914 to 1918, the ensuing isolation of Germany and economic and political problems of the Weimar Republic clearly reduced scientific output. Nevertheless, throughout the first two decades of the 20th century, many problems in atomic spectroscopy were attacked using quantum theory à la Bohr. The theory was given more solid mathematical footing by the work of Arnold Sommerfeld [173]. He generalised Bohr's circular stationary orbits, numbered by the main quantum number n, to elliptical ones, with additional quantum numbers l and m, which he called azimuthal and magnetic quantum numbers. They number the quantum states of orbital angular momentum and its orientation along an arbitrary axis. Spectroscopy thus defined a shell structure of atomic energy levels [178], with closely spaced levels sharing the same n. This structure was confirmed by the X-ray scattering experiments of Charles Barkla [138] and Henry Moseley [158]. Angular momentum states, which can be identified by the modulus of an orbit's angular momentum, can vary from $l = 0$ to $l = n - 1$. Their orientation, which can be thought of as the projection of the angular momentum vector onto an arbitrary axis, varies from $m = -l$ to $m = +l$. Sommerfeld's book "Atomic Structure and Spectral Lines" [181] concluded what is now called the "old" quantum theory. The angular momentum quantum numbers explain the line splitting in a magnetic field, the (normal) Zeeman effect. Paul Epstein [170] and Karl Schwarzschild [172] independently derived equations for the Stark effect in hydrogen.

However there remained substantial flaws in the old quantum theory. Perhaps the most important one was that it provided no means to calculate the intensities of the spectral lines, i.e. the probability that excitation and de-excitation happens. It also failed to explain the anomalous Zeeman effect, where spin comes in.

Spatial quantisation was clearly another quantum effect at odds with classical electrodynamics, in which an angular momentum vector can make any angle with an externally axis, e.g. defined by the direction of a magnetic field. While searching to confirm that it indeed existed, electron spin was accidentally discovered in the famous Stern-Gerlach experiment (see Focus Box 5.6). Otto Stern had worked with Einstein in Prague and Zürich and became assistant to Max Born at the Frankfurt Institute for Theoretical Physics after the first world war ended. Before the war, he had learned about atomic beams formed by effusion of heated substances under vacuum[6]. Walter Gerlach had worked with Wilhelm Wien during the war to develop wireless telegraphy. He joined the Frankfurt Institute for Experimental Physics, adjacent to Born's institute, in 1920. The two got together to develop a method to prove or

---

[6]For a review of the history of molecular beams see [423].

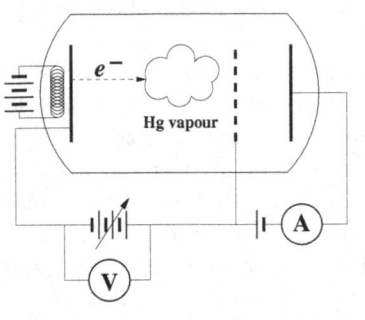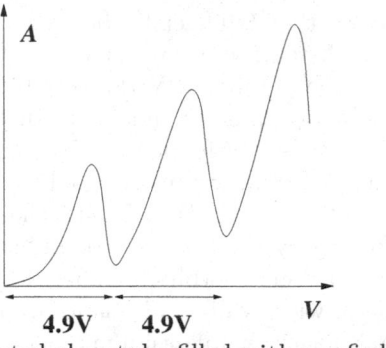

The Franck-Hertz experiment used an evacuated glass tube filled with rarefied mercury vapour, sketched above on the left. Electrons are emitted by the heated cathode on the left, and accelerated by a variable voltage V towards the middle grid. A small counter-voltage keeps low energy electrons passing the grid from reaching the anode on the right. The current A between grid and anode is measured by an ampere meter. Its dependence on the grid voltage is shown schematically in the right figure.

Electrons on average have several hundred elastic interactions with mercury atoms on their way to the grid, the electron beam is thus diffused in direction, but the electrons keep their energy. The current first increases with grid voltage. When the grid voltage approaches 4.9V, the current drops sharply. This indicates that electrons of kinetic energy 4.9eV interact with mercury atoms in an inelastic collision and lose most of their energy. At first Franck and Hertz thought that is energy was the ionisation level of mercury. It later became clear that this energy corresponds to the preferred transition of outer electrons ($6^1S_0 \rightarrow 6^3P_1$)[613]. When one increases the grid tension further, the anode current increases again up to a maximum at 9.8V. This means that now the electrons have enough energy for inelastic collisions half way to the grid, their kinetic energy is lost there instead of close to the grid. The remaining potential difference accelerates them again, up to the necessary kinetic energy for inelastic collisions close to the grid.

The zones of inelastic scattering can be made visible when using neon gas instead of mercury vapour. The neon excitation energy is approximately 18.5eV, a weighted average of several transitions [613]. When the neon atom falls back to its ground state, it emits visible light. The zone of inelastic scattering becomes visible as a faintly glowing slice close to the grid. When the accelerating voltage is increased, the zone wanders towards the cathode. At twice the grid voltage, when the zone has reached half the cathode-grid distance, a second glowing zone appears close to the grid and the anode current again reduces abruptly.

Focus Box 5.5: The Franck-Hertz experiment

disprove space quantisation, despite everyone's and especially Born's scepticism. Atomic beams are key to the test, because they have no net charge and are not subject to Lorentz forces, but interact with the magnetic field via their magnetic moment only. The critical experimental obstacle was to produce a sufficiently strong gradient in a magnetic field to make the small magnetic moment visible. In a feasibility study [191], Stern had concluded that even a gradient at the limit of feasibility would only result in a minute deflection of the beam.

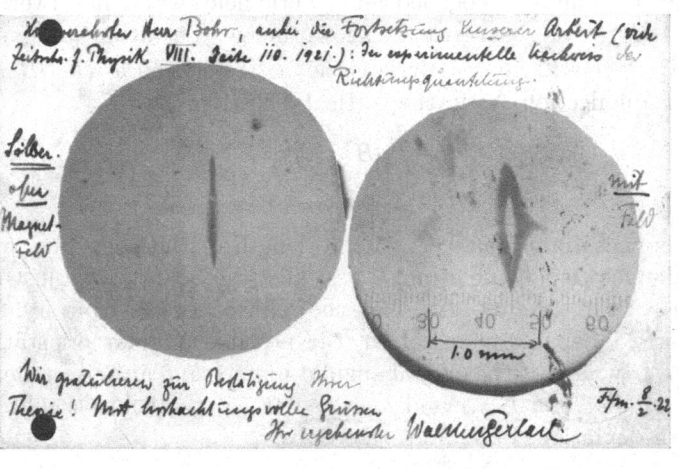

Figure 5.4 Gerlach's postcard to Niels Bohr dated February 8, 1922. It shows a photo of the split silver beam deposit, with a short statement congratulating Bohr on the confirmation of his theory of spatial quantisation. (Credit: Niels Bohr Archive P006)

After many tries and improvements to their apparatus and methods, in 1922 they succeeded to observe the magnetic deviation of silver atoms. A triumphing postcard, which is reproduced in Figure 5.4, was sent to Niels Bohr [542]. It was first believed that they had demonstrated the quantisation of orbital angular momentum direction, which is not possible since the ground state of silver atoms has none. It took new ideas by George Uhlenbeck and Samuel Goudsmit to find out that they really discovered the electron spin.

George Uhlenbeck was a man with wide interests in languages, history, mathematics and physics. Not being admitted to University because of a lack of Latin and Greek, he started his scientific education at the Delft Institute of Technology. When the law changed, he joined the prestigious University of Leiden in 1919, where he received a scholarship. After graduating in 1920, he attended courses with the charismatic Paul Ehrenfest and Hendrik Lorentz. After an episode as a private instructor in Rome, where he made friends with Enrico Fermi, he seriously hesitated between a career in physics or in history. He was advised to first finish his advanced physics studies and became

An atomic electron which has total angular momentum $\vec{J}$ carries a magnetic moment $\vec{\mu}$, proportional to the Bohr magneton μ_B:

$$\vec{\mu} = -\frac{e}{2m_e}\vec{J} = -\mu_B\frac{\vec{J}}{\hbar}$$

The second equation assumes that the angular momentum is quantised in units of $\hbar$ as predicted by the Bohr-Sommerfeld model. The atom passes horizontally through an inhomogeneous magnetic field, which has a vertical inhomogeneity $\partial B/\partial z$ as schematically shown below. The atom will experience a force $\mu(\partial B/\partial z)\cos\theta$, where θ designs the angle between $\vec{\mu}$ and $\partial\vec{B}/\partial z$. It will thus suffer a deflection Δz in the vertical direction:

$$\Delta z = \frac{\mu}{2m}\left(\frac{\partial B}{\partial z}\right)\frac{(\Delta x)^2}{v^2}\cos\theta$$

where m is the atomic mass, Δx is the length of the deflecting field region and v is the velocity of the atom. Classically, the beam would just be spread continuously from $-\Delta z$ to $+\Delta z$, since the value of $\cos\theta$ is arbitrary. According to Sommerfeld's space quantisation, the possible angular orientations have $\cos\theta = m_l/l$, where l is the orbital angular momentum and m_l is the magnetic quantum number. For $l = 1$ and $m_l = -1, 0, +1$, the beam should thus split in two with deviations $-\Delta z$ and $+\Delta z$ only.

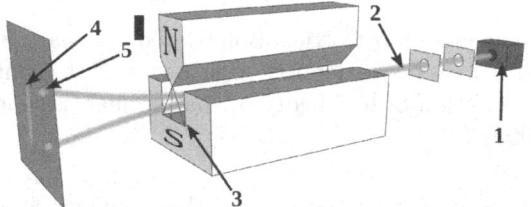

Starting in 1921, Stern and Gerlach set up a series of experiments schematically shown above. An oven (1) emits a beam of silver atoms at 1000°. It passes through a pair of collimators (2) in pencil shape and goes through a strongly inhomogeneous magnetic field (3). It is deposited on a plate, either as a straight line (4, classical case), or as two symmetric dots (5, quantised case). The latter was observed by Stern and Gerlach when photographically developing the plate after many unsuccessful tries in early 1922. They and their contemporaries thought the deflection was due to the spatial quantisation of orbital angular momentum. In reality, the total angular momentum in Ag atoms solely comes from the spin $\vec{S}$ of its outermost 5S electron. It has $l = 0$, thus $m_l = 0$ in Sommerfeld's notation. However, the total angular momentum is not zero, but $\vec{J} = \vec{L} + \vec{S} = \vec{S}$, with a component along the magnetic field of $m_s = \pm 1/2$. The spin causes a magnetic moment $\vec{\mu}_s = g\mu_b\vec{S}$ with the Landé factor $g \simeq 2$.

Focus Box 5.6: Stern-Gerlach experiment

Ehrenfest's *doctorandus*. His first assignment was to learn about atomic spectroscopy from Samuel Goudsmit, a master student in Leiden. In summer of 1925 they started to work together. According to Abraham Pais [466], "George was the more analytic one, better versed in theoretical physics, a greenhorn in physics research and an aspiring historian... Sem [Goudsmit] was the detective, thoroughly at home with spectra... known in the physics community and a part-time assistant to Pieter Zeeman in Amsterdam." Together they discovered that while the normal Zeeman splitting came from the orbital angular momentum of electrons, the abnormal Zeeman splitting must be due to an additional property of the electron, the spin. Spin is an angular momentum, originally thought to be due to a rotational degree of freedom of particles. However, since the electron has no size (as far as we know, see Chapter 9), it cannot be mechanically realised. Spin is an intrinsic property of the particle, like its charge and mass. Goudsmit and Uhlenbeck published their findings in a short note in 1925 [204], followed by a more substantial *Nature* paper in 1926 [205]. The latter was followed by a very interesting postscript by Niels Bohr, where he pointed out that their conclusion, although still incomplete, "... must be more welcomed at the present time, when the prospect is held out of a quantitative treatment of atomic problems by the new quantum mechanics initiated by the work of Heisenberg, which aims at the precise formulation of the correspondence between classical mechanics and quantum theory."

At first the interpretation as an effect of spin met difficulties because of the Landé factor [186] of 2 which relates the spin angular momentum to its magnetic moment. The factor was soon identified as a relativistic effect [215]. Wolfgang Pauli worked out the non-relativistic theory of spin, using Heisenberg's matrix quantum mechanics. He realised that the complicated numbers of electrons in closed shells can be reduced to the simple rule of one electron per state, if the electron states are defined using four quantum numbers instead of the three (n, l, m_l) in the Bohr-Sommerfeld atom. For this purpose he introduced a new two-valued quantum number, m_s, corresponding to spin orientation [203]. His conclusion required the Pauli exclusion principle, stating that no two electrons –more generally no two fermions– can be in the same quantum state. For example, if two electrons reside in the same orbital, then their n, l, and m_l values are the same, thus their spin projections must be different, with $m_s = +1/2$ and $-1/2$. When Paul Dirac derived his relativistic quantum mechanics in 1928, electron spin was an essential part of it.

5.4 CANONICAL QUANTA

The complex path from the quantum atom to a complete quantum mechanics involved many contributors. It has been detailed by a contemporary contributor, Bartel L. van der Waarden, reproducing and commenting original articles, in reference [399]. There were several "schools" involved in the process, which Victor F. Weisskopf described in the diagram reproduced in Figure 5.5. The arrows indicate the intensity of scientific exchange between the centres, the many multiple entries witness the mobility of scientists in the aftermath of the first world war. We do not need to go into the detail of this development, but we will jump to the canonical formulations of quantum mechanics right away.

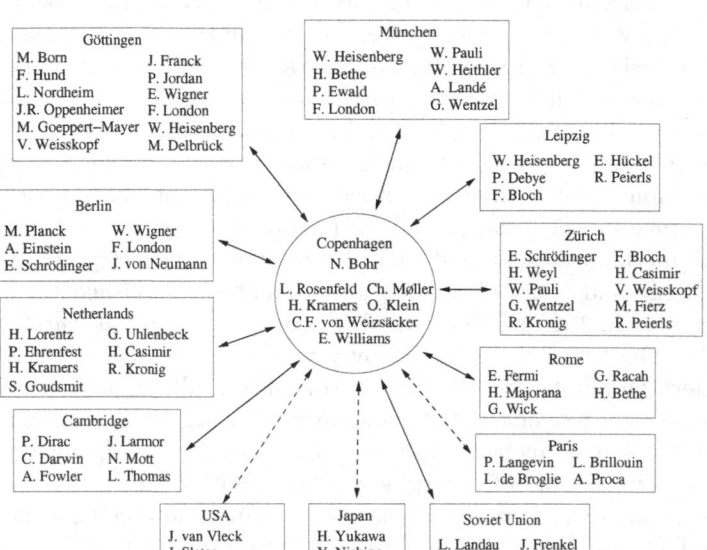

Figure 5.5 Schools of theoretical physics in the years of the foundation of quantum mechanics. Adapted from [441].

We have used plural here on purpose. There are indeed two such formulations, matrix mechanics by Werner Heisenberg, Max Born and Pascual Jordan; and wave mechanics by Louis de Broglie, Erwin Schrödinger and Peter Debye. Let us start with the matrix approach.

One of the principle problems with quantum mechanics à la Bohr is that it can describe stationary states but has trouble with time dependent processes. Thus the energy levels are predicted roughly right, absorbed and emitted radiation energies come out correctly, when using the Bohr prescription decoupling these from the energy of a single orbit. But how long does it take the electron to fall back to a lower energy state and how does it decide which one to choose? In other words, what are the intensities of emission lines? The problem with time dependence can be traced back to assuming classical motion of the electrons. Hans Kramers made a first attempt to calculate transition probabilities between quantum states in terms of a discrete-time Fourier transform of the orbital motion [195, 196], ideas which were extended in collaboration with Werner Heisenberg to a semiclassical matrix-like description of atomic transition probabilities [201]. The standard narrative of quantum mechanics

Classically, the emission and absorption frequencies of an oscillating system are determined by their eigenfrequencies. This property is preserved in the old quantum theory of the atom à la Bohr. However, the orbital angular momentum is quantised, such that $mvr = nh/(2\pi)$ (see Focus Box 5.4). This can be generalised to an orbital condition for momentum p and position q:

$$\int_0^T p\frac{dq}{dt}\,dt = \int_{q(0)}^{q(T)} p\,dq = nh$$

In the Kramer approach, emission and absorption is described via the time-discrete Fourrier transform of the orbital coordinate:

$$q(t) = \sum_{n=-\infty}^{\infty} e^{2\pi in\nu t} Q_n$$

The Q_n are complex coefficients, with $Q_n = Q_{-n}^*$. According to the correspondence principle, the classical and quantum frequency should be the same for large n. The energy difference of close orbits n and m, $(n - m) \ll (n, m)$, should then be $(E_n - E_m) \simeq h(n-m)/T$, where T is the classical period of the orbit. Heisenberg replaced this explicit description of the orbital motion (which is unobservable) by an equivalent description of the transition from n to m, which is observable via the radiated photon with frequency $(E_n - E_m)/h$. For small n and m, these are no longer multiples of any orbital frequency. Heisenberg noticed that the Fourier transform Q_{nm} of the transition amplitude has only one frequency:

$$Q_{nm} = e^{2\pi i(E_n - E_m)t/h} Q_{nm}(0)$$

This is an expression for the evolution in time of Q_{nm}, it can be read as an equation of motion for the emission or absorption. The equation shows again, that energy is the *vis viva* of evolution. The quantities described follow matrix algebra, such that a multiplication between e.g. position P and momentum Q matrix elements is defined as

$$(QP)_{nm} = \sum_{k=0}^{\infty} Q_{mk} P_{kn}$$

The multiplication is not commutative:

$$\sum_k (Q_{nk} P_{km} - P_{nk} Q_{km}) = i\hbar\delta_{nm}$$

This relation gives rise to Heisenberg's uncertainty principle (see Focus Box 5.8).

Focus Box 5.7: Matrix mechanics

has it that Heisenberg had the basic idea of a solution during a forced stay on the isolated island of Helgoland, where Max Born sent him in spring of 1925 because of a bad hay fever. Heisenberg reformulated all of quantum theory in terms of a version of the transition matrices [200]. The principle ideas behind Heisenberg's radical approach are sketched in Focus Box 5.7. Their foundation is Heisenberg's deeply rooted conviction that physics should only deal with observable quantities. By this we obviously denote quantities that are measurable in principle, not necessarily in practice. The orbital motions of electrons in an atom are not in this category, thus Heisenberg disposed of the concept. Instead, he based matrix mechanics solely on transitions between quantum states. Since quantum states can be numbered using quantum numbers, the transition from state i to state f thus involves a matrix element M_{if}, the square of which is the probability for the transition to happen. All possible matrix elements form a matrix, with index i running from 1 to the number of initial states, f from 1 to the number of final states. Consecutive transitions are described by multiplying matrix elements following matrix algebra, as Max Born and Pascual Jordan noted [198, 202]. Our modern view of the matrices is that they are operators acting on a state vector. They describe observables. If the state is an eigenstate of the operator (corresponding to an eigenvector of the matrix), it will project out an eigenvalue which corresponds to the result of a measurement. Wolfgang Pauli applied the new method to the hydrogen atom in 1926 [211] and showed that it reproduced the correct spectrum.

An immensely important result of matrix mechanics is Heisenberg's uncertainty principle. It states that a quantum system can only be the simultaneous eigenstate of two operators if these commute. In other words, among a pair of observables described by non-commuting operators, only one can have a fixed value. Examples of such pairs of mutually exclusive observables are momentum p and position r, where the uncertainty principle states that $\Delta r \Delta p \geq \hbar/2$. For energy E and time t it reads $\Delta E \Delta t \geq \hbar/2$. We go into details in Focus Box 5.8. The principle has been introduced by Heisenberg in 1927 [221], and formally derived by Earl H. Kennard [222] and Hermann Weyl [229].

The uncertainty principle describes a fundamental property of quantum systems, not a limitation of measurement. It has often been confused with an interaction between a macroscopic measurement device and the quantum system itself. That would blame its existence on the interference of measurement and quantum object called the "observer effect". This is partially due to Heisenberg himself, who used the latter as an example of a physical cause of the principle [237]. However, it is much more important than this. A particle at a defined space point does not *have* a defined momentum. A particle of fixed momentum *is* everywhere.

In parallel to the impressively fast progress with matrix mechanics, an alternative was developed, which did not give up on the concept of describing the electron's motion. In 1924, Louis de Broglie [199] introduced the idea that the relation between energy-momentum on one side and frequency-wavevector

We look into Heisenberg's uncertainty principle using position and momentum as examples. Their respective operators Q and P do not commute, i.e. $QP - PQ = i\hbar$. Let us assume that the particle is in an eigenstate ψ of the position operator Q for position q:

$$Q\psi = q\psi$$

Applying the commutator above we find:

$$(QP - PQ)\psi = (Q - q\mathbb{1})P\psi = i\hbar\psi$$

where $\mathbb{1}$ is the identity operator, $\mathbb{1}\psi = \psi$. Let us assume that the particle could simultaneously be in an eigenstate to P with eigenvalue p. We would then have:

$$(Q - q\mathbb{1})P\psi = (Q - q\mathbb{1})p\psi = (q\mathbb{1} - q\mathbb{1})p\psi = 0$$

in contradiction to the commutation relation above. This means that the particle cannot have a fixed position and a fixed momentum at the same time. Denoting by Δp and Δq the momentum and position range the particle has, one finds Heisenberg's uncertainty relation for position and momentum, $\Delta q \Delta p \geq \hbar/2$, or energy and time, $\Delta E \Delta t \geq \hbar/2$. Since $\hbar = 6.62607004 \times 10^{-34}$ Js is a very tiny action indeed, the uncertainty only manifests itself in quantum systems. You cannot use it as an excuse when caught speeding in your car. But is does manifest itself everywhere at the quantum scale.

Focus Box 5.8: Heisenberg's uncertainty principle

on the other side is valid for all particles, not only photons. He assigned a wavelength of $\lambda = h/p$ to the electron, or in fact any matter particle. The de Broglie wavelength is thus the ratio of Planck's constant to the momentum of a particle. By describing the electron motion as a standing matter wave, he reproduced Bohr's quantisation condition. The idea was strongly supported by Einstein, who thought that Heisenberg's matrix mechanics was too formal. In 1927, Clinton Davisson and Lester Germer demonstrated the wave nature of electrons by diffraction [219, 224]. Peter Debye commented that if electrons behave as waves, their movement should follow a wave equation.

Soon after, Erwin Schrödinger found that equation [213, 214], a differential equation for the evolution of a wave function $\psi(\vec{r}, t)$ in three dimensions. He found it by the analogy with wave and geometrical optics[7]. We sketch the properties of the wave function and the Schrödinger equation in Focus Box 5.9. In 1926, Schrödinger solved the problem of an electron's motion in the potential well created by the nucleus [212] and found the correct energy levels. Wave mechanics works, at least when particles are moving non-relativistically.

Both approaches, matrix mechanics and wave mechanics, have been shown to be completely equivalent by Paul Dirac in 1926 [208]. The reason why I prefer the wave formulation is that it can be applied to interactions between particles, which has always been the subject of my own research. This widening of the scope of quantum mechanics was initiated by Max Born, who had become a professor at University of Göttingen in 1921. He attracted many brilliant people to this small provincial town in northern Germany. The first was his friend and colleague, the experimentalist James Franck. His PhD students included Siegfried Flügge, Friedrich Hund, Pascual Jordan, Maria Goeppert-Mayer, Robert Oppenheimer and Victor Weisskopf. Among his many notable assistants, Enrico Fermi, Werner Heisenberg, Wolfgang Pauli, Léon Rosenfeld, Edward Teller and Eugene Wigner stick out. In 1926, Born wrote two articles about the quantum mechanics of scattering processes in Zeitschrift für Physik, an introductory one [207] with little mathematics by his standards, followed by a more elaborate one [206]. In these papers he developed the statistical interpretation of the wave function which we summarise in Focus Box 5.9. In short, the square of the wave function $\psi(\vec{r}, t)$ is the probability per unit volume to find the particle at position $\vec{r}$ at time t. In analogy to electromagnetic waves, where the intensity is the square of the amplitude, the wave function itself can thus be called a probability amplitude. Scattering modifies the wave function according to Huygens' principle (see Focus Box 7.5). In a particularly enlightening presentation of his ideas in Nature [217], carefully translated into English by Oppenheimer, Born states: "We free forces of their classical duty of determining directly the motion of particles and allow them instead to determine the probability of states." We will come back to all of this in more detail

---

[7]An enlightening derivation of Schrödinger's equation by Richard Feynman is explained in [483].

• The abstract *state function* Ψ contains all information about a physical state which can be obtained in principle. It is the equivalent of the state vector of matrix mechanics. In coordinates, it is represented by a complex *wave function* $\psi(q_i, s_i, t)$, depending on the classical degrees of freedom q_i (like position and momentum) and the quantum degrees of freedom s_i (like spin), as well as time t. The wave function is not observable for a single particle (therefore Heisenberg did not like it).

• A physical *observable* of the system, i.e. a property which can in principle be measured, is described by a linear hermitian operator Ω. Examples are the momentum operator $p_i \rightarrow -i\hbar\frac{\partial}{\partial q_i}$ and the energy operator $E \rightarrow i\hbar\frac{\partial}{\partial t}$.

• A system is in an *eigenstate* Ψ_n of the operator Ω, if $\Omega\Psi_n = \omega_n\Psi_n$, i.e. if the operator projects out an *eigenvalue* ω_n of the observable. These are quantised, so they can be numbered. For hermitian Ω, the eigenvalues ω_n are real numbers. In coordinates one has $\Omega(q, s, t)\psi_n(q, s, t) = \omega_n\psi_n(q, s, t)$. The eigenstates form an orthonormal basis, thus every possible state can be written as a weighted sum of eigenstates, $\psi = \sum_n a_n\psi_n$ with complex coefficients a_n. The squares of these coefficients are the probability densities to find the system in eigenstate n at time t, their space integrals are the corresponding probabilities. This statement is called the *Born rule*, the "fundamental interpretive postulate of quantum mechanics" according to Steven Weinberg [628, p. 29]. In particular, for a pure eigenstate n, $\rho = \psi_n^*\psi_n = |\psi_n|^2 \geq 0$ is the probability density to find it at t with properties q and s.

• Because of the orthonormality, one has $\sum_s \int (dq_i...)\psi_n^*(q_i, s, t)\psi_m(q_i, s, t) = \delta_{nm}$. A *measurement* of an observable delivers the nth eigenvalue of Ω with probability $|a_n|^2$. Its *expectation value*, i.e. the average weighted by probability, is thus $\langle\Omega\rangle_\psi = \sum_s \int (dq_i...)\psi^*(q_i, s, t)\Omega\psi(q_i, s, t) = \sum_s \sum_n |a_n|^2\omega_n$.

• The time evolution of a system is described by the *Schrödinger equation*, $i\partial\psi/\partial t = H\psi$ with the linear hermitian *Hamilton operator H*, the corresponding observable is energy. In a closed system, H does not explicitly depend on time, $\partial H/\partial t = 0$. The eigenstates of H are then stationary, the eigenvalues are conserved. Because of hermiticity, one has

$$\frac{d}{dt}\sum_s \int (dq_i...)\psi^*\psi = -i\sum_s \int (dq_i...)\left[(H\psi)^*\psi - \psi^*(H\psi)\right] = 0$$

Probability is thus a conserved quantity, $\frac{d}{dt}\int \rho\, d^3x = \frac{d}{dt}\int (dq_i...)\psi^*\psi = 0$.

• The simplest solution to Schrödinger's equation is a *plane wave* with amplitude ψ_0, propagating in empty space in the direction $\vec{k}$ with frequency $\omega = 2\pi\nu$ and wave number $k = 2\pi/\lambda$: $\psi(\vec{r}, t) = \psi_0 e^{-i(\omega t - \vec{k}\vec{r})}$.

Focus Box 5.9: A summary of non-relativistic wave quantum mechanics following Bjorken and Drell [379]

in Chapter 7. Born received the 1954 Nobel prize in physics for the statistical interpretation of quantum mechanics.

I personally agree with Einstein and Born that wave mechanics is preferable to matrix mechanics when thinking about quanta. I like the idea that particles are point-like objects which move like waves. The wave aspect just means to me that one must abandon the idea of a trajectory, a smooth curve in space-time which locates the particle. Instead the probability amplitude defines a distribution of positions in space and time, a field, where the particle can manifest itself when measured. A decisive conceptual difference between trajectories and waves is that the latter can interfere with each other, the former cannot. We discuss interference phenomena in Focus Box 5.10 using light as a familiar example.

5.5 QUANTUM NUCLEI

Some twenty years after the discovery of the atomic nucleus, most physicists believed that it was a bound state of protons and electrons. The number of protons would account for nuclear weight, the difference between the number of protons and electrons for the nuclear charge, which gives the atomic numbering scheme of the periodic table and determines their chemical properties. There are a couple of obvious problems with this idea. First, it is entirely unclear why a certain group of electrons would choose to be bound in the compact nucleus, while an equally large second group would fill the much larger atomic volume. No way an electromagnetic binding force between electrons and protons could provide such a feature. But there also was corroborating evidence. Nuclear beta decay emitted electrons, so shouldn't they be present in the nucleus? In 1919, Rutherford reported experiments which showed that bombarding nitrogen nuclei with energetic alpha particles kicked out protons [179]. We can take the grandmaster himself as a witness of public opinion. In 1920 he wrote in his Bakerian lectures [185]: "We also have strong reason for believing that the nuclei of atoms contain electrons as well as positively charged bodies, and that the positive charge on the nucleus represents the excess positive charge. It is of interest to note the very different role played by the electrons in the outer and inner atom. In the former case, the electrons arrange themselves at a distance from the nucleus, controlled no doubt mainly by the charge on the nucleus and the interaction of their own fields. In the case of the nucleus, the electron forms a very close and powerful combination with the positively charged units and, as far as we know, there is a region just outside the nucleus where no electron is in stable equilibrium. While no doubt each of the external electrons acts as a point charge in considering the forces between it and the nucleus, this cannot be the case for the electron in the nucleus itself." He speculated about the existence of a neutral nuclear bound state of a proton with an electron, which would be able to penetrate matter and even atoms.

But in the same period of time, more doubts appeared. Franco Rasetti had measured the spectra of di-atomic molecules like O_2 and N_2, and found that

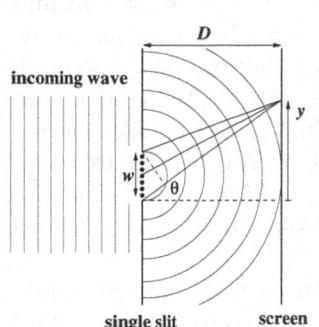

single slit   screen

Diffraction of a plane wave of light occurs when it meets an obstacle. On the left, we consider a single slit of width w at a distance D from a screen. When the light with wavelength λ meets the slit, according to Huygens' principle, every point in the aperture emits a cylindrical wavelet. The amplitudes of all these, originally in phase, add up to form the outgoing wave. We discuss diffraction of the wave in the Fraunhofer limit, i.e. for $w^2/(D\lambda) \ll 1$. What arrives at the screen at normal incidence, $\theta = 0$, is a bright strip with a fuzzy edge.

At any other angle, when we compare the phase of the wavelet emitted from the middle of the slit to that emitted at the lower edge, we find a phase shift of $\delta \simeq w\sin\theta$. All of the wavelets coming from points with distance $w/2$ have that phase shift, so they sum up constructively. After a minimum at $w\sin\theta = \lambda$, there is thus a secondary maximum at $w\sin\theta/2 = \lambda$. More maxima follow at regular intervals. Their displacement on the screen is $y_m = D\tan\theta \simeq D\sin\theta = m2D\lambda/w$ with an integer m. The intensity distribution is $I(y) = I_0 \sin^2\delta/\delta^2$ with $\delta \simeq \pi wy/(D\lambda)$. The diffraction pattern is shown in the upper figure on the bottom left.

When there is a double slit, with negligible width, we can treat both à la Huygens as point sources of in-phase monochromatic light. The geometry is shown on the right. The path difference between two rays is $d\sin\theta$, such that maxima occur at $y_m = mD\lambda/d$, all with equal intensity, as shown in the middle figure below.

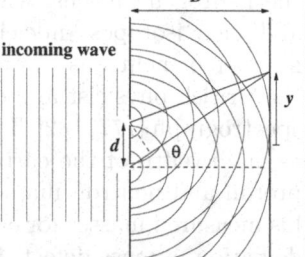

double slit   screen

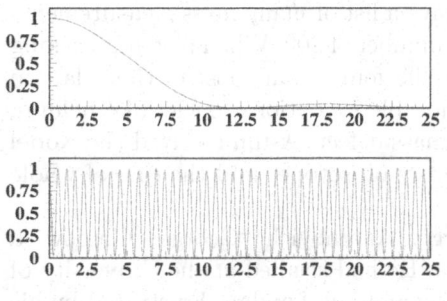

When the width of the slits is not negligible, we find an overall intensity distribution characterised by the slit width, but with regular strips, as shown in the lower figure. In the sketches above we have shown the wave character of the light as wavefronts, the particle character as straight rays. Indeed both aspects are necessary to understand the phenomena of diffraction and interference. Classical interference experiments with visible light were performed by Thomas Young in 1801 [7].

Focus Box 5.10: Diffraction and interference

both nuclei have integer spins. For oxygen, this fits with a nucleus made of 16 protons and 8 electrons. But for nitrogen, there ought to be 14 protons and 7 electrons, thus half-integer spin. The plot thickened when Walther Bothe and Herbert Becker observed a penetrating neutral radiation when beryllium nuclei were bombarded with high energy α particles [235]. They conjectured that the radiation was of electromagnetic origin. That became unlikely when Irène and Frédéric Joliot-Curie measured very high energy protons exposing hydrogen-rich substances to Bothe's radiation [247], which would be in contradiction to their own absorption measurements if the radiation were indeed made of photons.

The particle behind the new radiation, the neutron, was found by James Chadwick in 1932 [244]. He was a student and close collaborator of Rutherford and his deputy in the direction of the Cavendish Laboratory. To determine the mass of the new neutral particle, he measured the recoil energy when they bounced off light nuclei like hydrogen and nitrogen. The result was, that their mass was not at all zero but roughly equal to the proton mass. The numbering scheme of Mendeleev, with atomic number equal to the number of protons in a nucleus, was completed by the fact that the nuclear mass was roughly proportional to the number of protons plus neutrons[8]. Nuclear masses were found to be roughly integer numbers, when expressed in the conventional mass units of one sixteenth of the oxygen mass. A notable exception was the hydrogen nucleus, with 1.008 mass units. The masses of many elements and their isotopes –nuclei with the same number of protons and a different number of neutrons– were measured with better that permille precision by the British physicist and chemist Francis William Aston, with his novel mass spectrograph [177, 182]. The mass unit, today called atomic weight and defined as one twelfth of the carbon isotope mass ^{12}C, is roughly the average of the proton and neutron mass. When subtracting the closest whole number from his measured masses for each isotope, Aston found that there was a significant deviation, a mass defect. Dividing by the mass number, he obtained what he called "packing fraction". It is in fact roughly the binding energy per nucleon. In a Bakerian lecture in 1927 [216], he gave a list of many mass measurements with elements and isotopes up to a mass number of 200. When plotting packing fractions as a function of mass number, he found that most of them lay on a smooth curve. However, there were notable exceptions: nuclei like helium, carbon and oxygen have a much larger mass defect. Aston received the Nobel prize in chemistry in 1922 for the discovery of isotopes and the rule of whole numbers.

These measurements inspired Georgiy Antonovich "George" Gamow, a physicist of Soviet origin born in Odessa. He had studied at the University of Leningrad, together with Dmitri Ivanenko and Lev Landau. Frustrated by his thesis work, he went to Göttingen to work with Max Born and Fritz Houtermans. Léon Rosenfeld [499] remembered him as "a Slav giant, fair-haired and

---

[8]For a more detailed history of nuclear mass measurements see [557].

speaking a very picturesque German; in fact he was picturesque in everything, even his physics." Gamow spent a year in Copenhagen and visited Rutherford's laboratory in 1929. In a discussion session at the Royal Academy [236], he proposed a "simple model of the nucleus built from α-particles in a way very similar to a water-drop held together by surface tension." This liquid drop model was the first attempt to formulate a model of the nucleus compatible with quantum theory. While Gamow returned to the Soviet Union in 1931, the idea was quickly picked up by Heisenberg [249, 250, 255, 453]. He did not take α-particles but protons and neutrons as nuclear constituents, thus recognising the neutron as an elementary particle. In 1932 he wrote [249]: "If one wanted to take the neutron as composed of proton and electron, one would have to attribute Bose statistics and spin zero to the electron. It does not appear useful to elaborate such a picture any further. Rather the neutron shall be recognised as a fundamental constituent, which however is assumed to decay into proton and electron under suitable conditions, where conservation laws of energy and momentum may no longer be applicable." These "suitable conditions" were elaborated shortly afterwards by Enrico Fermi in his theory of β-decay, and energy-momentum conservation restored (see Section 9.2). The nature of the neutron as an elementary particle was further corroborated by Chadwick's and Maurice Goldhaber's observation of the deuterium split induced by gamma rays [258].

A long lasting version of the liquid drop model is the semi-empirical implementation due to Carl Friedrich von Weizsäcker, German physicist, philosopher and brother of the later president of the Federal Republic of Germany, Richard von Weizsäcker. In 1936, he published the model [266] we explain in Focus Box 5.11. Scattering experiments demonstrated that the nuclear force was not only much stronger than the electromagnetic one, but also the same for protons and neutrons [264, 265], confirming one of the basic assumptions of the liquid drop model. Weizsäcker's work was popularised in the monumental, almost 500 page review of nuclear physics by Hans Bethe and collaborators in 1936/7 [270, 274, 276], which summarised the status of the field.

The dynamics of nuclei when bombarded by neutrons and alpha particles was a subject of intense experimental and theoretical activity at the time. Ten years earlier, Rutherford had invented a satellite model where alpha particles orbited a core nucleus so that they could escape and give rise to α decay [223]. Gamow [225] as well as Ronald Wilfred Gurney and Edward U. Condon [227] had clarified, why α particles are able to leave the nuclear potential by quantum tunnelling, but cannot easily enter into it. Fermi and his group in Rome had bombarded uranium with neutrons and observed the creation of radioactive elements they thought to be trans-uranium nuclei [259]; their work laid the grounds for the discovery of nuclear fission.

The father of nuclear chemistry, Otto Hahn, received his education at the University of Marburg, Germany, and originally intended a career in industry. To improve his language skills, he went abroad and worked with William Ramsay at University College London and Rutherford in Montreal. Returning to Berlin to pursue research in radioactivity, he met the Austrian physicist

Figure 5.6 Physicist Lise Meitner and chemist Otto Hahn in their laboratory. (Credit: imago images/leemage)

Lise Meitner [678], who became his lifetime collaborator and friend. A photo of both in their Berlin laboratory is shown in Fig. 5.6. Meitner had obtained her doctorate in physics from the University of Vienna in 1905, only the second promoted woman in sciences at that university. In 1911, when Fritz Haber became the director of the physical chemistry institute of the newly created Kaiser Wilhelm Gesellschaft (today Max Planck Gesellschaft), he appointed Hahn as the head of its radioactivity institute. From 1918 onward, Meitner headed the institute's physics department until she was forced to emigrate in 1938. To follow up on the work of the Fermi group and similar results by Irène Joliot-Curie and Pavle Savić in Paris [275], Hahn and his fellow radiochemist Friedrich Wilhelm "Fritz" Strassmann tried to isolate the final state components of the observed reactions of slow neutrons with uranium nuclei [281]. On December 17, 1938, they succeeded in showing that a much lighter element, the alkali metal barium was one of the resulting nuclei [285, 286]. Nuclear fission induced by slow neutrons had been discovered. That this was indeed the mechanism for this very exoenergetic nuclear reaction was first realised by Meitner and her nephew Otto Robert Frisch [288, 284], who were by then working in Stockholm and Copenhagen, respectively. While spending Christmas 1938 together, they used the liquid drop model to explain how the absorption of a neutron could deform the drop such that it would split into two smaller drops (see Focus Box 6.1). This happens when the surface tension make this final state energetically favoured, as can bee seen from Focus Box 5.11. When Niels Bohr learned about this, he was all excited [394]: "Oh, what fools we have been! We ought to have seen that before." He notified the New Yorker John A. Wheeler in early January 1939 [394] and together they formulated a quantitative theory [282], again using the liquid drop model. In particular, they calculated the neutron energy required to induce fission in various nuclei.

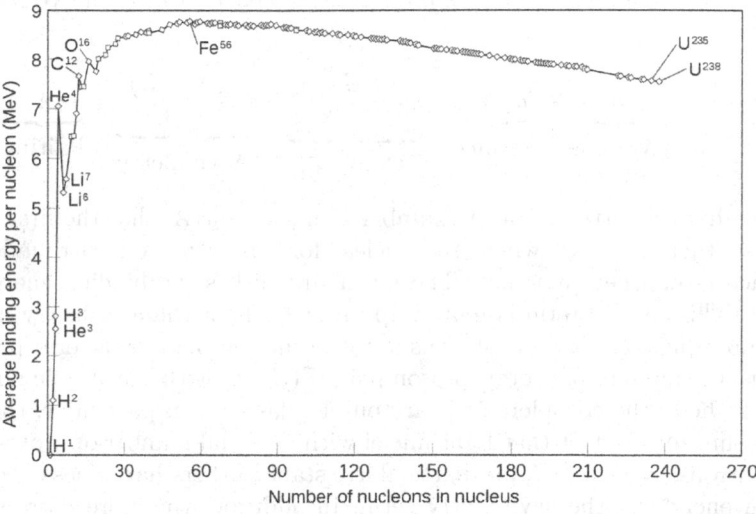

Figure 5.7 Average binding energy per nucleon as a function of the mass number A. The most tightly bound nuclei are found in the iron group and in the so-called magic nuclei. (Credit: Wikimedia Commons)

What the smooth curve of the liquid drop model completely misses are the "magic" nuclei, as Eugene Wigner called them [557], the ones with a number of neutrons or protons equal to 2, 8, 20, 28, 50, 82 and 126. For protons, this corresponds to He, O, Ca, Ni, Sn and Pb. Even more stable are the doubly magic isotopes, ^{4}He, ^{16}O, ^{40}Ca, ^{48}Ca, ^{48}Ni, ^{56}Ni, ^{100}Sn, ^{132}Sn and ^{208}Pb. An approach more close to quantum physics, describing the nuclear structure in analogy to the atom, is able to explain these exceptions. An early attempt to construct such a shell model of the nucleus appeared in 1932 by Dmitri Ivanenko [248], another member of the three "Leningrad musketeers" together with Lev Landau and George Gamow. Such an approach seems natural, but there are important differences to the physics of the atom. First of all, while the electromagnetic potential is well known since Maxwell, the nuclear potential is not. It is not created by a fixed outside source like in the atom, but for each nucleon by all the others present in the same nucleus. A quantitative formulation using a square-well potential was developed by Maria Goeppert-Mayer

The liquid drop model, invented by Gamow and turned into a semi-empirical model with adjustable parameters by von Weizsäcker, was the first nuclear model to describe the binding energy E_B and the radius R of nuclei. The latter is given by a compact packing of A incompressible nucleons, such that $R = r_0 A^{1/3}$ and the nuclear volume is roughly proportional to A, the atomic mass number. The radius parameter is $r_0 \simeq 1.25 \text{fm}$. The (negative) binding energy, $E_B = M(Z, A) - Z m_p - (A - Z) m_n$, is described by the Bethe-Weizsäcker formula:

$$E_B = \underbrace{-a_1 A}_{\text{Volume}} + \underbrace{a_2 A^{\frac{2}{3}}}_{\text{Surface}} + \underbrace{a_3 \frac{Z(Z-1)}{A^{\frac{1}{3}}}}_{\text{Coulomb}} + \underbrace{a_4 \frac{(N-Z)^2}{A}}_{\text{Asymmetry}} \pm \underbrace{a_5 A^{-\frac{1}{2}}}_{\text{Pairing}}$$

It is roughly proportional to the number of nuclei, $-a_1 A$, thus the first term. However, there is a core where the nuclear force is saturated, surrounded by a surface layer where it is not. The latter diminishes the binding energy by $+a_2 A^{\frac{2}{3}}$. This surface term is more important for light than for heavy nuclei. Coulomb repulsion between protons is taken into account by a term proportional to the number of proton-proton pairs $Z(Z-1)$ with the atomic number Z. Up to here, the considerations are purely classical. In particular, they do not account for the fact that light nuclei with an equal number of protons, Z, and neutrons, $N = A - Z$, are particularly stable; others have a less negative binding energy by the asymmetry term. In addition, there are more stable nuclei with both proton and neutron numbers even and few with both odd; this is described by the pairing term. All parameters a_i are positive. The asymmetry term is positive and reduces the binding energy when $N \neq Z$. The pairing term has a positive sign for odd N and Z, negative when both are even, and zero otherwise. The coefficients are determined by adjusting to measured nuclear masses. A recent result from 2228 experimentally measured atomic masses is [558]:

$$a_1 = 15.777 \pm 0.037 \; ; \; a_2 = 18.341 \pm 0.133 \; ; \; a_3 = 0.710 \pm 0.002$$
$$a_4 = 23.211 \pm 0.060 \; ; \; a_5 = 11.996 \pm 1.536$$

The formula gives rather good results for heavy nuclei, less so for light ones. The exceptionally stable "magic" nuclei are badly represented (see Figure 5.7). The formula is especially useful to understand nuclear fission and fusion. Fission splits a heavy nucleus ($A \gg 56$) into an asymmetric pair of lighter ones, towards more tightly bound nuclei (see Focus Box 6.1). Fusion unites two light nuclei ($A \ll 56$) to form a more tightly bound heavier one (see Focus Box 6.4).

Focus Box 5.11: Liquid drop model of the atomic nucleus

during her time in Chicago and at the Argonne National Laboratory in the 1940s, and published in 1949 [314, 317, 318]. Simultaneous to her, Otto Haxel, J. Hans D. Jensen and Hans Suess devised an equivalent model [315]. Their subsequent trans-Atlantic collaboration resulted in a comprehensive book on the subject [329]. The basics of the nuclear shell model are explained in Focus Box 5.12. Goeppert Mayer, Jensen and Wigner received the 1963 Nobel prize for the shell model of nuclear structure. She was only the second Nobel laureate in physics after Marie Curie. No other female laureate in physics was elected until Donna Strickland in 2018.

An attempt to unify the liquid drop model and the shell model was made in the 1950s by Aage Bohr, the son of Niels, Ben Mottelson and James Rainwater, who created the so-called collective model [341]. It also takes into account the interaction between nuclear protons and neutrons, which the shell mode treats separately. It combines collective and individual motion of nucleons and explains certain magnetic and electric properties which the shell model cannot account for. The three received the 1975 Nobel prize in physics for their work [426].

An ab-initio calculation of the nuclear potential had to wait for the field theory of Quantum Chromodynamics, QCD (see Section 9.2). However, applications of nuclear physics by far preceded a complete understanding of strong forces. The enormous energy stored in nuclei caused impressive efforts to liberate and use them for peaceful as well as military purposes. We will elaborate on the latter in Chapter 6.

5.6 QUANTA INTERPRETED?

I am certainly not equipped to give you an overview of the many interpretations of quantum mechanics which have appeared since its very beginnings. Nevertheless I believe that I cannot avoid to at least discuss my very naïve view of what "really" happens in a quantum system, so that you can disagree with me.

According to me, there are at least two aspects of quantum mechanics which are –or at least ought to be– in the focus of its interpretations and the discussion of its difficulties. The first one deals with the question of what the wave function "really" means. I have called it above an amplitude of probability. This probabilistic denomination has the same meaning as the amplitude in "real" waves like the electromagnetic one: it describes a quantity which evolves in space-time, as described by the quantum's equations of motion. Its square gives you, for each point in space-time, the probability for the quantum object to be there, just like the square of the electromagnetic wave amplitude gives you its intensity at every point in space-time. The difference is of course, that the amplitudes of an electromagnetic wave are observable electric and magnetic fields, while the probability amplitude is only observable via a product, with itself or with another wave.

The nuclear shell model is constructed in analogy to its atomic equivalent, but with a model potential. For a shell with principle quantum number n, there are n subshells with orbital quantum number $l = 0, 1, 2, \ldots, (n-1)$. For each one of them, there are $m_l = -l, -l+1, \ldots, 0, 1, \ldots, l-1, l$ projections of l on an arbitrary axis. These $2l + 1$ states are degenerate in energy. For each one, there are two spin orientations, with $m_s = \pm 1/2$. Without magnetic field, these are also degenerate in energy. A state is thus characterised by four quantum numbers, n, l, m_l, m_s. The degeneracy with respect to magnetic quantum numbers is removed by spin-orbit coupling. The notation used in nuclear physics for the principal quantum number is $n = N + l + 1$, where $N = $ is the number of nodes in the wave function. Energy levels are shown on the right [321].

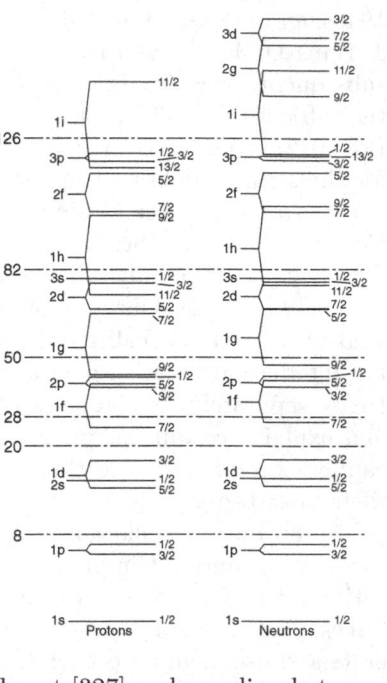

The potential can be modelled with a central part [327] and coupling between spin $\vec{S}$ and orbital angular momentum $\vec{L}$ [329]:

$$V_{tot} = V(r) - f(r)\vec{L} \cdot \vec{S} \; ; \; V(r) = \frac{-V_0}{1 + e^{(r-R)/a}}$$

with the radius parameter $R \simeq 1\text{fm}$. The spin-orbit coupling removes the energy degeneracy and the levels partially overlap. Large gaps in energy appear each time a shell is completed. This explains the "magic" numbers of protons and neutrons and in particular the "doubly magic" nuclei like ^{4}He, ^{16}O, ^{40}Ca, ^{48}Ca, ^{48}Ni, ^{56}Ni, ^{208}Pb etc., which are especially stable.

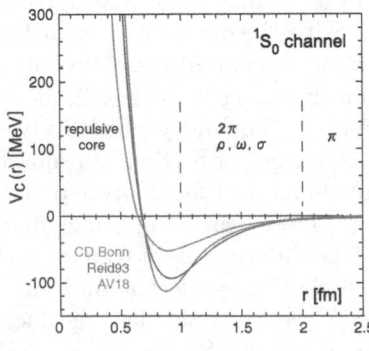

Modern versions of the nuclear potential are based on Quantum Chromodynamics (see Section 9.2), implemented on a lattice [572]. Potentials calculated this way are shown on the left [591]. They have a shallow minimum at a distance of about 1fm. The binding force at larger distances is transmitted by mesons [268]. At shorter distances there is a repulsive core which prevents nucleons from merging.

Focus Box 5.12: Shell model of the atomic nucleus

Naïve experimentalists like me may entertain a frequentist view of this probabilistic quantity. For a frequentist statistician, the probability of an event is the ratio between the number of times it occurs and the number of time it could have occurred. Thus, the phrase in the weather forecast, that the probability for rain tomorrow is 30%, makes no sense to a frequentist because there is only one tomorrow. For a Bayesian statistician, on the contrary, this statement makes perfect sense. For him, the probability just measures our degree of knowledge for what lies ahead.

For a frequentist, an observation of a quantum can be interpreted as the single instance of a repetitive measurement process. A large number of measurements under identical conditions would map out the probability of the quantum's presence as a function of space and time. Thus the question "where was the particle before I measured its position?" is meaningless. To this statement at least, the fathers of the Copenhagen interpretation of quantum mechanics, Born, Bohr and Heisenberg, would agree. Heisenberg wrote in his essay "The Development of the Interpretation of the Quantum Theory" [330]:

> The Copenhagen interpretation is indeed based upon the existence of processes which can be simply described in terms of space and time, i.e. in terms of classical concepts, and which thus compose our "reality" in the proper sense. If we attempt to penetrate behind this reality into the details of atomic events, the contours of this "objectively real" world dissolve–not in the mist of a new and yet unclear idea of reality, but in the transparent clarity of mathematics whose laws govern the possible and not the actual.

So he refuses to go further and discuss in what sense the wave function is "real". In his introduction to Heisenberg's "Physics and Philosophy" [359], Paul Davies sharpens these statements, saying that "the reality is in the observations, not in the electron."

But a question doesn't go away by declaring it illegal. Many, including me, are not content with abstaining from interpreting the wave function itself. Perhaps the most convincing (or least unconvincing) way of attacking the question of quantum motion originated from Louis de Broglie [218] and David Bohm [319] and is sometimes called the pilot wave interpretation. It interprets the wave function as a pilot which describes all possible paths a particle can take. Its evolution is governed by the quantum equations of motion, Schrödinger, Klein-Gordon or Dirac depending on velocity and spin of the particle. The particle takes a definite path which is uncovered in the measurement, a notion that would appeal to a Bayesian statistician. The theory is thus explicitly non-local for coherent states, but each particle has a definite position at any moment, in as much as the uncertainty relation allows. Bohm writes [319]:

> In contrast to the usual interpretation, this alternative interpretation permits us to conceive of each individual system as being in a

precisely definable state, whose changes with time are determined by definite laws, analogous to (but not identical with) the classical equations of motion. Quantum-mechanical probabilities are regarded (like their counterparts in classical statistical mechanics) as only a practical necessity and not as an inherent lack of complete determination in the properties of matter at the quantum level.

The wave nature of quantum states is of course not limited to motion, but also applies to stationary states, like atomic ones. What distinguishes a particle on a trajectory and a wave particle is interference. As far as this is concerned, there is no difference between electromagnetic waves [7] (which we are familiar with) and matter waves [219, 224] (which we are less familiar with). A double slit experiment can be realised with photons as well as electrons, and even leaving half an hour between the passage of subsequent photons or particles makes no difference to the appearance of an interference pattern in the observed intensity (see Focus Box 5.10). The reason is to be sought in the simple solutions to the equation of motion, i.e. plain waves (or their superposition). Plain waves with fixed momentum are everywhere[9], with fixed energy they are there for all times. There is thus an inherent non-locality in quantum phenomena. Once a system of particles is created, its coherence remains, until destroyed by an external intervention. This is exposed by interference experiments and by apparent paradoxes like the Einstein-Podolsky-Rosen one [263]. And it is put to fascinating use for example in quantum cryptography [536].

That there is destruction of coherence during the measurement process is also compatible with the fact that reactions happen or not, at random, but guided by a calculable probability. This in my view is the second question which one should focus on. When an interaction occurs, the Copenhagen interpretation claims that the wave function irreversibly "collapses" into point-like particles. The guide-wave interpretation does not see that happen, since the point-like nature of quanta is always preserved. Interactions can happen or not, at random and again in a wave-like manner, since interference between concurrent processes occurs. When they happen, they destroy coherence, otherwise the guide waves continue undisturbed and the particles follow them. This is as close as my naïve notion of point-like particles moving like a wave gets to the thinking of great men.

There is an interesting interpretation of the fact that quantum processes constantly cause boolean bifurcations, random yes-no decisions following calculable probabilities. In so-called "many-worlds" interpretations of this fact, this leads to a constant splitting of our universe into mutually exclusive histories. All of these are thought to be embedded in a "multiverse", including

---

[9]Note that "everywhere" in the quantum sense is to be taken with respect to the quantum scale of 10^{-15} m. Likewise a free particle is a bold abstraction, empty space does not exist, at least not for long (see Section 9.3).

an infinite number of possible evolutions. This may seem a little far-fetched, but to my knowledge there is no experimental fact contradicting this view.

I can only skim the surface of the interpretation problems quantum physics faces, only as deeply as my limited insight takes me. One of my favourite contemporary physicists, Steven Weinberg, has much more to say about this subject, both at the informal level [641, 645] and at the formal one [627]. The discussion has been going on since quanta were first interpreted and I cannot see an end to it. We will thus come back down to Earth rather violently and discuss the role of science and scientists in the two World Wars of the 20th century.

FURTHER READING

Hans Kangro, *Vorgeschichte des Planckschen Strahlungsgesetzes. Messungen und Theorien der spektralen Energieverteilung bis zur Begründung der Quantenhypothese*, F. Steiner, 1970.

James D. Bjorken and Sidney D. Drell, *Relativistic Quantum Mechanics*, McGraw-Hill 1998.

Carlos A. Bertulani, *Nuclear Physics in a Nutshell*, Princeton University Press 2007.

Finn Aaserud and Helge Kragh (Edts.), *One Hundred Years of the Bohr Atom*, Scientia Danica, Series M, Volume 1, Det Kongelike Danske Videnskabernes Selskab 2015.

Steven Weinberg, *Lectures on Quantum Mechanics*, Cambridge University Press 2015.

Luigi E. Picasso, *Lectures in Quantum Mechanics*, Springer 2016.

Shan Gao, *The Meaning of the Wave Function: In Search of the Ontology of Quantum Mechanics*, Cambridge University Press 2017.

CHAPTER 6

War time physics

> It's my hypothesis that the individual is not a pre-given entity which is seized on by the exercise of power. The individual, with his identity and characteristics, is the product of a relation of power exercised over bodies, multiplicities, movements, desires, forces.
>
> Michel Foucault, *Power/Knowledge*, 1980 [439]

T HIS CHAPTER discusses how nationalism infected society in the beginning of the 20th century, not sparing physicists. How the "great" war, World War I, severed many precious links between German and essentially all other scientists. How ideology polluted physics and caused an unprecedented brain drain in Europe during Nazi reign. I cover the concentrated efforts on the nuclear bomb on both sides of the Atlantic during the early 1940s. In particular, I look into the complicated relation between Heisenberg and Bohr during the German occupation of Denmark. And I briefly discuss how cold war affected physics and physicists.

6.1 A "WAR OF THE MINDS"?

When reading about science and scientists in World War I, one is often confronted to the view that it essentially was a chemists' war. This is of course due to the devastating use of chemical warfare, associated with the name of Fritz Haber, who became the founding director of the Kaiser-Wilhelm Institute for Physical Chemistry and Electrochemistry (now the Fritz-Haber-Institute of the Max Planck Gesellschaft) in 1911. This and the Institute for Chemistry were the first ones of the Kaiser-Wilhelm-Gesellschaft (KWG, now Max Planck Gesellschaft, MPG), in recognition of the importance of research for the rapidly growing chemical industry in Germany. The idea behind the KWG was to bring together academia and industry under the wings of the state, for the benefit of all three. Haber was a good choice, since he had already established close collaboration with Carl Bosch from the chemical company BASF when they together scaled up the catalytic synthesis of nitrates to an industrial level. The resulting fertilisers contributed a lot to feeding people, while nitrate-based explosives killed many.

On August 1, 1914, Germany declared war on Russia and the Kaiser ordered mobilisation of the troops. The same day, nine British scholars, including J.J. Thomson and William Ramsay, published a plea in The Times arguing against a British involvement [531]. Following Germany's invasion of the neutral Belgium in August 4, Britain entered what was to become World War I. The use of chemical weapons was banned by the Hague Convention of 1907, to which Germany was a signatory. Notwithstanding this clear proscription, the use of gas shells was in the focus of the German Ministry of War. In 1916, Haber was named head of its chemistry section and led the teams developing chlorine and other lethal gasses, notably sulfur mustard (LOST) and phosgene with the industrial chemists Wilhelm Lommel and Wilhelm Steinkopf. Haber personally attended the first release of gas grenades during the second battle of Ypres in spring of 1915. The medical officer of the French army's 1st African Battalion described what happened [608]. The sky darkened with a yellow-green cloud. Asphyxiating vapours caused burning throats, pain in the thorax, shortness of breath and coughing-up blood. This eye witness survived, but few other soldiers were that lucky; of the 15,000 French soldiers in the sector, 5000 were dead and a similar number was made prisoner. With Haber as an advisor, a special troop for gas warfare was formed. He drafted physicists, chemists and other scientists to this unit, like Otto Hahn, Gustav Hertz and James Franck, who himself became a victim of a gas attack in 1917.

An even deadlier gas was phosgene, developed by French chemists led by Victor Grignard (Nobel prize in chemistry of 1912) and first used by France in 1915 [515]. Around 37,000 tons of phosgene were manufactured during the war, out of a total of 190,000 tons for all chemical weapons, making it second only to chlorine. The German and French army were practically the only parties deploying it. Less notorious than LOST, phosgene nevertheless caused the vast majority of victims in World War I chemical warfare.

The second notable intervention of scientists in World War I technology was the detection of submarines. The German U-Boot was hard to detect by the rudimentary hydrophones the allied navies possessed. Paul Langevin in France, a student of Pierre Curie, had suggested to use supersonic sound [540] instead of audible signals. Urgency was created when the British ocean liner "Lusitania" was sunk by the German Navy in May 1915 [635], killing about 1200 people, including more than 100 U.S. citizens. In July, the British Admiralty set up the Board of Invention and Research and the Royal Navy set up its own Anti-Submarine Division. In the U.S., a Naval Consulting Board was set up in 1915 under the leadership of Thomas Alva Edison, but with little success. After the American entry into the war in 1917, British and French experts on underwater sound were called to a meeting with their American counterparts, followed by a meeting in Paris a year later. However, supersonic echo-ranging, sonar, was still in research by the end of the war. A more immediate success was met by the subterranean detection of telegraphic communications across the trenches [180].

In the U.S., even before its entry into the war, the National Academy of Sciences founded the National Research Council [620] with the aim to coordinate fundamental and applied research for defence purposes. It was a decentralised body under the chairmanship of the astronomer George E. Hale, who had excellent international contacts. Robert Millikan served as its vice-chairman. In May 1917, a group of British and French physicists, including Rutherford, informed their colleagues of their own research, primarily on submarine detection. Also there the success was very limited and did not have a significant influence on the outcome of the war. But the war certainly helped to organise governmental support and direction of science.

A very disturbing involvement of scientists in the beginning of World War I is their contribution to war propaganda in late 1914, seen from within as a "War of the Minds" [164]. Fritz Haber was a war enthusiast of the first hour, signing with other scientists, writers and artists the so-called "Manifesto of the Ninety-Three" in October 1914 [488]. This manifesto had the signature of 15 otherwise respectable scientists, including the six Nobel laureates Planck, Röntgen, Ostwald, Wien, Fischer and Baeyer. Taken away by a surge of patriotism, without having first hand information to support their claims and sometimes without even knowing the exact text, the signatories formulated their "protest to the civilized world against the lies and calumnies with which our enemies are endeavoring to stain the honor of Germany in her hard struggle for existence" [166]. A notable absentee from the list of signatures is Albert Einstein, who had arrived in Berlin in March 1914 to head the Kaiser-Wilhelm-Institute for physics, which was to be created in 1917. Einstein had been German, stateless, Austrian-Hungarian and Swiss in his life and was profoundly uninterested in the notion of nations [168]:

> I am far from keeping my international conviction secret. How close a man or a human organisation is to me only depends on my judgement of his desires and abilities. The state to which I belong as a citizen does not play the faintest role in my feelings; I regard the affiliation to a state as a business matter, rather like the relation to a life insurance.

With only three others[1], Einstein signed a "Manifesto to the Europeans" [495] in mid-October 1914, which stated that "no war has ever so intensively interrupted the cultural communalism of cooperative work as this present war does." The clear-sighted authors predicted:

> The struggle raging today will likely produce no victor; it will leave probably only the vanquished. Therefore it seems not only good, but rather bitterly necessary, that educated men of all nations marshall their influence such that –whatever the still uncertain end of war may be– the terms of peace shall not become the wellspring of future wars.

---

[1] It was redacted by the physiologist Georg Friedrich Nikolai, and signed by Einstein, the astronomer Wilhelm Julius Förster and the philosopher Otto Buek.

In the middle of nationalist fervour, this stayed the standpoint of a small minority.

The "Manifesto of the Ninety-Three" backfired completely, causing outraged reactions outside Germany. British scholars reacted with a countermanifesto, signed by 117 scholars including Bragg, Rayleigh and Thomson, published as a "Reply to the German Professors" in The Times [161]. It concluded:

> We grieve profoundly that, under the baleful influence of a military system and its lawless dreams of conquest, [Germany] whom we once honoured now stands revealed as the common enemy of Europe and of all peoples which respect the law of nations.

Both sides claimed that war had been forced on them, reflecting the view of their respective governments. The Würzburg professor Wilhelm Wien (see Section 5.1) went a step further and solicited colleagues to always balance the number of citations of British authors with an equal number of German papers. Stefan L. Wolff has meticulously analysed the history of Wien's ridiculous initiative [531] using the abundant correspondence of physicists during that time. He found that, in addition to opposing the perceived "Engländerei" of German researchers, it was motivated by an aversion against the domination of physics research by the liberal Berlin scientists. Soon, the initiative was taken over by more radical folkish-nationalist physicists like Philipp Lenard and Johannes Stark, trying to construct an antinomy between German and foreign physics. Opposition to the initiative was minor but notable. Friedrich Paschen, known for his contributions to atomic spectroscopy, and Emil Warburg, the president of the Physikalisch-Technische Reichsanstalt refused it vigorously. Max Planck, a rather moderate nationalist, argued that science and politics had to be kept strictly separate. Arnold Sommerfeld must have felt the antiscientific nature of the initiative, since he recommended to keep it secret. In any case, this attack against the global nature of science marks the infection of part of the German science community by views alien to scientific thinking. It culminated in the Third Reich in a movement in favour of a would-be "German physics".

World War I left about 8.5 million dead and more than 20 million wounded[2]. Of these, about 90,000 were killed by chemical weapons and many more left impaired. In German public opinion, the Nobel prizes attributed to German scientists shortly after the end of World War I, 1918 in physics (attributed in 1919) for Max Planck and the 1919 double prize for Johannes Stark in physics and Fritz Haber in chemistry, were taken as a "German victory" [489]. But at the end of the war, German scientists were excluded from international conferences for ten years [611]. German, a former language of sciences, was completely replaced by English.

---

[2]See https://www.britannica.com/event/World-War-I/Killed-wounded-and-missing

6.2 "GERMAN" PHYSICS

Bigotry, nationalism, racism and other non-scientific intrusions into science of course did not suddenly appear in 1914 out of nothing. Scientists are not living in a political vacuum, they are citizens and often engage in the political and social questions of their time. A good example may be Max Planck [519], respectful towards hierarchies, a supporter of monarchy during World War I and signatory of the "Manifesto of the Ninety-Three". He however maintained a respectable conservative attitude, refused any boycott of foreign institutions and lobbied for the access of women to higher education.

In the period between the two world wars, Planck actively participated in the reconstruction of intellectual life in Germany. Indeed, the first World War left Universities bankrupt. In 1920, Planck helped create the "Emergency Association of German Science", together with Fritz Haber and the former Prussian Minister of Culture Friedrich Schmidt-Ott. Its objective was to provide the academic community with a central institution of fund raising, both from governmental sources and from industry. This organisation continued to exist through the Third Reich as "Deutsche Forschungsgemeinschaft" (DFG) and was re-established in 1945. Planck intervened in favour of fundamental research and opposed the folkish-nationalist and racist views of Lenard and Stark. In the 1920s Stark published a book on "The present crisis in German Physics" [190] where he constructs an antinomy between an overrated theoretical physics, to which he attests a lack of intuitiveness, and an underrated experimental physics. Max von Laue dismantled his arguments in a review in 1923 [193], where he argued a difference in essence between the two branches of science, instead of a difference in value. He also rejected that it concerns only German physics since relativity and quantum physics shook up established convictions in all countries. He ended with the devastating conclusion: "All in all, we would have wished that this book had remained unwritten." Stark left academia in 1920, Lenard resigned from the German Physical Society in 1925.

Lenard and Stark had an early association to the Nazi party. While Adolf Hitler was locked up in the fortress of Landsberg and wrote his programmatic book "Mein Kampf", they published an essay entitled "Hitlergeist und Wissenschaft" [197]. In this pamphlet, they claim emergency[3]:

> Do not be deceived, the Aryan-Germanic blood, the carrier of its unique spirit, is already in the process of disappearing rapidly; a racially alien spirit has been busily at work for over 2,000 years. The exact same force is at work, always with the same Asian people behind it that had brought Christ to the cross, Jordanus Brunus to the stake, and that shoots at Hitler and Ludendorff with machine guns and confines them within fortress walls: It is the fight of the dark spirits against the torchbearers in an endeavor to eliminate their influence on earth.

---

[3]The English translations of documents in this section come from the anthology of Klaus and Ann M. Hentschel [487].

In this pathetic rhetoric, they pledge adherence to the national-socialist movement, "with Hitler beating the drum." In the same spirit, Lenard in his four-volume textbook "German Physics" [271] claims that relativity and quantum theory were Jewish inventions, to be replaced by superior Aryan-Germanic theory. In reality he revives the older conflict between theoretical and experimental physicists. In the introduction he writes: "The meaning of all well-tested natural laws is simple; they can thus be formulated not only using equations, but also using normal words. All undisputed scientific knowledge can be presented without a significant amount of mathematics." The textbook was discussed in a series of articles in the press 1936. The student Willy Menzel welcomed it in the Nazi propaganda newspaper "Völkischer Beobachter", Heisenberg called it "erroneous and misleading" in a subsequent article. Stark in response again dug up his old claim of the priority of experiment versus theory, especially "Jewish" theory. He made the outrageous claim that theory had nothing to do with the discovery of the positron, X-rays and quanta, and concluded: "Theory must be brought back within bounds from its presumptuous position." Anti-modernist and antisemitic prejudice was thus reunited by physicists disconnected from progress. In 1933, when Hitler came to power, Stark was made president of the PTR against the advice of experts. He also became president of the DFG in 1934.

The mounting anti-semitism started to attack several great scientists, including Albert Einstein. In 1933, when Hitler became chancellor of the German Reich, Einstein decided not to return from a visit in the U.S. In May 1933, Planck met Hitler and tried to intervene in favour of his Jewish colleagues. To no avail. The alignment of German science with national-socialist power proceeded. Planck finally gave up all official functions in 1938. The Nazi government and party tried to steer academic appointments, but with limited success. Also the movement for "Aryan physics" of Lenard and Stark only had a handful of second rate adherents, mostly among their own students and collaborators.

The prosecution of Jews in Germany, however, took full swing, when on November 9, 1938 a first organised pogrom by SA paramilitary forces and civilians ransacked Jewish homes, hospitals, businesses and schools. More than 250 synagogues were destroyed all over Germany. Tens of thousands of Jewish men were imprisoned in concentration camps. The events were widely reported around the world. The correspondent of The Telegraph wrote on November 11 [160]:

> An officially countenanced pogrom of unparalleled brutality and ferocity swept Germany today. Beginning in the early hours of this morning and continuing far into tonight, it puts the final seal to the outlawry of German Jewry. Mob law ruled in Berlin throughout this afternoon and evening and hordes of hooligans indulged in an orgy of destruction. I have seen several anti-Jewish outbreaks in Germany during the last five years, but never anything as nauseating as this.

The exodus of German Jews had started as soon as Hitler came to power. It now accelerated[4]. By 1935, when the so-called Nuremberg racial laws became applied, about 65,000 Jewish citizens had left, including some 1600 scholars, according to a list established by the "Notgemeinschaft deutscher Wissenschaftler im Ausland". The Nuremberg laws formalised the policy to remove Jews, socialists, communists and other disagreeable activists from civil service. By 1936, this had led to the dismissal of more than 1600 university teachers, about 20% of the total teaching staff, including 124 physicists, again according to the "Notgemeinschaft". Until 1942, about 330,000 more Jews were expelled, leaving all of their belongings behind. After the take-over of Austria in 1938, 150,000 Austrian Jews joined the forced expatriates.

The racial cleansing of Universities mostly affected the physics faculty in Berlin and Göttingen due to their size and liberal orientation. According to Hentschel [487], it basically cut the scientific output in Germany by half as determined by bibliometry. Moreover, he finds that "the number of papers by future emigrés is positively correlated with the novelty of the topics and negatively with the conservativeness of a journal." The loss is thus larger than the pure numbers suggest, since the more innovative physicists left. Also technology suffered. Between 1930 and 1940, the number of patents almost halved. Schrödinger, Franck and Haber left their positions in protest against Nazi policy, Sommerfeld went to Oxford in 1939.

Preferred host countries of the expatriates were Britain with about 10% and especially the U.S. with almost 50%. Relief organisations were created in many countries immediately after mass emigration started. In addition, influential individuals like Rudolf Ladenburg, Wigner, Einstein and Weyl did their very best to help displaced colleagues. Despite the difficulties of many emigrés to become integrated in their host countries, these clearly profited from the influx of brain power, both quantitatively and qualitatively. Especially the U.S. with its rapidly expanding R&D sector for nuclear physics presented many opportunities to newly arriving physicists. Notwithstanding strict security screening, many of them joined the weapons program after the U.S. entered into the war in 1941 (see Section 6.4).

In the period between 1939 and 1941, the policy of the Nazi government escalated from enforced emigration to mass murder [569]. In January 1942 at the Wannsee Conference held near Berlin, Nazi officials formulated the plan for a "final solution to the Jewish question", which culminated in the Holocaust, killing two thirds of the Jewish population in Europe. Bruno Tesch, a former student of Fritz Haber, was instrumental in the extermination at industrial scale. The Hamburg company Tesch & Stabenow which he co-founded delivered the granular prussic acid (HCN) cartridges under the brand name Cyclon B to the gas chambers. By 1945, only about 25,000 Jews in Germany survived, out of an initial population of more than 560,000. Tesch and his deputy Weinberger were sentenced to death by a British court marshal and executed in

---

[4]Hentschel [487] has compiled the available statistics on emigration in general and on that of scholars in particular. The numbers quoted here are based on his findings.

1946. Prominent leaders of Nazi Germany who planned and carried out the Holocaust and other war crimes were trialed in a series of military tribunals in Nuremberg shortly after the war. The prosecution of concentration camp personnel still continues to this day.

6.3 HITLER'S NUCLEAR BOMB?

Directly after the discovery of nuclear fission by Hahn and Strassmann and its explanation by Meitner and Frisch (see Focus Box 6.1), a group of scientists formed in Germany, officially called "Working group for nuclear physics" and informally known as "Uranverein", to look into technical applications of the enormous energy stored in nuclear matter. The Ministry of Education with its Reichs Research Council (Reichsforschungsrat, RFR) as well as the Ministry of War were notified about the potential technical and military application of nuclear power. In parallel, Siegfried Flügge, an assistant to Otto Hahn, speculated about applications in a published paper [283], taken up by the press [485]. The early Uranverein had members from Göttingen (Georg Joos, successor of Franck, and Wilhelm Hanle), Heidelberg (Bothe and Wolfgang Gentner), Leipzig (Gerhard Hoffmann, successor of Debye, and Robert Döpel), as well as Berlin (Geiger). It was a rather loose working group, there was no effort by political authorities to centralise this research. The work was interrupted after a few months since some of its members were drafted for military service. Meanwhile, Bohr and Wheeler worked out the theory of nuclear fission using the liquid drop model [282].

A new effort of collaboration was started after the beginning of hostilities in September of 1939. This second Uranverein now included Flügge, Heisenberg from Leipzig and Carl Friedrich von Weizsäcker (see Focus Box 5.11) from the KWI for Physics in Berlin, Walther Gerlach from Munich and the physical chemist Paul Harteck from Hamburg. Kurt Diebner, advisor of the Army Ordnance Office (Heereswaffenamt, HWA), represented the military. At about the same time, the KWI for Physics in Berlin was taken out of the Kaiser-Wilhelm-Gesellschaft and subordinated to the military, with Diebner as its administrative director replacing the Dutch Peter Debye who went to Cornell. It was especially supported by Erich Schumann, who also taught courses on acoustics and explosives, his areas of research, at Berlin University. Schumann was the doctoral advisor to Wernher von Braun, who had obtained his doctorate in 1934. Another strong political supporter was Abraham Esau, president of the Reichs Research Council.

According to a 1939 memorandum by Heisenberg [287] and his post-war account [307], it was clear to the Uranverein members from the beginning that two applications of uranium fission were possible: a nuclear reactor using slow neutrons and moderately enriched uranium; or a nuclear explosive using uranium almost pure in the isotope ^{235}U and fast neutrons. In fact, fission by slow neutrons is not usable for explosives. With all other parameters the same, the energy yield in a fission by thermal neutrons is a factor 10^8 less than by fast neutrons. The physics of nuclear fission is explained in Focus Box 6.1.

The entrance of neutrons into a nucleus is not hindered by the Coulomb barrier, they can become absorbed, $(A, Z) + n \rightarrow (A + 1, Z)^*$. The final state nucleus is in an excited state and can undergo fission into two more tightly bound lighter nuclei. An example is the induced fission of uranium: $^{235}\text{U} + \text{n} \rightarrow {}^{137}\text{Ba}^* + {}^{97}\text{Kr}^* + 2\text{n}$. The nuclei in the final state are typical examples, many others are possible. The excited nuclei of the final state can cool off emitting further neutrons, the average total number is about 2.2.

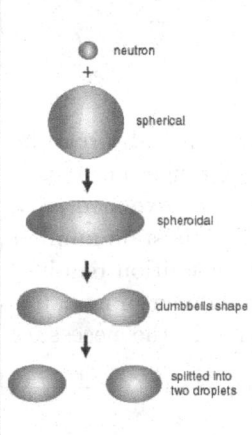

The binding energy is roughly 7.5MeV per nucleon for ^{235}U and 8.4MeV for the two daughter nuclei (see Figure 5.7). Fission thus liberates the enormous energy of almost 1MeV per nucleon, or more than 200 MeV per uranium nucleus. So why doesn't it occur spontaneously for most nuclei? The answer is the same as for e.g. α decay: there is an energetic barrier to be crossed which is schematically shown below. When a nucleus is excited, e.g. by an incoming neutron, its shape is deformed from the spherical ground state as shown on the left. In the simplest case, there is an elongation along a single axis, by a distance ϵ. Three cases can occur [567, p. 320]:

(a) For light nuclei, deformation increases the potential energy steeply, as shown by the dotted curve. The nucleus will normally return to its spherical ground state by getting rid of the excess energy, emitting e.g. neutrons or photons. (b) For heavier nuclei, the barrier is shallower as shown by the solid curve. (c) For nuclei above $Z^2/A \simeq 50$, the barrier disappears (dashed curve) and fission occurs spontaneously. The height of the barrier, E_a, is called the activation energy.

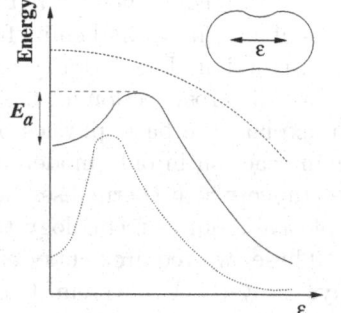

It mainly depends on the interplay between surface energy (which increases with ϵ) and Coulomb energy (which decreases with ϵ). The simplest deformation is into an ellipsoïd with elongation ϵ, changing the surface and Coulomb terms of the liquid drop model (see Focus Box 5.11) like this:

$$a_2 A^{\frac{2}{3}} \rightarrow a_2 A^{\frac{2}{3}} \left(1 + \frac{2}{5}\epsilon^2\right) \quad ; \quad a_3 \frac{Z^2}{A^{\frac{1}{3}}} \rightarrow a_3 \frac{Z^2}{A^{\frac{1}{3}}} \left(1 - \frac{1}{5}\epsilon^2\right)$$

However the quantum mechanical terms also play a role. The isotope ^{235}U is an even-odd nucleus, ^{238}U a more tightly bound even-even one. This explains why fission of the latter is only induced by energetic neutrons with an energy above 6 MeV, while the lighter isotope is fissioned by neutrons of any energy.

Focus Box 6.1: Nuclear fission

The daughter nuclei of the fission process are in an excited state and fall back to the ground state by emitting neutrons and photons. If there are enough neutrons and if they have the right energy range, a chain reaction can be caused as sketched on the right. ^{235}U produces on average 2.2 neutrons per fission. The ratio $k = n_{i+1}/n_i$, the effective number of neutrons n produced in stage $i+1$ divided by the number of neutrons from the previous stage i, can be used to characterise the process.

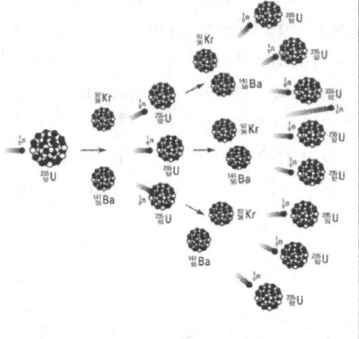

The effective number of neutrons only counts those which are not absorbed by other materials or escape, and which have the right energy range. For $k < 1$, the process is called sub-critical and the reaction will die out exponentially. Critical conditions, $k = 1$, correspond to a steady state with a constant number of neutrons and a constant energy production: this is the condition required for a nuclear reactor. In the over-critical state, $k > 1$, the number of neutrons and fission reactions grows exponentially with time. This is the necessary condition for a nuclear explosion.

Focus Box 6.2: Nuclear chain reaction

The rare component of natural uranium, ^{235}U, can be fissioned with neutrons of any energy and emits further neutrons, thus potentially establishing a chain reaction. The majority component of natural uranium, ^{238}U has a high absorption cross section for fast neutrons and thus hinders the chain reaction. We explain the basic physics of nuclear chain reactions in Focus Box 6.2. A chain reaction in only moderately enriched uranium requires a moderator, a substance which thermalises neutrons without absorbing them. Usage as an explosive requires technology to separate the two isotopes of uranium.

These two requirements defined the strategy for pursuing applied nuclear physics by the Uranverein. Concerning the reactor line of research, a search for an efficient moderator concentrated on two substances: heavy water D_2O and pure carbon in the form of graphite. The absorption cross sections for the two materials were measured in 1940, by Döpel and Heisenberg in Leipzig for heavy water, and by Bothe and collaborators in Heidelberg for graphite. The pure graphite used for absorption measurements by the latter had residual impurities of boron and cadmium, resulting in much too high absorption [293, 302]. It was thus decided to concentrate the efforts on moderation with heavy water.

Heavy water is a natural component of water, though only at a level of 0.01%. The main supplier at the time was a company in Norway close to Rjukan, which was running a hydroelectric power plant delivering 108 MW. Part of the power was used for nitrogen separation and fertiliser production. Another part was used for water electrolysis. In the residual water after electrolysis, the proportion of heavy water is enriched. After the German invasion of Norway in April 1940, the plant was requisitioned and improved by Harteck. Its peak production rate was about 200 litres of D_2O per month. The plant was sabotaged by partisans and bombarded by allied forces several times, in 1943 it had to cease production. The required uranium was ordered from the Auer company in Oranienburg, run by the SS. It used natural uranium coming from mines in Jáchimov, close to Karlsbad in the occupied Sudetenland. However, purification experiments by Harteck and others with gaseous UF_6 failed.

In 1940, Weizsäcker [292] pointed out that in a nuclear reactor also the isotope ^{239}U would be produced. In its decay chain, elements 93 and 94, neptunium and plutonium, would appear, with fission properties similar to ^{235}U. There was no experimental proof, since this would have required a working reactor, but the conclusion was considered likely [307]. Indeed plutonium, ^{239}Pu, can be used to construct a nuclear bomb. This possibility of breeding nuclear fuel was used in the Manhattan Project.

In summer of 1941, about 150 litres of heavy water were available and an experimental set-up, named L-IV, was built in Leipzig to measure the multiplication rate for neutrons in layers of uranium, separated by heavy water in a spherical geometry. Using metallic uranium, Robert and Clara Döpel together with Heisenberg measured a multiplication rate greater than one [296] in 1942, indicating that a scaled-up version of the reactor would sustain a chain reaction and produce energy. The L-IV exploded in June 1943, when air leaked in during an inspection, creating the world's first nuclear accident.

The results were reviewed in 1942, in meetings with the minister of Science and Education, Bernhard Rust, and the minister of Armaments and War Production, Albert Speer. It was decided to continue research on the existing low level. The declared goal was to construct a nuclear reactor capable of driving machines, like in U-boots or other marine vessels. The KWI for Physics in Berlin was returned to the Kaiser-Wilhelm-Gesellschaft and Heisenberg appointed as director. Under difficult war-time conditions, the research led to the construction of an experimental reactor in Berlin Dahlem and a subsequent model in an underground facility in Haigerloch close to the Black Forest. In early 1945, the latter came close to a critical multiplication of neutrons without actually reaching it. On April 22, 1945, Haigerloch was occupied by U.S. troops. The reactor was confiscated, dismantled and later destroyed.

The sequence of events briefly sketched above is basically undisputed by historians and contemporary witnesses. What is the subject of sometimes passionate debate is the motivation of German physicists in the Uranverein and the reason for their failure to pursue the construction of a nuclear bomb. Heisenberg and Weizsäcker emphasise their reluctance to work on nuclear weapons due to lack of resources and time. They advance that despite the already difficult military situation, the military in 1942 still believed in a soon-to-come Final Victory ("Endsieg"). For themselves, they state their belief that the war would have fatally ended anyway before a nuclear bomb would have been available, and that non-military applications would be more important after the war ended.

The Dutch-American Samuel Goudsmit, proponent of electron spin together George Uhlenbeck (see Section 5.3), developed a very different point of view. He was the chief scientific advisor of the Alsos mission, an intelligence exercise part of the Manhattan Project, created after the allied invasion of Italy in 1943. Its aim was to collect intelligence related to German nuclear research, but also chemical and biological weapons. Their assignment was to seize personnel, documents and material to evaluate the state of German research and prevent it from falling under concurrent control. Alsos personnel closely followed the progressing front line and sometimes even crossed into enemy territory. They reached the left-overs of Uranverein in Hechingen, where the KWI for Physics had relocated. Among others, Alsos took Heisenberg, Hahn, von Laue and Weizsäcker into custody and interned them in the Farm Hall estate in England, where their conversations were surreptitiously taped. In many articles published in 1946 and 1947, as well as his book "Alsos" [458], Goudsmit concludes that the Germans did not get anywhere close to creating a weapon. He basically claims that his German colleagues had not understood how a nuclear bomb would function at a basic level, let alone at an engineering level. Heisenberg took this as an attack on his understanding of nuclear physics, and tried to counter it with an article in Nature [308]. An exchange of letters followed, where each one stood his ground, until both men agreed that it would be better to only discuss physics.

While incompetence and unwillingness of the protagonists and the Nazi authorities are claimed to be the main drivers for the Uranverein's failure, it is clear that Germany did not have and did not want to create the necessary resources for a serious attempt on nuclear weapons. The funds made available to the Uranverein's activities, estimated to be about 2M$, were at a per mil level of the U.S. effort in the Manhattan Project. At the height of Germany's activities in 1942, no more than about 70 scientists worked on nuclear applications. The only German development effort during World War II which came anywhere near the Manhattan Project was the rocket program led by Wernher von Braun in Peenemünde. In 1942, it employed an estimated 2000 scientists and twice as many workers and costed at least half a billion dollars.

The historical discussion on the motivations of the German nuclear scientists often converges on a single person and a single incident, Heisenberg's 1941 visit to Niels Bohr in Copenhagen, occupied by German troops since April 1940. The visit is to be seen in the context of the propaganda lecture tours of Heisenberg, von Weizsäcker and others to occupied countries [470]. Heisenberg used such an occasion to see Bohr in September 1941. On the advice of Weizsäcker, he brought up German nuclear research and the possibility of a bomb based on the principle of nuclear fission. Bohr was deeply shocked and angry, and stopped the meeting abruptly. Heisenberg and von Weizsäcker claimed after the war that Heisenberg was merely seeking Bohr's advice on what to do, insinuating that he was trying to withhold crucial information from Nazi authorities [402]. But Bohr took it as a threat by a representative of an enemy country, trying to squeeze information about allied activities out of him and even ask for his collaboration. Heisenberg's point of view was reflected in Robert Jungk's very popular book "Brighter than a Thousand Suns" [333] published after the war, based on interviews with von Weizsäcker and Heisenberg himself. When Bohr learned about it in 1957, he drafted a letter to Heisenberg in response, but never sent it [532]. In the draft, he described what he recalled from the meeting, his anger and disappointment.

Since there are no witnesses to this encounter, which destroyed the old friendship and close working relationship between the two men, it has inspired artistic interpretations. Poetic accounts are to be discovered in a short story by Per Olov Enquist [494] and a novel by Jérôme Ferrari [619]. The most remarkable re-enactment for me is Michael Frayn's award winning play "Copenhagen" [518], which first ran in London in 1998 and on Broadway in 2000. It triggered a lively discussion among scholars, e.g. in the New York Review of Books [523, 524, 528, 533, 534, 538, 539, 547, 637]. In 2002, it also prompted Niels Bohr's family to release the unsent letters to Heisenberg which Bohr had drafted between 1957 and 1962 [532].

There are many books describing the German effort for a nuclear reactor and nuclear explosives during World War II; some I have consulted are listed in the Further Reading section of this chapter. If you wonder which books to read, I recommend that you consult the very fair comparative review of Klaas Landsman [535]. Some of the authors subscribe to Heisenberg's and von Weizsäcker's post-war claim that they, for lack of collusion with Nazi authorities, deliberately held back crucial information such that there was no serous bomb project. Their line of arguments is that they actively supported the reactor project, with mainly post-war applications in mind. In an interview, Edward Teller (see Section 6.4) went so far as to say that Heisenberg sabotaged the bomb effort [661].

Goudsmit's point of view of German incompetence, supported e.g. by Paul L. Rose [507], is often criticised, documentation and witnesses at hand. Heisenberg and his Uranverein wrote at least two reports to their authorities, which show that they clearly understood the difference between a controlled chain reaction in a reactor and an uncontrolled one in a bomb [662], differences in the required fuel and moderator materials as well as the process itself. Von Weizsäcker even tried to get a patent on plutonium breeding from a nuclear reactor and its usage for explosives [662]. An argument in favour of incompetence seems to be the reaction of German internees in Farm Hall [525], when they heard about the nuclear bombing of Hiroshima and Nagasaki. It appears that Heisenberg had never seriously calculated the critical mass (see Focus Box 6.3) of pure ^{235}U or plutonium which it would take to construct a nuclear bomb. Only at Farm Hall did he get a correct estimate, even though all necessary information had long been available to him [626].

Goudsmit generalised his findings [458]: "I think the facts demonstrate pretty conclusively, that science under fascism was not, and in all probability

Producing more neutrons in each step of a fission reaction is not sufficient to establish a chain reaction. As the neutrons diffuse through the material, they must have a sufficient chance to cause another fission before being absorbed or lost through the outer boundary. If absorption is negligible, the number of produced neutrons is proportional to the volume, the number of lost ones to surface. There is thus a critical size, where both are equal. For larger sizes, the number of fissions grows exponentially, for smaller ones is dies out. For a known shape, this size corresponds to a mass of fissionable material, the critical mass.

In pure ^{235}U, neutron absorption is negligible and neutrons of about 1 MeV kinetic energy are fast enough to cause a chain reaction in a material of sufficient size, before the assembly blows up thermally. In a large amount of active material, the number of neutrons per unit volume, $N(\vec{x}, t)$, follows a diffusion equation, with diffusion coefficient D:

$$\frac{\partial N}{\partial t} - D\nabla^2 N = \frac{\nu - 1}{\tau} N$$

This is a continuity equation for N and its current density $\vec{j} = -D\vec{\nabla}N$, with a source term on the right describing the number of neutrons produced per unit time, using the average number ν of neutrons produced per fission and the mean time τ between fission reactions. A stationary solution, corresponding to reactor conditions, is defined by $\partial N/\partial t = 0$. For an exponential growth in fission reactions, we must have $N(\vec{x}, t) = N_1(\vec{x})e^{\nu' t/\tau}$, where ν' is the effective number of neutrons amenable to fission reactions and τ is the average time between two of those. We assume a spherically symmetric active region, such that $N_1(\vec{x}) = N_1(r)$. In addition, the neutron density and current must be continuous at the outer boundary. We then find for the spatial part of the neutron distribution:

$$\nabla^2 N_1(r) + \frac{\nu - 1 - \nu'}{D\tau} N_1(r) = 0 \; (r < R) \quad ; \quad \nabla^2 N_1(r) = 0 \; (r > R)$$

The solution is $N_1 = \sin{(\pi r/R)}/r$, provided that $\nu' = (\nu - 1) - \pi^2 D\tau/R^2$. For an exponential growth, ν' must be greater than zero. There is thus a critical radius $R_c = \pi l$ with the so-called diffusion length $l = \sqrt{D\tau/(\nu - 1)}$. Using numerical values available at the time for the required constants [299, 626], one gets a critical radius of about 13cm and a critical mass of almost 200kg, an overestimate by roughly a factor of 3. This is due to the fact that the diffusion equation to be valid requires a size much larger than the diffusion length, which is not the case. The full calculation is most easily done by numerical simulation. Enrico Fermi, Stanislaw Ulam, Nicholas Metropolis and John von Neumann [310, 461, 464] developed the so-called Monte-Carlo method for this purpose. It is used in particle physics today to simulate the passage of particles through active and passive materials.

Focus Box 6.3: Critical mass

could never be, the equal of science in a democracy." That totalitarian regimes are unable to organise a coherent scientific effort, even for military purposes, unfortunately turned out to be wrong. Shortly after his book first appeared in 1947, the Soviet Union conducted its first test of a nuclear bomb.

6.4 THE MANHATTAN ENGINEER DISTRICT

Landsman summarises the conditions which had to be satisfied for a successful crash program to develop and produce a nuclear bomb [535]:

1. There was a strong initial drive by a small group of physicists to get the program off the ground.

2. From a certain point in time, there was unconditional support from he Government.

3. Practically unlimited industrial resources and man power were available.

4. There was an unprecedented concentration of brilliant scientists working on the project.

In Germany the first condition was met, but none of the others were. Heisenberg always emphasised the lack of conditions 2 and 3. In addition, the last condition could not have been satisfied because of the massive forced exodus of first class physicists from Germany and occupied countries. On the contrary, all conditions were clearly met in the U.S. in the context of what is called the Manhattan Project[5], soberly described in the report "Atomic Energy for Military Purposes" by Henry DeWolf Smyth [305], a massive secret effort to develop nuclear weapons with the official code name "Manhattan Engineer District".

Allied efforts to use nuclear energy for weaponry started on both sides of the Atlantic shortly after the discovery of nuclear fission. Just before the first hostilities between Germany and Poland began in September 1939, the Hungarian émigré Leo Szilard and Albert Einstein wrote a letter to President Roosevelt[6] alerting him to the possibility that Germany would use nuclear fission for military purposes. President Roosevelt created a governmental committee coordinating and modestly funding nuclear research, both on a reactor and on explosives. In 1940, Alfred Nier from Minnesota and his colleagues at Columbia verified the prediction by Bohr and Wheeler that the isotope ^{235}U was responsible for fission with fast neutrons [290, 291]. They concluded [291]: "These experiments emphasize the importance of uranium isotope separation on a larger scale for the investigation of chain reaction possibilities in uranium."

---

[5]The chronology of the Manhattan Project and many primary documents can be found on the websites of the U.S. Department of Energy, https://www.osti.gov/opennet/manhattan-project-history/, and the National Security Archive, https://nsarchive2.gwu.edu/NSAEBB/NSAEBB162/index.htm

[6]https://www.osti.gov/opennet/manhattan-project-history/Resources/einstein_letter_photograph.htm#1

In Britain, Marcus Oliphant in Birmingham had attracted the émigrés Rudolph Peierls and Otto Frisch to his University. Alarmed by the possibility that Germany would make an effort towards military use of nuclear fission, Frisch and Peierls wrote a two-part paper, a technical and a non-technical one collectively called the "Frisch-Peierls Memorandum"[7], to explain the physics and rough parameters of a nuclear bomb. The original is kept in the Bodleian Library of Oxford University. Due to only approximately known properties of Uranium, they came up with an overly optimistic estimate of the critical mass, roughly one pound. The memorandum made it into governmental channels and an advisory committee, called MAUD, was formed to coordinate research on nuclear explosives at British universities. Experimental data confirmed the memorandum's punch line and convinced sceptics like James Chadwick that a nuclear bomb was feasible. The final report of the MAUD committee in 1941 had a considerable impact on both sides of the Atlantic [614, Section 3.7]. On the initiative of Gregory Breit, the National Research Council set up a self-censorship committee in 1940, to withhold publications on nuclear issues until after the war [429]. All relevant research papers were only published after the war, but quoting their original submission date.

The Japanese attack on the U.S. fleet in Pearl Harbor of December 7, 1941, catapulted the United States into the war. The following month President Roosevelt secretly gave his tentative approval to proceed with the construction of a nuclear bomb. By early 1942, a series of military defeats in the Pacific caused the pace to be accelerated, following two lines of research and development, a uranium and a plutonium line. A bomb made of pure ^{235}U required isotope separation at an industrial level. Enrichment seemed feasible by gaseous diffusion through fine meshes, which Frisch and Peierls had already envisaged; by gas centrifuge; or by spectroscopic separation. A bomb made of plutonium seemed a quicker route, since ^{239}Pu can be chemically separated from other elements. However, it must be bred in a nuclear reactor by the reaction chain ^{238}U + n $\to$ ^{239}U $\to$ ^{239}Np e$^-\bar{\nu}_e$ $\to$ ^{239}Pu 2e$^-$2$\bar{\nu}_e$. The two beta decays have half lives of 23 minutes and 2 days, respectively. Plutonium itself has a long lifetime, more than 24,000 years. This line of research was led by Arthur Compton at the new Metallurgical Laboratory (Met Lab) at the University of Chicago, today the Argonne National Laboratory. Enrico Fermi's work at Columbia and Glenn T. Seaborg's work at Berkley were moved to Chicago. A review in May 1942 found that progress had been made on all paths to uranium enrichment as well as plutonium breeding. However, there was no clear front-runner. Since there was no time to be lost if the bomb was to be available for war-time usage and since there was no room for wrong decisions, it was decided to follow all paths in parallel. At the same time, large parts of the project were put in the hands of the Army Corps of Engineers, which set up the Manhattan Engineer District (MED) to manage the project,

---

[7]See transcripts in http://www.atomicarchive.com/Docs/Begin/FrischPeierls.shtml and http://www.atomicarchive.com/Docs/Begin/FrischPeierls2.shtml

under the leadership of general Leslie R. Groves. A site at Oak Ridge, Tennessee was selected for the construction of production plants, concentrating on spectroscopic and gaseous diffusion and dropping the centrifuge process. J. Robert Oppenheimer was appointed head of the bomb research and design laboratory to be built at Los Alamos, New Mexico. In December 1942, President Franklin Roosevelt gave his final authorisation to proceed with construction of the nuclear bomb.

On December 2, 1942, researchers in Chicago headed by Enrico Fermi achieved the first self-sustaining chain reaction in a graphite-moderated uranium reactor. Based on this success, DuPont company constructed an experimental production reactor, and a pilot chemical separation facility at Oak Ridge. Met Lab scientists designed the water-cooled plutonium production reactors for the Hanford site in the U.S. state of Washington. Starting in 1943, three production reactors and corresponding chemical separation plants were built, with the first one operational in late September 1944. Los Alamos received its first plutonium from Hanford in early February 1945. Neutrinos produced by this reactor were used by Frederic Reines and Clyde Cowan to directly demonstrate the existence of the electron neutrino ν_e in 1953 (see Section 9.1).

The isotope separation at Oak Ridge went into production rather slowly due to the immense technical difficulties with both electromagnetic and gaseous diffusion technique at industrial dimensions. Both were eventually successful, as well as the liquid thermal diffusion technique brought in as a back-up. In the end, it took the combined efforts of all three facilities to produce enough enriched uranium for the one and only uranium bomb produced during the war.

Meanwhile at the Los Alamos site, close to Santa Fe in New Mexico, scientists and engineers were busy turning theory into a viable bomb design. Newcomers were given a series of lectures by Robert Serber and Oppenheimer [299]. The problem is how to almost instantly bring together sub-critical pieces of fissionable material to exceed the critical mass. Efforts for the uranium bomb, led by Seth Neddermeyer, concentrated on the rather simple gun-type design, where a piece of uranium is shot into a uranium target by ordinary explosive. Plutonium, on the other hand, is likely to melt before the two pieces could come together. Von Neumann proposed an alternative ignition method, where the fissionable material is compressed by a surrounding explosive to reach critical mass density. The engineering for this method is much more complex and required a test of the technology.

In June 1944, allied troops landed in France and the war in Europe entered into its final phase. In parallel, the Alsos mission had determined that Germany was nowhere near mastering nuclear weapons technology. Germany as a target for a nuclear bomb thus ceased to be of primary interest and Japan came into focus, with the aim to end the war in the Pacific. In the first months of 1945, the facilities at Oak Ridge and Hanford produced enough enriched uranium and enough plutonium for at least one bomb each. At Los Alamos, bomb designs were finalised and ready for testing.

With the construction of a nuclear bomb at arm's length, a debate started on whether and how to use it. In spring of 1945, the Secretary of War of the Truman administration formed a so-called Interim Committee to examine the question. It was composed of representatives of the administration, the military, industry and science. Robert Oppenheimer, Enrico Fermi, Arthur Compton, and Ernest Lawrence served as scientific advisors. The debate turned around the question whether it was a sufficient threat to Japan to announce and launch a demonstration bomb on uninhabited territory, or if a target in Japan was to be hit in an unannounced attack. The aim was to stop the war in the Pacific either way. The committee rejected the first alternative as too dangerous, because of the possibility of enemy preparation and intervention, but also the risk that the demonstration could fail. It was thus decided to recommend the launch on a military complex surrounded by workers' homes in a Japanese city. The committee also recommended that after the war ended, nuclear technology should be controlled by an international organisation, which would have the right to inspect installations on-site. In the absence of such an organisation, it advocated to try and preserve American superiority as long as possible. The conclusions were challenged by scientists at Met Lab, including James Franck, Glenn Seaborg and Leo Szilard. They argued that a preemptive strike on Japan without a prior threatening demonstration would inevitably cause an arms race and that international control of the proliferation of nuclear arms was of primordial importance. However, the Interim Committee stuck to their point of view, which was communicated to President Truman during the Potsdam conference.

To verify the technology of implosion ignition required for a plutonium bomb, a test was conducted in mid-July 1945 in the New Mexico desert, under the code name "Trinity". It was attended by several hundred people, including Leslie Groves, James Chadwick, Enrico Fermi, Richard Feynman, Robert Oppenheimer and John von Neumann. The test was preceded by exploding a demonstration device for the implosion mechanism, with almost 100t of conventional explosive mixed with radioactive material. In the early morning of July 16, the full nuclear device was fired. A photograph taken from 20 miles away is shown in Figure 6.1. Richard Feynman witnessed the test from behind a truck windshield [438]. He remembered[8]:

> So I look back up, and I see this white light changing into yellow and then into orange. The clouds form and then they disappear again; the compression and the expansion forms and makes clouds disappear. Then finally a big ball of orange, the center that was so bright, becomes a ball of orange that starts to rise and billow a little bit and get a little black around the edges, and then you see it's a big ball of smoke with flashes on the inside of the fire going out, the heat.

---

[8]The quote comes from a transcript of Feynman's lecture at Santa Barbara in 1975, which you can listen to at https://www.youtube.com/watch?v=hTRVlUT665U

All this took about one minute. It was a series from bright to dark, and I had seen it. I am about the only guy who actually looked at the damn thing, the first Trinity test. Everybody else had dark glasses, and the people at six miles couldn't see it because they were all told to lie on the floor. I'm probably the only guy who saw it with the human eye.

Finally, after about a minute and a half, there's suddenly a tremendous noise – BANG, and then a rumble, like thunder – and that's what convinced me. Nobody had said a word during this whole thing. We were all just watching quietly. But this sound released everybody – released me particularly because the solidity of the sound at that distance meant that it had really worked.

Figure 6.1 Amateur photograph of the Trinity test explosion of the first nuclear bomb on July 16, 1945. (Credit: Photo courtesy of Los Alamos National Laboratory)

President Truman was promptly informed of the successful test and used this new strengthening of the U.S. position in the negotiations about a post-war order for Europe. He did not know that the Soviet Union had information about the U.S. program since 1941 and was well aware of the situation. The final order for a nuclear attack on Japan was given on July 25, in a directive drafted by Leslie Groves. It ordered the Army Air Force to attack Hiroshima, Kokura, Niigata, or Nagasaki (in that order of preference) as soon after August 3 as weather permitted.

The first bomb, a uranium gun-type nicknamed "Little Boy", was dropped on Hiroshima on August 6, 1945. It detonated half a kilometre above a Japanese army training field. It contained little more than the critical mass of ^{235}U, with an effective explosive power of 15k tons of TNT, i.e. 63 TJ. This actual yield was only 3% of the theoretical energy yield; most of the ^{235}U was dispersed in the explosion before contributing to the chain reaction. The bomb caused widespread death and destruction in the densely populated city of Hiroshima. An estimated 70,000 people died as a result of initial blast, heat, and radiation effects[9]. By the end of the year, some 30,000 more were killed by radioactive fallout and other collateral effects. There is no official statistics of the long-term death toll due e.g. to increased probability of cancer, but it may amount to similar numbers.

The second launch, with a plutonium implosion-type bomb nick-named "Fat Man", took place over Nagasaki on August 9, 1945. The target had been chosen by the aircraft commander because of technical difficulties and bad weather, even though it was low on the pre-established list of targets. The bomb exploded again half a kilometre above the city, in the middle between two war industries, Mitsubishi-Urakami Torpedo Works and Mitsubishi Steel and Arms Works. Despite a partial evacuation, there were almost 200,000 people living in Nagasaki. The estimated yield was 21 kt of TNT, thus 40% larger than that of "Little Boy". Because of the topography of the area and launch site outside of the city, the destructions were less than at Hiroshima, but nevertheless substantial[10]. About 40,000 people died immediately, 70,000 within a year and twice as many within the next five years. In the area immediately under the detonation, the death rate was comparable to that at Hiroshima.

As a consequence of these two attacks, Japan offered surrender to the Allies on August 10, provided that Emperor Hirohito be spared. On August 12, the U.S. accepted the surrender, on the condition that the Emperor would retain only a ceremonial role. World War II had ended due to nuclear power.

Although collaboration across science, the military and industry, piloted by government, was first implemented in World War I, the Manhattan Project pushed back all previous boundaries. At its height, it employed more than 130,000 people and costed an estimated 2 billion dollars in 1945 [360] (equivalent to ten times that much in today's US$). In Los Alamos alone, it united the *crème de la crème* of nuclear physicists of the time. The DoE web site[11] lists Hans Bethe, James Chadwick, Albert Einstein, Enrico Fermi, Richard Feynman, James Franck, Klaus Fuchs (see Section 6.5), Joseph Rotblat, Glenn Seaborg, Robert Serber, Leo Szilard, Edward Teller, Eugene Wigner and Herbert York, no less than eight Nobel laureates. This count does not include the numerous other contributors at Universities and visitors from abroad like Niels and Aage Bohr. It is thus clear that Landsman's fourth condition was more than fulfilled, a concentration of brain power that could have been equalled nowhere in the world.

It is not possible here to seriously estimate the total death toll of World War II [586]. In Europe, many tens of millions died under the reckless attack of Nazi Germany, including six million Jews and other minorities, victims of the Holocaust. The loss of life in the Pacific war was equally horrific, especially in China, where an estimated 15 to 20 million people died during the Japanese

---

[9]The death toll estimations come from the U.S. Department of Energy, https://www.osti.gov/opennet/manhattan-project-history/Events/1945/hiroshima.htm
[10]The death toll estimations come from the U.S. Department of Energy, https://www.osti.gov/opennet/manhattan-project-history/Events/1945/nagasaki.htm
[11]https://www.osti.gov/opennet/manhattan-project-history/People/people.htm

occupation. The total death toll is often quoted to be of the order of 65 million people, with about 27 million casualties in the Soviet Union alone. Thus only a small but significant percentage is due to the use of nuclear weapons, which luckily enough has so far stayed the only one.

6.5 COLD WAR PHYSICS

Tensions between the allies agains the Nazi regime appeared immediately after the end of World War II[12]. The strict security measures in the Manhattan Project succeeded to keep the bomb program from being spied on by German and Japanese enemy intelligence, such that only rumours leaked. Soviet intelligence, however, was more successful, probably also because of people with sympathy for communism in Britain and the U.S. The Soviet government learned about the nuclear bomb plans already in 1941, a year ahead of the formation of the Manhattan Engineer District. The information allegedly came from within the so-called "Cambridge Five" spy ring in Britain, whose members were Guy Burgess, Donald Maclean, Kim Philby, Anthony Blunt and John Cairncross[13]. At the same time the sudden drop of American and British publications on nuclear research caught the attention of Soviet physicists, but interestingly enough not German ones, probably since they though they were far ahead anyway. Soviet intelligence made intensive attempts to penetrate via key laboratories like the University of California at Berkeley and the Chicago Met Lab, attempts mostly stopped by FBI and internal counterintelligence. However, one central figure, the British physicist Klaus Fuchs, made it into the heart of the Manhattan Engineer District, working with the Theoretical Division at Los Alamos. After returning home he continued to inform the Soviet Union about British bomb research and development. After his uncovering in 1950, he served nine years in prison in the U.K., moved to East Germany and became one of the GDR's scientific leaders. Additional engineering information was funnelled to Soviet channels by Julius and Ethel Rosenberg, née Greenglass, and her brother David Greenglass. The Rosenbergs were executed in 1953 despite an international campaign in their favour.

Notable scientist spies in the Canadian part of the Manhattan Project were Alan Nunn May, a collaborator of Chadwick, and Bruno Pontecorvo from Enrico Fermi's group. Both worked on the water-moderated nuclear reactor at Chalk River, Ontario, and returned to Britain after the war. Following the arrest of Klaus Fuchs, Pontecorvo's Soviet handlers became worried that he would be uncovered, and in 1950 Pontecorvo covertly went to the Soviet Union with his family. Pontecorvo continued his work as a physicist, and made important contributions e.g. to neutrino physics (see Section 9.1), all

---

[12]Many facts and figures in this Section come from the Atomic Archive series of web pages, http://www.atomicarchive.com/History/coldwar/index.shtml and Odd Arne Westad's seminal work [646].

[13]For an interesting making-of the Cambridge Five, see the book by Davenport [651], who identifies British societal tendencies of the time which according to him lead to the long-term rise of populism and Brexit. See also the review of his book by John Banville [663].

the while continuing to deny that he had been a spy. Many other sources in the Manhattan Project remained known only by their codenames from intercepted messages.

In 1946, the U.S. government made a proposal to put all nuclear energy activities under international control, including on-site inspections, called the Baruch plan after its author[14], and based on a comprehensive report by Secretary of State Dean Acheson and David Lilienthal [306]. The plan was promptly rejected by the Soviet Union and not adopted by the United Nations either, just like the Soviet counter-proposal of universal nuclear disarmament. The U.S. resumed nuclear bomb testing immediately afterwards, since 1947 under the civilian authority of the Atomic Energy Commission (AEC) [360]. The AEC was originally chaired by Lilienthal, and scientifically advised by a sub-committee headed by Oppenheimer. Both men notably opposed the development of a nuclear fusion weapon, popularly know as a hydrogen or thermonuclear bomb, leading to their removal from the AEC.

The first two nuclear tests after World War II –code-named Crossroads– were conducted over the Bikini Atoll in the Pacific, using plutonium fission bombs of the Nagasaki type. One bomb detonated over water, the other under water, with derelict battle ships as test targets. They were largely publicised as yet another demonstration of the unprecedented power of nuclear armament and U.S. dominance.

Meanwhile the Soviet effort to gain access to this new weapon had been jump-started by the successful intelligence during the Manhattan Project. The project thus progressed rapidly under the leadership of Igor V. Kurchatov[15]. Following detailed information obtained through Klaus Fuchs, a copy of the Fat Man design, nick-named Joe-1 in the West, exploded at the Semipalatinsk test site in Kazakhstan end of August 1949. The arms race predicted by the Met Lab scientists was on. But no other Soviet test followed for two years.

During the Manhattan Project, Edward Teller had already come forward with the idea to use nuclear fusion for military purposes. Bombs using nuclear fusion are also called thermonuclear weapons or (misleadingly) hydrogen bombs. Given the quick success of the Soviets with a fission bomb, Teller's idea gained momentum. The reasons are threefold. First, the energy gain per nucleon in the fusion of light elements into e.g. the doubly magic ^{4}He is even greater than that of fission, as shown in Focus Box 6.4. Second, the raw materials for nuclear fusion, like deuterium and tritium, are relatively abundant and readily available in normal water. Third, in contrast to fission bombs, where there is a lower and an upper limit on the fissionable mass, the total energy yield in a fusion bomb is scalable: the more nuclei are fused, the more energy is released. The proposal to pursue this path met substantial resistance, not only by the AEC but also by individual Nobel laureates like Fermi and Isidor I. Rabi, who criticised the plan as immoral and technically unfeasible [360,

---

[14]http://www.atomicarchive.com/Docs/Deterrence/BaruchPlan.shtml
[15]The Russian National Research Center for nuclear and particle physics is named after Igor Kurchatov, see http://eng.nrcki.ru

When you look at the graph of the binding energy per nucleon in Figure 5.7, you notice that the rise at low masses is much steeper than the slow decrease at large masses. This steep rise is used by nuclear fusion in stars to liberate binding energy. The yield per nucleon is larger that the gain in fission, but since the nuclei have few constituents, the gain per nucleus is smaller. Light nuclei are, however, readily available as raw material.

To fuse two nuclei, Coulomb repulsion must be overcome. The maximum potential V_C, when two nuclei (A, Z) and (A', Z') just touch, is the order of a few MeV:

$$V_C = ZZ' \frac{e^2}{A^{1/3} + A'^{1/3}}$$

To overcome it by thermal motion of the nuclei, a substantial density and a very high temperature, $T = V_C/k \simeq 10^{10}$ K, is required. Typical temperatures in the interior of the Sun are more like 10^7 K, but the tail of the Maxwell temperature distribution is sufficiently long to allow fusion. Raw material in the form of hydrogen plasma is abundantly available. The principle energy source for light stars is the so-called proton-proton chain:

$$
\begin{aligned}
{}^1\mathrm{H} + {}^1\mathrm{H} &\rightarrow {}^2\mathrm{H} + \mathrm{e}^+ + \nu_e + 0.42\mathrm{MeV} \\
{}^1\mathrm{H} + {}^2\mathrm{H} &\rightarrow {}^3\mathrm{He} + \gamma + 5.49\mathrm{MeV} \\
{}^3\mathrm{He} + {}^3\mathrm{He} &\rightarrow {}^4\mathrm{He} + 2\,{}^1\mathrm{H} + 12.86\mathrm{MeV}
\end{aligned}
$$

The large energy liberated in the last step is due to the exceptionally high binding energy of the doubly magic ^{4}He nucleus. The total process burns four hydrogen nuclei and generates almost 25MeV energy. The positron produced on the way annihilates with plasma electrons and contributes to the energy balance. So do the gamma rays. The Sun has enough fuel to shine for another $\simeq 10^9$ years. A second important source of energy for heavy stars is the CNO cycle, which can be summarised as $3(^4\mathrm{He}) \rightarrow {}^{12}\mathrm{C} + 7.27\mathrm{MeV}$. Heavier elements are synthesised by fusing additional helium nuclei, those beyond iron by fusing neutrons, e.g. in the merger of binary objects.

When constructing a nuclear fusion bomb, similar conditions of density and temperature must be reached. There is thus usually a two-staged process: a fission detonator, itself an implosion-type uranium or plutonium bomb, generates the pressure and temperature to cause a pallet of deuterium and tritium to fuse via the reaction $^2\mathrm{H} + {}^3\mathrm{H} \rightarrow {}^4\mathrm{He} + \mathrm{n} + 17.6\mathrm{MeV}$. The pellet usually consists of lithium deuterate, $^6\mathrm{LiO_2D_2}$, which can supply both the deuterium by thermal dissociation and the tritium by fast neutron fission.

For the last decades, there is a world-wide effort to use controlled nuclear fusion for energy production. The latest example is the ITER project [689] under construction in the south of France. It will use magnetic confinement in a Tokamak to contain a plasma in its own magnetic field and an external toroidal one. Another confinement method is the Stellarator [504], which uses a more complex magnetic field and relies less on the dynamics of the plasma itself. A German-American demonstrator project is the Wendelstein 7X.

Focus Box 6.4: Nuclear fusion

p. 380ff]. However, in 1950 President Truman decided that the development of a fusion bomb was to go ahead in view of the intensifying cold war. The opponents Lilienthal and Oppenheimer were subsequently removed from the AEC.

In the end of 1952, a first fusion bomb constructed according to the two-stage design by Teller and Ulam was detonated on an uninhabited atoll of the Marshall Islands, with a stunning yield of more than 10 megatons TNT equivalent, about 42×10^{15} J. The explosion wiped out the island housing the installation, its mushroom shaped cloud was more than 150km wide and reached to an altitude of 40km. The device was not a transportable bomb, since it used liquid tritium at $-250°$C as fusion material and required a cooling plant. It was ignited by a fission bomb, creating the necessary temperature and pressure conditions for nuclear fusion. Subsequent models used a similar two-stage design, but much more compact with solid lithium deuterade as fuel. The second test in spring 1954 on the Bikini atoll, with a yield of almost 16 megatons TNT equivalent, was the most powerful U.S. bomb ever detonated. It left a giant crater and blasted millions of tons of radioactive debris into the atmosphere. The inhabitants of neighbouring islands and the crew of a Japanese fishing boat suffered from severe exposure to radiation. Short term radiation sickness and long term effects like thyroid cancer were caused by the exposure.

The Soviet answer did not take long. Based on initial information from Klaus Fuchs, Andrei Sakharov designed a fusion bomb with layers of fusion fuel and uranium. In summer of 1953, a test of such a device in Siberia yielded about 400 kt TNT equivalent energy. However, the design was not really scalable and thus abandoned in favour of a two-stage design à la Teller-Ulam. In November 1958, the first Soviet megaton bomb exploded at the Semipalatinsk test site. In a series of further tests, a maximum yield of almost 60 megatons was achieved in 1961.

Meanwhile, the U.S. was under the influence of anti-communist paranoia during the McCarthy era. In 1953, Oppenheimer saw his security clearance suspended by the AEC, on a charge of long term association with the Communist Party. He demanded a security hearing in front of the AEC which lasted several weeks and heard 40 witnesses [537]. General Groves, Bethe, Teller and Rabi testified in his favour. Nevertheless, the majority of the AEC, with the notable exception of Smyth, did not recommend that he be given further access to sensitive information. After this devastating decision, Oppenheimer retired from public life and continued his research as the director of the prestigious Princeton Institute of Advanced Studies.

Nuclear weapons were not limited to the U.S. and the Soviet Union for long. Britain followed in 1952 with a fission bomb explosion in Australia, in 1957 with a fusion bomb test over Christmas Island in the Pacific. France followed in 1961 with a fission test in Algeria, and in 1968 with a fusion bomb in the South Pacific, and continued atmospheric testing until 1974. In 1995, France even resumed nuclear testing under global protest. However, France has now joined the other major nuclear powers in ratifying the Comprehensive Nuclear Test Ban Treaty (CTBT)[16] and the Treaty on Non-Proliferation of Nuclear Weapons (NPT)[17].

Despite all efforts to prevent further proliferation of nuclear weapons, China, India, Pakistan and lately North Korea joined the nations officially claiming the possession of nuclear armament. Israel is suspected to have a covert nuclear weapons program but does not acknowledge it. Other countries in the Middle East like Iran are regularly accused to aim at producing highly enriched ^{235}U for military purposes. The Cold War strategy of Mutually Assured Destruction (MAD), which strongly deterred first use of nuclear armament by guaranteeing that nuclear retaliation would lead to total destruction of the aggressor, led to an unimaginable nuclear arsenal. That the MAD strategy worked was convincingly demonstrated in the Cuban missile crisis. Starting in 1962 a large number of Soviet military personnel arrived on the island to install nuclear ballistic missile sites, promptly detected by U-2 spy planes. The largest rockets installed had a range of 1200 miles, enough to reach large cities in the south and east of the U.S. The threat was countered by the Kennedy administration with a naval blockade on October 22, 1962. The U.S. made clear it would not allow further delivery of Soviet arms to an island 90 miles off its coast and demanded that all weapons already in place be dismantled and brought back to the Soviet Union. Through threat and intense parallel diplomacy, an agreement was reached and the crisis resolved 13 days later [646, p. 305 ff]. The whole world felt that it had been on the brink of a nuclear catastrophe. My parents, like many Germans encouraged by the Adenauer government, stockpiled water and food according to a list distributed by civil defence, in the basement that had already served as a makeshift shelter in World War II. Switzerland continues to maintain a whole system of public bomb shelters beyond the end of the Cold War[18] in 1989.

The nuclear arms threat did not stop. For 2018, the Stockholm International Peace Research Institute reported a total of "approximately 13,865 nuclear weapons, of which 3750 were deployed with operational forces" [674]. Proliferation control is made difficult by the dual use of enriched uranium. Nuclear power plants need uranium enriched in ^{235}U to about 3 or 4%. There are about 450 of those operating today, a number grossly constant since 1990; they supply about 10% of the world supply in electricity[19]. Despite three major nuclear accidents with core meltdown [632], nuclear energy stays an important component of the energy budget in many countries. The NPT thus recognises the "inalienable right" of all nations to develop nuclear technology for peaceful purposes. But the technology that enriches uranium for power plants is of course also capable of producing weapon-grade fissionable material, making

---

[16]https://www.ctbto.org/fileadmin/content/treaty/treaty_text.pdf
[17]https://www.un.org/disarmament/wmd/nuclear/npt/
[18]For Westad's perception of the end of the cold war, see [647].
[19]https://www.world-nuclear.org/information-library/current-and-future-generation/nuclear-power-in-the-world-today.aspx

uranium enrichment the Achilles' heel of non-proliferation. The International Atomic Energy Agency (IAEA) in Vienna does its best to "establish and administer safeguards"[20] preventing military use of fissionable material and other nuclear technology.

FURTHER READING

W. Heisenberg, *Physics and Beyond: Encounters and Conversations*, Allen and Unwin, 1971.

Samuel A. Goudsmit, *ALSOS*, American Institute of Physics, Reprint, 1985.

Mark Walker, *German National Socialism and the Quest for Nuclear Power, 1939-1949*, Cambridge University Press, 1989.

Thomas Powers, *Heisenberg's War: The Secret History of the German Bomb*, Knopf, 1993.

Paul Lawrence Rose, *Heisenberg and the Nazi Atomic Bomb Project: A Study in German Culture*, University of California Press, 1998.

David Irving, *The Virus House*, Focal Point Classic Reissue, 2010.

Bruce Cameron Reed, *The History and Science of the Manhattan Project*, Springer, 2014.

Jon Agar, *Science and the First World War*, UCL Lunch Time Lectures, The Guardian, June 26, 2014, https://www.youtube.com/watch?v=oA1hMahtSKQ

Odd Arne Westad, *The Cold War: A World History*, Hachette Book Group, 2017.

---

[20]https://www.iaea.org/sites/default/files/statute.pdf

CHAPTER 7

Quantum fields

The third aspect of my subject is that of science as a method of finding things out. This method is based on the principle that observation is the judge of whether something is so or not. All other aspects and characteristics of science can be understood directly when we understand that observation is the ultimate and final judge of the truth of an idea. But "prove" used in this way really means "test," in the same way that a hundred-proof alcohol is a test of the alcohol, and for people today the idea really should be translated as, "The exception tests the rule." Or, put another way, "The exception proves that the rule is wrong." That is the principle of science. If there is an exception to any rule, and if it can be proved by observation, that rule is wrong.

Richard P. Feynman, *The Meaning of It All*, 1999 [512]

IN THIS CHAPTER I describe how Quantum Field Theory puts together quantum mechanics and relativity. The shift from atomic states and transitions towards particle reactions is an important aspect of this development. I will thus also discuss what quantum fields are, how they move and interact. Biographies of the prominent protagonists, Max Born, Enrico Fermi and Richard P. Feynman, are quoted in the Further Reading section.

7.1 LIMITS

Quantum mechanics certainly has philosophical problems which we glossed over in Section 5.6. But more important are the physical limits of its applications. Its equation of motion, the Schrödinger equation (see Focus Box 5.9), is clearly not covariant under Lorentz transformations and cannot be used to describe systems in relativistic motion. The need for a relativistic extension is thus evident. The equation's second problem is that it does not allow the creation or disappearance of particles. There is a continuity equation, which describes the fact that the probability density in an arbitrary volume can only change when probability "flows" in or out through the volume's boundaries. We demonstrate this fact in Focus Box 7.1. This means that the number of particles in a box can only change when particles enter or leave the box. The total number of particles in a system is conserved.

The Schrödinger equation implies the conservation of probability. We demonstrate this for a free particle:

$$i\frac{\partial}{\partial t}\psi \ + \ \frac{1}{2m}\vec{\nabla}^2\psi = 0$$

We multiply the equation with ψ^*, and its complex conjugate with ψ:

$$i\psi^*\frac{\partial}{\partial t}\psi + \psi^*\frac{1}{2m}\vec{\nabla}^2\psi = 0 \quad ; \quad -i\psi\frac{\partial}{\partial t}\psi^* + \psi\frac{1}{2m}\vec{\nabla}^2\psi^* = 0$$

Taking the difference of the two equations, we get:

$$(\psi\frac{\partial}{\partial t}\psi^* + \psi^*\frac{\partial}{\partial t}\psi) \ - \ \frac{i}{2m}(\psi^*\vec{\nabla}^2\psi - \psi\vec{\nabla}^2\psi^*) = 0$$

The first term denotes the time variation of the probability density ρ, the second is the divergence of the probability current density $\vec{j}$:

$$\rho = \psi\psi^* \quad ; \quad \vec{j} = -\frac{i}{2m}(\psi^*\vec{\nabla}\psi - \psi\vec{\nabla}\psi^*)$$

The two quantities are related by a continuity equation $\partial\rho/\partial t + \vec{\nabla}\vec{j} = 0$. What it means is most evident when we consider free particles with momentum $\vec{p}$ and energy E. Their wave function is a plain wave:

$$\psi = \sqrt{N}e^{-i(Et-\vec{p}\vec{x})}$$

The probability density ρ and the current density $\vec{j}$ of such a wave are:

$$\rho = \psi^*\psi = Ne^{-i(Et-\vec{p}\vec{x})}e^{i(Et-\vec{p}\vec{x})} = N$$
$$\vec{j} = \frac{iN}{2m}\left(e^{i(Et-\vec{p}\vec{x})}i\vec{p}e^{-i(Et-\vec{p}\vec{x})} - e^{-i(Et-\vec{p}\vec{x})}(-i\vec{p})e^{-i(Et-\vec{p}\vec{x})}\right) = \rho\frac{\vec{p}}{m}$$

The current density is thus the product of density and velocity, in close analogy to the classical electromagnetic current of a particle cloud. The equation does not allow the number of particles to change. When their local density diminishes, particles have flown out, when it increases, particles have flown in.

Focus Box 7.1: Continuity equation for non-relativistic particles

We know that this is not true, and so did Schrödinger. Photons can be absorbed and emitted by charged particles. And even those can be created in pairs, like electrons and positrons, and also annihilate into light. It is thus quite possible to create particles and make them disappear, when the conservation of quantum numbers like the electric charge is respected. Schrödinger proposed to replace the particles density ρ by the charge density $Q\rho$ and the corresponding current density $Q\vec{j}$. The conservation of probability then implies the conservation of electric charge Q. We will find back this "trick" in what follows, but for an even more important reason.

7.2 MOVING FIELDS

The prototype of a relativistically moving particle is the photon. Consequently, the classical Maxwell equations are already covariant under Lorentz transformations. The motion of the photon field consisting of $\vec{E}$ and $\vec{B}$ has the form of a wave equation, as shown in Focus Box 5.9. In analogy to this, Louis de Broglie [199] proposed in the beginning of the 1920s to also describe the motion of matter particles like the electron by a wave equation (see Section 5.4). Interference experiments with electrons indeed showed that matter also propagates in waves [219, 224]. One thus cannot maintain the classical concept of a trajectory –a smooth curve describing the position as a function of time– at distances comparable to the atomic size.

It remained unclear for quite a while, how the concept of matter waves was to be applied to anything else but a free particle. And thus the next step in the development of quantum mechanics was rather the matrix concept of Heisenberg, Born, Jordan and Pauli in the middle of the 1920s (see Focus Box 5.7). Since the observation of quantum mechanical systems is only possible via interactions delivering observables, the matrix mechanics focussed on radiative transitions in the atom and other interactions, not on the motion of particles. We come back to the focus on observables in Section 7.3.

Erwin Schrödinger revived wave mechanics in 1925 in his famous series of four publications [213, 214]. In his first delivery, he presented the equation of non-relativistic motion named after him. In the fourth, he describes a relativistic version also based on considerations of energy-momentum conservation. Schrödinger at first abandoned this route since it does not describe the energy levels in atoms correctly, unless one introduces spin. Spin is important here because in relativistic motion its magnetic interaction is not negligible. In parallel, Oskar Klein [210] and Walter Gordon [209] published the same idea, the relativistic equation of motion for scalar fields is named after them [455]. The Schrödinger equation follows from the non-relativistic energy-momentum conservation by operator substitution, i.e. quantisation. Likewise, the Klein-Gordon equation follows from its relativistic equivalent, as shown in Focus Box 7.2. It is a second order partial differential equation for fields described by a scalar wave function ψ. It appeared less attractive than Schrödinger's equation, which is linear in time, because it requires more initial conditions.

It was thus forgotten for a while until Wolfgang Pauli and Victor Weisskopf brought it back to public attention in 1934 [262]. The equation has solutions with positive and negative energy eigenvalues. While we are familiar with negative potential energies from the classical physics of bound systems, negative kinetic energies for free particles make no sense. Moreover, the signs of energy and probability are the same, thus negative energies correspond to negative probabilities. This makes even less sense, since probabilities are real numbers between 0 and 1 by definition. The problem can be fixed using the same "trick" of re-interpreting probability density as charge density à la Schrödinger. That now links the sign of the particle charge to the sign of its energy, a very significant fact.

The senseless negative probabilities may well have motivated the British theoretician Paul Dirac to search for an equation of motion linear in time. Since the relativistic energy conservation law has roots of negative energy due to its quadratic form, one may suspect that the negative probabilities in the Klein-Gordon equation are due to its quadratic form. A promising ansatz would thus be a linear wave equation analog to Schrödinger's, but linear in time *and* space, as detailed in Focus Box 7.4. The solutions of this equation must simultaneously obey the Klein-Gordon equation to ensure energy-momentum conservation. For scalar fields and scalar operators this is not possible. The operators are rather 4×4 matrices and the wave function has four components. Dirac's equation of motion for these fields thus consists of four coupled partial differential equations for four fields $(\psi_1, \psi_2, \psi_3, \psi_4)$. The total field ψ is called a spinor[1], since the four components correspond to states with positive and negative energy, each with spin "up" and spin "down". All have the same mass, thus propagate in the same way when no force is present.

The probability density of particles propagating according to Dirac's equation is positive definite. The problem of negative probabilities is thus solved, the conservation of the electromagnetic current density is no longer obtained by a trick, but directly follows from the conservation of the probability current. However, since Dirac particles must also respect the Klein-Gordon equation, negative energies are still with us. They must be interpreted.

In the 1930s, Dirac proposed to interpret the solution with negative energy as if the vacuum were filled with an infinite "sea" of electrons with negative energy. Normally all states with negative energy would thus be filled. If an energy of more than $2m_e c^2$ is transferred to such a state, it would transport an electron from negative to positive energy. Simultaneously, a "hole" with positive charge would be created in the vacuum, in analogy to holes created in the valence band of a semiconductor when an electron is lifted to the conduction band. This interpretation has numerous problems. Among others the electrostatic energy of the vacuum would be infinitely large and it is unclear how the hole would move through the sea of electrons. Dirac's proposal, however, has the important significance to associate negative energies with antiparticles, for

---

[1] Although a Dirac spinor has four components, it is not a four-vector. Its four components behave like scalars under Lorentz transformations.

The Klein-Gordon equation for the evolution of a scalar relativistic field follows from relativistic energy momentum conservation by operator substitution (see Focus Box 5.9):

$$E^2 - \vec{p}^2 - m^2 = 0 \quad \rightarrow \quad \left(\vec{\nabla}^2 - \frac{\partial^2}{\partial t^2} - m^2\right)\psi = 0$$

Here and in what follows, we use what is called "natural units", explained in Focus Box 7.3, so that the constants $\hbar$ and c no longer appear. The Klein-Gordon equation is manifestly covariant, which becomes obvious by using the four-vector gradient $\partial_\mu \equiv \partial/\partial x^\mu = (\partial/\partial t, \vec{\nabla})$:

$$\left(\partial_\mu \partial^\mu + m^2\right)\psi = 0$$

Using the same approach as for the Schrödinger equation, one derives a continuity equation for the following probability and current density:

$$\frac{\partial}{\partial t}\underbrace{\left[i\left(\psi^*\frac{\partial\psi}{\partial t} - \psi\frac{\partial\psi^*}{\partial t}\right)\right]}_{\rho} + \vec{\nabla}\underbrace{\left[-i\left(\psi^*\vec{\nabla}\psi - \psi\vec{\nabla}\psi^*\right)\right]}_{\vec{j}} = 0$$

One may be surprised that the probability density now contains a derivative of the wave function. With the probability current density it forms a four-vector $j^\mu = (\rho, \vec{j}) = i(\psi^*\partial^\mu\psi - \psi\partial^\mu\psi^*)$, which obeys a covariant continuity equation, $\partial_\mu j^\mu = 0$. For plane waves, $\psi = \sqrt{N}e^{i(\vec{p}\vec{x} - Et)}$, the components of j^μ are:

$$\rho = i\left(\psi^*\frac{\partial\psi}{\partial t} - \psi\frac{\partial\psi^*}{\partial t}\right) = 2EN \quad ; \quad \vec{j} = -i\left(\psi^*\vec{\nabla}\psi - \psi\vec{\nabla}\psi^*\right) = 2\vec{p}N$$

The current density is proportional to the four-momentum, $j^\mu = 2p^\mu N$, its timelike component is proportional to energy. This is due to the requirement of relativistic invariance of the probability, $\rho\, d^3\vec{x}$. After all, the number of particles is independent of the reference frame. The volume element transforms like $d^3\vec{x} \rightarrow d^3\vec{x}/\gamma$ because of Lorentz contraction in the direction of motion. To obtain invariance, the probability density must thus transform like $\rho \rightarrow \gamma\rho$. Since $\gamma = E/m$, ρ must thus be proportional to energy. A unit volume thus contains $2E$ particles with our normalisation.

The energy eigenvalues for a free relativistic particle are obtained when one inserts a plane wave function into the Klein-Gordon equation: $E = \pm(\vec{p}^2 + m^2)^{\frac{1}{2}}$. There are thus solutions with positive and negative energy. And since the probability density is proportional to energy, the probability is no longer positive definite. One must re-interpret the continuity equation for a four-vector of electromagnetic current density, $j^\mu = iQ(\psi^*\partial^\mu\psi - \psi\partial^\mu\psi^*)$. Proportional to the electric charge Q, j^0 becomes the charge density, which can indeed be positive as well as negative. Solutions with positive and negative energy are distinguished by the charge of the particle. And the continuity equation describes the conservation of electric charge.

Focus Box 7.2: Klein-Gordon equation

Relativistic quantum systems are characterised by two natural constants, Planck's constant of action $\hbar$ and the speed of light c:

$$\hbar \equiv \frac{h}{2\pi} \simeq 1.055 \times 10^{-34} \text{J s} \quad ; \quad c \simeq 2.998 \times 10^8 \text{m/s}$$

Their dimensions are $[\hbar] = ML^2/T$ and $[c] = L/T$, where M denotes mass, L length and T time. The natural system of units makes both constants dimensionless and equal to unity, $\hbar \equiv c \equiv 1$; it defines $\hbar$ as the unit of action and c as the unit of velocity. Both constants thus disappear from all equations. This permits e.g. to measure all masses (M), momenta (Mc) and energies (Mc^2) with the same units:

$$[E, M, p] = \frac{ML^2}{T^2} = \text{GeV} = 10^9 \text{ eV} \simeq M_p$$

The base unit electronvolt (eV) is defined as the energy gained by a unit charge particle traversing a potential difference of 1 V. A billion of these units, 1 GeV, corresponds roughly to the mass of the proton and thus forms the natural scale of high energy physics processes. When calculating ordinary quantities, one obviously need to convert results back into SI units. Useful conversion constants are given in the table below.

Quantity	Conversion factor	Natural units	SI units
Mass	1 kg = 5.61×10^{26} GeV	GeV	GeV/c^2
Length	1 m = 5.07×10^{15} GeV^{-1}	GeV^{-1}	$\hbar c$/GeV
Time	1 s = 1.52×10^{24} GeV^{-1}	GeV^{-1}	$\hbar$/GeV

In the system of natural units, the electric charge is also dimensionless, $[e] = [\sqrt{\hbar c}] = [1]$. It often appears in the form of the fine structure constant α, which is the electrostatic energy of two electrons at unit distance, divided by the mass of the electron:

$$\alpha = \frac{\frac{1}{4\pi} \frac{e^2}{\hbar/mc}}{mc^2} = \frac{e^2}{4\pi\hbar c} \simeq \frac{1}{137}$$

Focus Box 7.3: Natural units

We construct a relativistic equation of motion which has first order time and space derivatives, the Dirac equation: $i\partial\psi\partial t = (\vec{\alpha}\vec{p} + \beta m)\psi$. Energy and momentum are conserved if its solutions also respect the Klein Gordon equation (see Focus Box 7.2), $-\partial^2\psi/t^2 = (\vec{p}^2 + m^2)\psi$. Comparing coefficients we get an algebra for the operators $\vec{\alpha}$ and β:

$$(\vec{\alpha}\vec{p} + \beta m)^2 = \vec{p}^2 + m^2 \ ; \ (\sum_i \alpha_i p_i + \beta m)(\sum_j \alpha_j p_j + \beta m) = \sum_k p_k^2 + m^2$$

$$\alpha_1^2 = \alpha_2^2 = \alpha_3^2 = \beta^2 = \mathbb{1} \ ; \ \alpha_i\alpha_j + \alpha_j\alpha_i = 0 \ ; \ \alpha_i\beta + \beta\alpha_i = 0 \ \ (i \neq j)$$

The α_i and β anti-commute, they cannot be simple scalars. Their algebra is respected by hermitian and traceless 4×4 matrices with eigenvalues ± 1. The wave function ψ thus also has four components, $(\psi_1, \psi_2, \psi_3, \psi_4)$, it is called a spinor. Multiplying the Dirac equation by β from the left, we get a compact notation for the Dirac equation:

$$i\beta\frac{\partial\psi}{\partial t} = -i\beta\vec{\alpha}\vec{\nabla}\psi + m\beta\beta\psi \ \ \rightarrow \ \ (i\gamma^\mu\partial_\mu - m)\psi = 0$$

We are defining a four-vector of operators, $\gamma^\mu \equiv (\beta, \beta\vec{\alpha})$. The algebra of its components follows from that of the $\vec{\alpha}$ and β matrices:

$$(\gamma^0)^2 = \beta^2 = \mathbb{1} \ ; \ (\gamma^k)^2 = \beta\alpha^k\beta\alpha^k = -\mathbb{1} \ ; \ \gamma^\mu\gamma^\nu + \gamma^\nu\gamma^\mu = 2g^{\mu\nu}$$

The "square" of a spinor is $\psi^\dagger\psi$ with the hermitian adjoint spinor $\psi^\dagger = (\psi_0^*, \psi_1^*, \psi_2^*, \psi_3^*)$. Using the same method as for the Schrödinger and Klein-Gordon equation, we derive a continuity equation:

$$\partial_\mu(\bar{\psi}\gamma^\mu\psi) = 0$$

with the Dirac adjoint spinor $\bar{\psi} \equiv \psi^\dagger\gamma^0$. The term in parentheses is a four-vector current of the probability density:

$$j^\mu = \bar{\psi}\gamma^\mu\psi \ ; \ j^0 = \bar{\psi}\gamma^0\psi = \rho \ ; \ \vec{j} = \bar{\psi}\vec{\gamma}\psi$$

Its time-like component is positive definite, the problem of negative probabilities is thus solved. The electromagnetic current density is: $j^\mu = Q\bar{\psi}\gamma^\mu\psi$. For a free electron with energy E and momentum $\vec{p}$ it is $j^\mu = -eN(E, \vec{p})$, the one of a positron with the same energy-momentum is $j^\mu = +eN(E, \vec{p}) = -eN(-E, -\vec{p})$. A positron thus evolves the same way as an electron with opposite four-momentum.

Focus Box 7.4: Dirac equation

the first time. Some take this as a prediction for the existence of the positron, discovered two years later in cosmic rays (see Section 8.1). The hole theory itself did not survive.

The Genevan Ernst Stückelberg[2] [297] and the American Richard P. Feynman [311] found a much more convincing explanation of negative energy states. Their ansatz is to identify negative energies with particles evolving backwards in time, or equivalently antiparticles evolving forward in time. The current density of both is the same. Thus the emission of a positron with positive energy is identical to the absorption of an electron with negative energy. One can also say that particle solutions with negative energy can always be replaced by antiparticle solutions with positive energy and inverted momentum direction, as shown in Focus Box 7.4.

This has more dramatic consequences than it seems at first glance. Look at the space-time sketch of Figure 7.1. In the left diagram, the incoming electron scatters twice off the potential symbolised by a cloud, first at position (1), then at (2). In the right diagram, the time order is reversed. The intermediary electron now moves backward in time. We replace it by a positron moving forward. Thus an electron-positron pair is first created at (1), the resulting positron is annihilated with the incoming electron at (2), only the electron is left in the final state. According to Heisenberg's uncertainty relation, if the energy is sufficiently fixed, the time order of (1) and (2) cannot be known, the two processes are indistinguishable. At the same time, they only differ in that an electron moving backward in time is replaced by a positron moving forward, which changes nothing according to Feynman and Stückelberg. Their interpretation of negative energy solutions thus naturally leads to particle creation and disappearance processes required by observation.

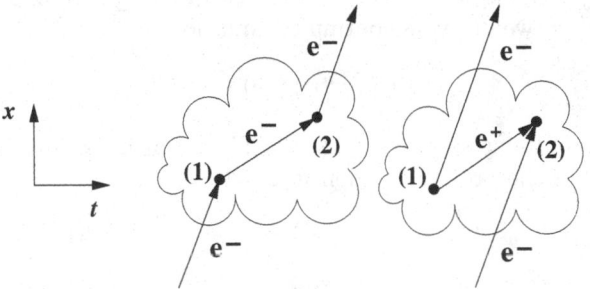

Figure 7.1 Space-time diagram for the scattering of an electron off a potential symbolised by a cloud. On the left, the electron simply scatters twice at positions (1) and (2). In the right diagram, the creation of an electron-positron pair at (1) is followed by an annihilation of the positron with the incoming electron at (2). Because of Heisenberg's uncertainty, the two processes are indistinguishable at small scale.

---

[2]His full name was Baron Ernst Carl Gerlach Stückelberg von Breidenbach zu Breidenstein und Melsbach.

7.3 INTERACTING FIELDS

Knowing how quantum fields move and how potentials are created by charges and currents, it remains to understand how they interact with each other. Relativistic quantum field theory[3] provides that understanding. But an additional conceptual step is required: not only observables, the matter fields and potential fields themselves become operators. This way the focus shifts from the motion of a particle in a potential towards the interaction of quantised fields. Matter particles are the quanta of fields. Force fields (or rather potential fields) have quanta like the electromagnetic potential, which is quantised in photons as seen in the photoelectric effect (see Section 5.2). We thus come back to the particle nature of light, as already discussed by Newton in his "Opticks".

The form of quantised electrodynamics we know today has been devised by Richard P. Feynman in the 1940s. Matter and force fields are operators. The matter field is an operator for the creation or annihilation of a particle at a given space-time point. With Freeman Dyson [323] we can also call them emission and absorption operators. I find this more to the point, since the space-time point we are talking about is typically an interaction vertex. An incoming particle is absorbed at that point, a final state particle is emitted. Probability amplitudes result when such operators are applied to the space which contains all states. While wave functions have thus changed their meaning, we can still take them as amplitudes of probability waves. We will continue to go back and forth freely between operators, their representations and their eigenvalues. In particular, the meaning of the conserved current density stays the same. It plays an important role in scattering theory, as we will see shortly.

The central notion of a scattering theory based on these concepts is called an invariant amplitude if one likes wave mechanics or a matrix element if one prefers matrix terminology. It is the probability amplitude for the transition of an initial state ψ_i into a final state ψ_f, and it is written as $\mathcal{M}_{fi}$. Consequently, its square $|\mathcal{M}_{fi}|^2$ is the probability density for this process. The square of the invariant amplitude, suitably normalised, enters into the calculation of an observable, which measures the probability for the process in question. In atomic physics, this observable is the intensity of a spectral line. For particle reactions, it is the cross section σ or the decay width Γ. The concept of the cross section is a geometric one, sketched in Figure 7.2 for a reaction $a + b \rightarrow c + d$. We associate a fictitious surface σ to each target particle b. Whenever a projectile a hits such a surface, the reaction takes place, otherwise it doesn't. The reaction rate is thus the product of the flux of incoming particles a, i.e. their number per unit surface and per unit time, the number of target particles exposed to this beam, and the cross section. This simple relation supposes that the surface density of target particles is small enough, such that the fictitious surfaces do not overlap and their total surface presented to the incoming beam

---

[3] A detailed history of quantum field theory, with complete documentation, is found in [479, p. 1-38].

is simply their sum. The cross section thus has the dimension of a surface, but of course no such surface exists. Instead, σ measures the probability for the process to take place. We explain the relation between invariant amplitude and cross section, Fermi's golden rule [420], in Focus Box 7.6.

The product between incoming projectile flux and the number of target particles is a characteristic of each experiment, called the luminosity. It is the maximum rate of reactions which would be observed if each projectile would interact with a target particle. The cross section is thus the ratio of the actual reaction rate and its maximum:

$$\frac{\#\text{reactions}}{\text{time}} = \underbrace{\frac{\#\text{projectiles}}{\text{time} \times \text{surface}} \times \#\text{targets}}_{\text{luminosity}} \times \text{cross section}$$

It is indeed a probability in strange units, measuring the strength of a reaction independent of experimental details. For particle waves, the cross section is discussed in Focus Box 7.5 using Huygens' principle.

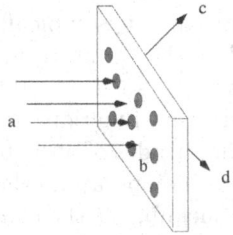

Figure 7.2 Geometrical concept of the cross section for a particle reaction $a + b \rightarrow c + d$. Each target particle b is associated with a fictitious surface σ. The reaction takes place whenever a projectile a hits that surface.

In an experiment, the cross section is extracted from the reaction rate, by dividing by the only experiment-dependent factor, the luminosity. It is measured in the enormously large units of barn, $1\text{b} = 10^{-28}\text{m}^2$. Cross sections one encounters for reactions among elementary particles are rather of the order of nanobarn, $1\text{nb} = 10^{-9}\text{b}$, or picobarn, $1\text{pb} = 10^{-12}\text{b} = 10^{-40}\text{m}^2$. For neutrinos they are even as small as 10^{-42}m^2 (at 1 GeV). Instead of just counting the total number of reactions, one can classify them according to observables like the scattering angle, i.e. the space angle Ω between an incoming particle a and an outgoing reaction product c. In that case, one measures the differential cross section $d\sigma/d\Omega$.

In case of a decay $a \rightarrow c + d + \ldots$, i.e. of a single particle in the initial state, the corresponding notion quantifying the strength of the interaction is the decay width Γ, as explained in Focus Box 7.7. Its inverse is the lifetime τ_a of the mother particle a. The width has the dimension of an energy and corresponds to the width of the mass distribution of the final state. The momenta and energies of the final state particles add up as four-vectors. The square of the

Let us formulate Huygens' principle for an incoming plane wave with frequency ω and wave vector $\vec{k}$, $\psi_i(x,t) = \sqrt{N}e^{-i(\omega t - \vec{k}\vec{x})}$, impacting on a point-like scattering centre. What happens is sketched below.

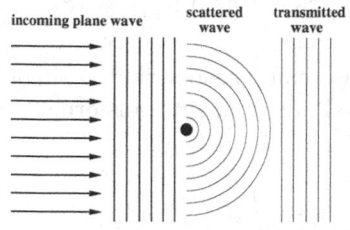

A slightly diminished plane wave goes through undisturbed, and a spherical wave with small amplitude originates from the scattering centre. Let us normalise the incoming plane wave simply with $N = 1$, such that $\psi_i = e^{-ik_\mu x^\mu}$ is a wave with probability density $\rho = \psi^2 = 1$. Its intensity I is proportional to this density and the velocity $v = \omega/k$, $I = \rho v = \omega/k$.

According to Huygens' principle, the outgoing wave is a superposition of the transmitted wave with amplitude $(1 - f)$ and the scattered wave with amplitude f:

$$\psi_f = (1 - f(\theta, \phi))\, e^{-ik_\mu x^\mu} + f(\theta, \phi)\frac{e^{-ik_\mu r^\mu}}{r}$$

The first term describes the transmitted wave. The factor $1/r$ in the term for the scattered wave, inversely proportional to the distance r from the scattering centre, is necessary to preserve the normalisation. At any point and time, the total intensity will be the square of this wave function and thus have three terms: the square of the transmitted wave, the square of the scattered wave and an interference term between the two. Ignoring interference for the moment, the intensity scattered into a volume element characterised by the solid angle element $d\Omega$ and the radial thickness dr is $\left| f(\theta, \phi)\frac{e^{-ikr}}{r} \right|^2 (r^2 d\Omega) dr = |f(\theta, \phi)|^2 d\Omega dr$. The flux scattered into $d\Omega$ is then $|f(\theta, \phi)|^2 d\Omega \frac{dr}{dt} = |f(\theta, \phi)|^2 v\, d\Omega$. The ratio of this flux to the incoming one is the probability of scattering, the cross section:

$$\frac{d\sigma}{d\Omega} = |f(\theta, \phi)|^2$$

We can generalise this finding for an arbitrary initial state. The free state before the scattering can be written as a superposition of eigenstates ψ_n of the free Hamiltonian (energy) operator: $\psi_a(\vec{x}, t) = \sum_n a_n(t)\psi_n(\vec{x})e^{-iE_n t}$. The final state ψ_c can also be written in this way, but with other coefficients. Now take the action of the potential as a small perturbation at time $t = 0$ and assume a pure initial state with a single coefficient $a_i = 1$ and all others zero. According to Huygens' principle, the incoming amplitude will be a little reduced, while a small additional amplitude a_f is excited by the scatter. When the potential is know, it can be calculated. Indeed, as shown above, a_f, correctly normalised, is nothing else but the invariant amplitude $\mathcal{M}_{fi}$, the probability amplitude for this process.

Focus Box 7.5: Huygens' principle for quanta

Consider a reaction, where a fixed electromagnetic potential A^μ, like that of a heavy nucleus, converts a scalar initial state ψ_a into a final state ψ_c.

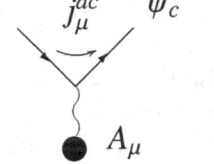

To first order, we can consider that a single action of the local potential A^μ transforms ψ_a into ψ_c:

$$\mathcal{M} = -i \int d^4x \, j_\mu^{ac} A^\mu$$

The electromagnetic current density, $j_{ac}^\mu = -ie\,(\psi_c^* \partial^\mu \psi_a - \psi_a \partial^\mu \psi_c^*)$, which couples to the electromagnetic potential, A^μ, is the same as the one we had found for free scalar particles (see Focus Box 7.2), except that the incoming and outgoing wave functions are no longer the same. The argument of the integral is the probability density to find ψ_c among all the possible final states produced by A^μ at the interaction vertex, $x = (t, \vec{x})$. The integration over space-time takes into account that the interaction can take place wherever A^μ reigns, i.e. anywhere for a static potential. The cross section can then be calculated by properly normalising the probability of the reaction, represented by $|\mathcal{M}|^2$. One obtains Fermi's Golden Rule:

$$d\sigma = \frac{|\mathcal{M}|^2}{F} dQ$$

The incident flux F depends on the relative velocity $|\vec{v}_a - \vec{v}_b|$ of projectile and target:

$$F = |\vec{v}_a - \vec{v}_b|\, 2E_a 2E_b = 4\sqrt{(p_a p_b)^2 + m_a^2 m_b^2}$$

The second equation is valid for $\vec{v}_a || \vec{v}_b$ and shows that this quantity is an invariant under Lorentz transformations. So is the phase space factor dQ, which takes into account the number of possible final state configurations:

$$dQ = (2\pi)^4 \delta^{(4)}(p_c + p_d - p_a - p_b)\frac{d^3 p_c}{(2\pi)^3 2E_c}\frac{d^3 p_d}{(2\pi)^3 2E_d}$$

It contains a δ-term ensuring energy-momentum conservation as well as the number of quantum states for all reaction products. The factors $N_i = 2E_i$ come in because of our normalisation of the wave function (see Focus Box 7.2).

Focus Box 7.6: Fermi's golden rule

total mass of all decay products is the square of their total energy minus the square of their total momentum, $m^2 = (\sum E_i)^2 - (\sum pi)^2$. The invariant m corresponds to the mass of the mother particle a. For a finite lifetime, this is not a fixed number, but distributed around the mean value, m_a. The width of the mass distribution Δm_a is inversely proportional to its lifetime, due to Heisenberg's uncertainty, $\Delta m_a \tau_a \geq \hbar$. The shorter the lifetime of a particle, the wider the mass distribution around its mean value.

For a generic decay of the mother particle a into at least two daughter particles, $a \to c + d + \ldots$, the connection between decay width Γ_a and invariant amplitude $\mathcal{M}$ is:

$$d\Gamma = \frac{1}{2E_a} |\mathcal{M}|^2 \frac{1}{(2\pi)^2} \delta^4(p_a - p_c - p_d - \ldots) \frac{d^3 p_c}{2E_c} \frac{d^3 p_d}{2E_d} \cdots$$

For a two-body decay in the rest system of the mother particle, we have:

$$\Gamma(a \to c + d) = \frac{|\vec{p}_f|}{32\pi^2 m_a} \int d\Omega \, |\mathcal{M}|^2$$

with the momentum $|\vec{p}_f| = |\vec{p}_c| = |\vec{p}_d|$ in the final state. The integration goes over the solid angle element $d\Omega = \sin\theta d\theta d\phi$ with the polar angle θ and the azimuth ϕ. The decay width is measured via the decay rate, the fraction of mother particles, $\Delta N_a / N_a = -\Gamma \Delta t$, which decays in a laps of time Δt. Thus a constant fraction decays per unit of time, the remaining number of mother particles diminishes exponentially, with $N_a(t) = N_a(0) e^{-\Gamma t}$. The invariant mass of the decay products, $m = \sqrt{(E_c + E_d)^2 - (\vec{p}_c + \vec{p}_d)^2}$, i.e. the mass of the mother particle, follows a distribution around the mean value m_a with width Γ_a.

Focus Box 7.7: Decay width

What we have quoted so far is the first term of a perturbative expansion, since we assumed that the initial and final state are free particles, which interact with the potential at a single space-time point. After its pioneer Max Born [217], this first order term is also called the Born approximation. For electromagnetic interactions, we calculate it in Focus Box 7.8. The progress of the perturbation series is counted by the order of the coupling constant between the current and the potential, e for electromagnetic forces. Further accuracy of the calculation is achieved when higher order terms are included, with a second interaction, a third one and so on. Provided, of course, that higher order terms are successively smaller and smaller perturbations, such that the series converges. The conditions under which this is indeed the case deserve a more in-depth discussion which we defer to Chapter 9.

We now have to go one step further and describe how the electromagnetic potential is generated and how it reaches the space-time point where it interacts with the current density. The generation of fields and potentials is described by Maxwell's equations. Or rather, Maxwell's equations describe the *change* of fields and potentials. But that is all we need. Once a current density has caused an electromagnetic wave, it will propagate according to wave mechanics. We can thus describe the interaction as one between two currents, one that emits the photon, the other which absorbs it. And again this is an *inter*action, since the roles can be interchanged. Figure 7.3 represents such an electromagnetic scattering by exchange of a photon. Diagrams of this type are called Feynman diagrams after their inventor [457]. They represent the process visually, but they are also precise prescriptions for the calculation of the corresponding invariant amplitude. One can interpret them as space-time diagrams of what happens during the interaction, but of course within the limits of Heisenberg's uncertainty. As you can see in Focus Box 7.8, the conversion of interacting current densities into a probability amplitude involves an integral over space-time. It is really an integral over all possible paths which the particles can take from the initial to the final state of a reaction. Thus Feynman called his method a space-time approach [311]. After the integration, all that is left are energies and momenta of the particles involved. We no longer care, nor can we possibly know where or when the interaction happened. For us, the straight lines with arrows in a Feynman diagram indicate matter particle momenta, wiggly lines indicate the momenta of force quanta, photons so far.

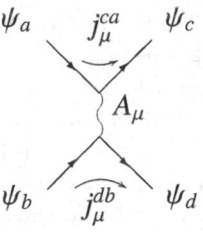

Figure 7.3 Feynman diagram of an electromagnetic scattering $a + b \to c + d$ in Born approximation.

From the annotated Feynman diagram one sees that each element of the graph corresponds to an operator, or a probability amplitude; all enter multiplicatively into the invariant scattering amplitude. The straight lines correspond to matter fields and symbolise their field operators. The current densities $\psi_{a,b}$ annihilate the incoming fields at the interaction vertices, $\psi_{c,d}$ create the outgoing fields. The wiggly line represents the exchanged photon, A_μ, with its four-momentum q. We note that matter fields interact with the potential, and not with the electromagnetic field. We will discuss this further in Chapter 9. The scattering amplitude can be interpreted as the probability amplitude for the complete process. Its square is then the probability density, integrated over space-time it becomes the probability and determines the cross section.

The correspondence between the elements of a Feynman diagram and the factors entering into the invariant amplitude for a process are called Feynman rules. How they can be derived for a give field theory is fixed by simple

We calculate the invariant amplitude for the process in Figure 7.3, using the result obtained in Focus Box 7.6. We have to evaluate the four-potential A^μ as generated by the current density of the target, j^μ_{db}. The Maxwell equations being covariant under Lorentz transformations, we can also use them in the relativistic domain. They are also invariant under gauge transformations of the form $A_\mu \to A_\mu + \partial_\mu\chi$ with an arbitrary scalar function $\chi(x)$, because these transformations do not change the field tensor. We can thus change the gauge for the four-vector of the electromagnetic potential such that $\partial^\mu A_\mu = 0$. This is called the Lorenz gauge, named after the Danish physicist and mathematician Ludvig Valentin Lorenz (not to be confused with the Dutch Henrik Antoon Lorentz). In this gauge, Maxwell's equations take the form of a wave equation:

$$\partial_\mu A^\mu = j^\nu$$

A transition current density j^μ_{db} for fields, which describe free particles before and after the interaction, produces the electromagnetic potential:

$$A^\mu = -\frac{1}{q^2} j^\mu_{db}$$

The four-momentum transfer, $q = p_c - p_a = p_b - p_d$, is the difference between outgoing and incoming four-momentum of the projectile, which is transferred to the target. The invariant amplitude for this scattering process is then:

$$\mathcal{M} = -i \int d^4x\, j^\mu_{ab} \frac{1}{q^2} j^{ab}_\mu$$

Diagrammatically the amplitude is represented by the Feynman diagram 7.3. For fermions, the scattering amplitude in terms of spinor currents has some more complex math, but the underlying idea is exactly the same.

Focus Box 7.8: Electromagnetic scattering in Born approximation

rules, almost like in a cookbook. One can indeed say: field theory = Feynman rules [486]. Here they define quantum electrodynamics, QED [457]. The electric charge e in the current density, which enters at each vertex, determines the order of magnitude for the probability amplitude. The total scattering amplitude is thus proportional to e^2, the cross section is proportional to $\alpha^2 = e^4/(4\pi)^2 = 1/137$. The photon propagator $1/q^2$ expresses the probability amplitude that the two current densities exchange a photon of four-momentum q.

At each vertex, the four-momentum is strictly conserved. That means, on the other hand, that the exchanged photon has to have strange properties. The square of its four-momentum, i.e. the square of its mass, is not zero, as it would be for a free photon. This deviation from its rest mass is a property of all particles, which connect two vertices in a Feynman diagram, i.e. which transmit forces. They are therefore called virtual particles. Their propagator has a pole at the "true" mass of the free particle. For the photon, it lies at $q^2 = 0$. That the amplitude becomes very large for this value of the four-momentum transfer is irrelevant for QED. It corresponds to the limiting case of an interaction without momentum transfer, an interaction which does not really take place. That the probability for no interaction is much larger than that for a noticeable one is in agreement with our perturbative approach to interactions in general.

We notice again one of the conceptually largest differences between the quantum and the classical approach to interactions. In classical physics an interaction, however small, always takes place, when a non-zero potential or force is present. In quantum processes, nature has a choice to let the interaction happen or not. And indeed no interaction is the default.

We also notice that matter fields interact with the potential, not with forces. From your course of classical physics you may have taken away the impression that the liberty to gauge potentials gives them a somewhat fuzzy meaning. After all one can add a constant to the electrical or gravitational potential and the predicted trajectory of particles remains unchanged. Likewise one can add a divergence-free vector field to the magnetic vector potential and nothing changes in the classical case. In quantum systems, this is no longer true, particles interact with the potential even in the absence of a force field. The first to point out this astonishing effect were Werner Ehrenberg and Raymond E. Siday [313]. The experimental proof has been proposed by Yahir Aharonov and David Bohm [342, 349], first implemented by Robert G. Chambers [346], by Gottfried Möllenstedt and W. Bayh [364] shortly after. The mechanism is that the electromagnetic potential changes the phase of the matter field. This phase by itself is unobservable since the probability is given by the square of the field, where the phase drops out. But interference experiments allow to observe phase changes. Focus Box 7.9 shows you how.

That all scattering processes are wavelike in nature, and thus described by Huygens' principle, is indeed proven when interference occurs. An important interference phenomenon occurs when more than one process can mediate a reaction between identical initial and final states. That is for example the

The electromagnetic potential changes the phase of charged matter fields. For simplicity we will consider a non-relativistic matter particle, where the same happens. Let the particle move in an electric potential $V(t)$, which depends on time:

$$i\frac{\partial \psi}{\partial t} = H\psi = [H_0 + V(t)]\,\psi$$

The operator H_0 describes the free motion, V is treated as a perturbation. The solutions will have a time dependent phase $\psi = \psi_0 e^{-iS(t)}$:

$$i\frac{\partial \psi}{\partial t} = i\frac{\partial \psi_0}{\partial t}e^{-iS} + \psi_0\frac{\partial S}{\partial t}e^{-iS} = H_0\psi_0 e^{-iS} + \frac{\partial S}{\partial t}\psi_0 e^{-iS} = \left[H_0 + \frac{\partial S}{\partial t}\right]\psi$$

A comparison of coefficients shows that $\partial S/\partial t = V(t)$. Thus on a closed path, the field will undergo a phase shift of $\Delta S = \oint V(t)dt$. The relativistic answer is completely analogous to this, but for a four-potential A_μ:

$$\Delta S = \oint V dt - \oint \vec{A}d\vec{x} = \oint A^\mu dx_\mu$$

In the absence of an electrical potential, the phase shift is thus proportional to the magnetic flux included in the path. It is observable if one lets the phase-shifted field interfere with the original field. This is the idea of what is called an Aharonov-Bohm experiment, a double slit experiment as shown on the right.

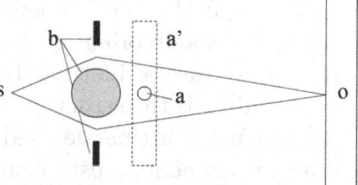

Electrons from the source **s** are brought to interference through a double slit **b**. In the shadow of the round middle screen, a long thin solenoid **a** is placed. Outside the solenoid, the magnetic field is zero, but the magnetic potential decreases inversely to the distance to the solenoid. Thus it cannot be zero everywhere, regardless of its gauge. The interference pattern is observed on the screen as the intensity as a function of the displacement **o**. The intensity of the interference stripes is modulated by the image of the single slit, as shown in Focus Box 5.10. Thus if there were a magnetic force on the electron, e.g. by a stray field **a'** of the solenoid, the envelope *and* the interference strips would shift alike. If there is no field, only the strips shift under a constant envelope. This is exactly what is observed. Chambers in his pioneering implementation of the experiment used a ferromagnetic whisker to implement the changing potential. Möllenstedt and Bayh used a very fine solenoid electromagnet. In a modern implementation, Akira Tonomura [460] and collaborators used a superconductor to confirm the result.

Focus Box 7.9: The Aharonov-Bohm experiment

case, when two different virtual intermediate particles can transform the initial into the final state, such that no observable allows to distinguish the two contributions, not even in principle. The ability to distinguish works regardless of whether the observable is actually measured or not, or even can be measured with existing devices. It suffices that it exists in principle. Indistinguishable processes can interfere with each other at the amplitude level. If we have two amplitudes, $\mathcal{M}_1$ and $\mathcal{M}_2$ describing the two contributions, the square of the total amplitude will be $|\mathcal{M}|^2 = |\mathcal{M}_1^2 + \mathcal{M}_2^2 + \mathcal{M}_1\mathcal{M}_2|$, an interference term appears. It can increase or decrease the total probability of the process, depending on whether the interference is constructive or destructive. A prominent example is the interference between the exchange of photons and Z bosons, electroweak interference, which we will discuss in Chapter 9.

In summary we note that in the framework of a field theory, detailed properties of particle reactions can be calculated with the tools of perturbation theory. Using a well defined prescription, these properties can be measured, such that an experimental check of the theory is feasible, as Feynman requires in the initial quote of thus chapter. In particular, it can be shown by example that the theory is not valid or limited in its application range. The parameters of the theory, like masses and coupling constants can be measured experimentally. Results of different experiments become comparable. Detailed properties are not only absolute or relative cross sections, measuring the strength of a reaction, but also their dependence on kinematic quantities like energy or scattering angle, spin orientation etc. As far as decays are concerned, not only the total decay rate or lifetime, but also their branching into different final states is calculable. If the structure of the interaction is known, scattering and decay experiments can thus be used to investigate the structure of matter. Rutherford's experiment to establish the existence of the atomic nucleus is a classic example. We will see in Chapter 9 how an analogous method established the existence of quarks.

The elements of a quantum field theory are operators, which describe the propagation of quantised fields through spacetime, and operators for their pointlike interactions. Feynman diagrams allow to describe particle reactions. Not only do they illustrate what happens, each element also has an associated mathematical expression, which can be interpreted as a probability amplitude. Together they form the set of Feynman rules which define a quantum field theory. Their multiplication and integration over space and time give the probability amplitude for a reaction, its square then is the probability.

Quantum electrodynamics is an early prototype of such a theory and extremely successful. And it is successful despite the fact that in calculating processes beyond the Born approximation, i.e. with more than one photon exchanged, divergent terms appear. A sophisticated method, called renormalisation, allows to absorb these infinities into measured quantities, which are by definition finite. We will come back to this method and its foundations in Chapter 9. Among other consequences, renormalisation leads to the astonishing fact that coupling constants like the electric charge are not really constant, but depend on distance. Applying renormalisation, perturbative quantum electrodynamics makes impressively precise predictions, which impressively precise experiments can confront. To give but one example, the magnetic moments of electron and muon have been calculated and measured to a relative accuracy of a few 10^{-13} and 10^{-10}, respectively.

One would thus think that the success of quantum electrodynamics ought to have led to an immediate effort to apply field theoretical methods to other forces. Foremost to weak interactions, which manifest themselves at low energies mainly by radioactive decay, but also to strong interactions including nuclear forces. This way, the triumph of the Standard Model of particle physics would have started in the 1950s. Instead, quantum field theory fell into deep disrespect until well into the 1970s. The reason was not only that the treatment of divergences by renormalisation was viewed by many as a dirty trick, designed to sweep basic inconsistencies of the method under the rug. There is also a conceptional problem, which has to do with the behaviour of the theory at very small distances. It is called the Landau pole problem [324, 325, 326, 386] after the Russian physicist Lev Landau, a very influential figure during that time and leader of a Russian school of particle physicists. The existence of a Landau pole is a common feature of field theories which are asymptotically free, i.e. which have a vanishing scattering amplitude at large distances. Reversely, it leads to a divergence of the coupling constant, the electrical charge for quantum electrodynamics, at a large but finite energy scale (or at a small but finite distance), independently of its value at "normal" energies. This pole has no practical importance for quantum electrodynamics, because it occurs at unbelievably small distances, even smaller than the so-called Planck scale, which itself is already some 10^{-20} times smaller than the proton radius. Nevertheless the existence of this pole was taken as a sign that there was something fundamentally inconsistent and incomplete –thus wrong according to the introductory quote of our chapter– with quantum field theory.

Philosophically speaking it can also be argued that quantum mechanical states are not by themselves observable, but only transitions between them are, as Heisenberg kept pointing out. It is thus intellectually attractive to demand a theory, which predicts the amplitudes $\mathcal{M}$ directly without reference to an underlying dynamics, solely on the basis of symmetry and other abstract requirements. Such arguments led to the renaissance of the matrix method, based on the S-matrix of John A. Wheeler [278]. Leading theorists like Steven Weinberg shared these doubts [383]: "It is not yet clear whether field theory will continue to play a role in particle physics, or whether it will ultimately be supplanted by a pure S-matrix theory." The hope was that very general requirements on the S-matrix, like unitarity and consistency, together with symmetry criteria, would completely define it. A perturbative approach would then also be unnecessary, the S-matrix would depend analytically on kinematic quantities. These hopes led to a strange intellectual atmosphere. Frank Wilczek, one of the fathers of Quantum Chromodynamics (see Section 9.2) later recalled [491] that "strict quantum field theory was considered to be naive, in rather poor taste, an occasion for apology." However, no unique S-matrix exists, rather there are many theories which are non-perturbatively consistent, each with its own S-matrix. Thus in the beginning of the 1970s, it was realised that the S-matrix approach was less a theory than a tautology. Francis Low said in the discussion following his summary talk at the International Conference on High-Energy Physics in Berkeley (1966) [391]: " I believe that when you find that the particles that are there in S-matrix theory, with crossing matrices and all the formalism, satisfy all these conditions, all you're doing, I think, is again showing that the S-matrix is consistent with the world the way it is; that is, the particles have put themselves there in such a way that it works out, but you have not necessarily explained that they are there."

The second principle taken for granted in such theories was that elementary hadrons, particles like proton, neutron and pion subject to strong interactions, have no constituents. All these hundreds of particles were supposed to be equally fundamental, each was to be a dynamical bound state of all the others. Following one of its main advocates, Geoffrey Chew, this axiom was called "nuclear democracy" [389]. While some like Werner Heisenberg adhered to this ideology until the mid 1970s, it has been refuted convincingly by the discovery of quarks and the description of their dynamics by quantum chromodynamics, the quantum field theory of strong interactions. We come back to these developments in Chapter 9.

S-Matrix theory nevertheless scored important points. It helped to understand e.g. that there are relations between current densities which enter into matrix elements –or rather into invariant amplitudes in my preferred way of thinking. The usage of such relations is called current algebra [374]. It allows, among other things, to derive so-called sum rules, specific integrals over differential cross sections, which relate to the properties of hadrons. Since they do not make reference to specific dynamics, they stay valid also in a field theory which does have definite dynamics. We will come back to all of this in Chapter 9. But before that we have to go into details about experimental techniques for particle physics and how their development fuelled progress.

FURTHER READING

Richard P. Feynman, *Surely you're joking, Mr. Feynman!*, W.W. Norton, 1985.

James Gleick, *Genius: The Life and Science of Richard Feynman*, Pantheon Books, 1992.

Steven Weinberg, *The Quantum Theory of Fields*, Vol. 1, Cambridge Univ. Press, 1995.

Nancy T. Greenspan, *The End of the Certain World: The Life and Science of Max Born*, John Wiley & Sons, 2005.

Jim Ottaviani and Leland Myrick, *Feynman*, First Second, 2011.

Gino Segré and Bettina Hoerlin, *The Pope of Physics: Enrico Fermi and the Birth of the Atomic Age*, Henry Holt & Co., 2017.

CHAPTER 8

Enabling technologies

> The effect of a concept-driven revolution is to explain old things in new ways. The effect of a tool-driven revolution is to discover new things that have to be explained. In almost every branch of science ... there has been a preponderance of tool-driven revolutions.
>
> Freeman Dyson, *Imagined Worlds*, 1998 [505][1]

I N THIS CHAPTER I will discuss the role that particle acceleration and detection technology plays in the development of the Standard Model of particle physics. I will concentrate on successes (as usual) but exceptionally also cover failed major projects, because there are interesting lessons to be learned.

8.1 COSMIC RAYS

In the 1930s and 1940s quantum electrodynamics emerged as the only robust theory of particles and their interactions. It not only correctly describes the transitions in atoms, but the whole wealth of electromagnetic interactions between charged matter particles and photons. In the following two decades, faith in quantum field theory was lost, but the methodology of extracting and interpreting information from experiments was preserved. The main observables were and still are cross sections and decay rates.

During the same period of time, experimental techniques developed with impressive speed, as if physicists had only waited for technology to become available to turn it into new particle physics experiments. This way, during quite a while, experimental particle physics kept a considerable lead over particle theory. It is arguable whether the development of disruptive theoretical ideas has since driven the staggering progress in particle physics [362] or whether it was progress in particle acceleration and detection techniques which provided the driving force [543]. However what can be safely said is that their mutual stimulation brought about today's situation where predictions

---

[1]Copyright © 1997 by the President and Fellows of Harvard College. Quoted by permission.

of theory are routinely verified or falsified by experiment with unprecedented accuracy. In turn, the findings of experiments –or lack thereof– strongly constrain the directions of further theory development.

As far as particle reactions are concerned, the only particle sources in the beginning of the 20th century were radioactive decays and cosmic rays[2]. Charles-Augustin de Coulomb had already observed in 1785 using a torsion electrometer, that gas filled capacitors discharge slowly despite careful electrical insulation. One called this phenomenon "air electricity" and put it in the same category as other electrical phenomena in the atmosphere, like lightning. At the end of the 19th century, Julius Elster and Hans Geitel [643], two German school teachers, friends and collaborators, showed that the discharge was due to the ionisation of air molecules in the capacitor volume, caused by charged particles. Around the same time, Becquerel discovered radioactivity (see Section 2.2). So the question was whether the charged particles came from Earth, from the atmosphere itself, or from outside. Research to answer this question was vigorously pursued by the Catholic priest Theodor Wulf SJ, using the sensitive and transportable string electrometer he had invented. This device is explained in Focus Box 8.1.

Wulf's plan was to measure the rate of discharge in an ionisation chamber as a function of altitude with respect to sea level. The rate is roughly 1 Volt per hour (see Focus Box 8.1), so the measurement takes patience and constant conditions. Measuring the rate at various places [130, 136], including the top of the Eiffel tower in Paris –at that time the world's highest building– did not reveal a conclusive altitude dependence. Domenico Pacini and collaborators [144] made similar measurements below the surface of Lake Bracciano and concluded that that the radiation was probably not emitted by the Earth's crust.

The proof for an extraterrestrial origin of the ionising radiation required balloon flights. The first ones were made in 1910 by Albert Gockel [132], a German working at the newly founded University of Fribourg, Switzerland. He used a balloon on loan from the Swiss Aero-Club in Zürich and a rather unreliable electrometer he had also borrowed. This was due to the fact that his research on air electricity was not at all supported by his boss, who thought that he should rather concentrate on solid state physics for the benefit of Swiss industry [587, 593], an argument against fundamental physics still heard today. The data collected during a short flight over Lake Zürich showed strong statistical and systematic variations. Nevertheless they confirmed that the radiation did not emerge from the Earth's surface. Gockel was first to call it "cosmic radiation", but missed a fundamental discovery. Two lessons are to be learned from his failure: if discovery is what you are after, use only first class instruments; and do not listen to your boss.

---

[2]The literature covering the history of cosmic rays is rather extensive, thanks to the centenary of the discovery of their extraterrestrial origin in 2011. See e.g. the articles of the special edition of Astroparticle Physics in 2014 [610].

An ionisation chamber is a closed volume filled with gas, normally air. Its interior forms a capacitor, the voltage on its plates can be read from the outside using an electrometer, like the bifilar one invented by Theodor Wulf shown below [589]. Wulf patented the device in 1906 (DRP 181284).

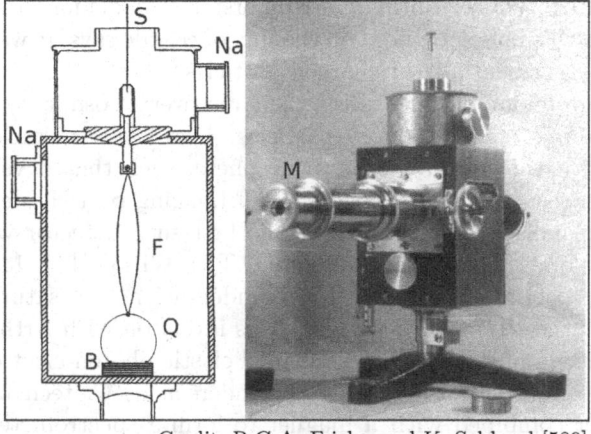

Credit: R.G.A. Fricke and K. Schlegel [589]

A cut through the instrument is shown on the left. The lower end of thin metal filaments (F) was fixed to a quartz string (Q), using its elasticity as a restoring force. The string was mounted on an amber insulator (B). An external source can be connected to S and the case grounded. Two sodium dryers (Na) reduced the humidity on the inside of the device. The distance of the two filaments was measured through a microscope (M). A photograph of the instrument is shown on the right. It was manufactured by the Günter&Tegetmeyer (G&T) company. A similar device [589], modified according to suggestions by Victor Hess, was used by him on his balloon flights.

The rate of voltage decrease dU/dt in the ionisation chamber measures the rate of ionisation density $d\rho/dt$:

$$\frac{dU}{dt} = \frac{1}{C}\frac{dQ}{dt} = \frac{V}{C}e\frac{d\rho}{dt}$$

With the capacitance C and the volume V of Wulf's set-up, a voltage decrease of 1 V/h corresponds to an ionisation rate of close to 1 ion/cm^3/s.

Focus Box 8.1: Electrometers and ionisation chambers

The experimental proof of the extraterrestrial origin of cosmic rays is thus rightly credited to the Austrian Victor Francis Hess [142]. In a series of long balloon flights up to 5300m altitude and with a first class electrometer manufactured to his standards (see Focus Box 8.1), he measured the intensity of the ionising radiation systematically. His rather precise data showed that starting at an altitude of about 800m the intensity increases quadratically with altitude. His measurements were repeated with an even higher accuracy by Werner Kolhörster [151] for altitudes reaching 8000m. The extraterrestrial origin of the radiation was thus established, the term cosmic rays generally accepted. In 1929, Walter Bothe and Kolhörster [230] demonstrated the influence of the Earth's magnetic field on the flux of cosmic rays. It was thus shown that they consist essentially of charged particles.

During about four decades after their discovery, cosmic rays stayed the dominating source of high energy particles. They cause particle showers by interaction with the nuclei of the atmosphere, such that a variety of long lived particles reaches the Earth's surface. Imaging particle detectors allow those to be visualised and identified. The first such detector was the cloud chamber, developed by the meteorologist C.T.R. Wilson [141, 145]. The track of an ionising particle is visualised by condensation of a saturated vapour. Wilson shared the 1927 Nobel prize for this invention with Arthur H. Compton, the American physicist who discovered elastic photon-electron-scattering (see Focus Box 5.3). In 1932, Carl D. Anderson at Caltech [243], using a cloud chamber combined with a magnet to form a spectrometer, discovered the positron, the anti-electron predicted by Paul Dirac's equation (see Focus Box 7.4). Figure 8.1 shows the spectrometer as well as the cloud chamber photo of a positron. Its charge $+e$ is determined through the curvature of its track and its specific energy loss, measured through the density of droplets, as well as the energy lost when traversing a lead plate. The energy loss of charged particles in a medium by excitation and ionisation of atoms is explained in Focus Box 8.2. The year after, Giuseppe Occhialini and Patrick Blackett [253] at the Cavendish Laboratory of Cambridge confirmed the existence of this first antiparticle. Anderson and Hess shared the Nobel prize in physics of 1936 for their respective discoveries.

The second important imaging detector of the time was the photographic emulsion, a glass plate covered by a light-sensitive layer. Similar to light on a photographic film, ionising radiation leaves a track of grains, which can be made visible by developing the emulsion. It can then be analysed with a microscope. The grain density is proportional to the specific energy loss by ionisation and thus proportional to the square of the particle charge.

A non-imaging device for counting the passage of ionising particles was invented by Rutherford's collaborator Hans Geiger in 1908 [126]. Its principle of operation and that of other gaseous detectors is explained in Focus Box 8.6. In its original version it worked only for the heavily ionising α particles with charge $+2e$. When Geiger moved back to Germany, he and Walter Müller at University of Kiel perfected the device [226] such that it was also sensitive to β particles with a single charge $-e$, thus four times less ionisation. It consists

The dominating source of energy loss for charged particles in matter are multiple interactions with atomic electrons. The energy lost by an incident particle interacting with a single atomic electron can be calculated classically in analogy to Rutherford scattering (see Focus Box 4.5), except that the recoil of the struck electron is not at all negligible. In the early 1930s, Hans Bethe [233, 322] calculated the average energy $-dE$ lost by a projectile of charge ze, mass $M \gg m_e$ and velocity $\beta = v/c$ in a thickness dx of a material characterised by its volume density n_e of atomic electrons, according to quantum mechanics and special relativity:

$$-\frac{dE}{dx} = \frac{e^4}{m_e c^2 4\pi\epsilon_0^2} \frac{n_e z^2}{\beta^2} \left[\ln\left(\frac{2m_e c^2 \beta^2}{I(1-\beta^2)} \right) - \beta^2 - \frac{\delta}{2} \right]$$

The term δ, added later, corrects the result for the polarisation of material atoms due to the incoming charge. This polarisation effectively reduces the charge of distant atoms and thus the energy loss, i.e. $\delta > 0$. The ionisation constant I regroups the global ionisation properties of the atom, i.e. the different excitation levels weighted by the probability to excite them. It depends on the atomic charge Z and is empirically found to be:

$$\frac{I}{Z} = \begin{cases} 12 + 7/Z & \text{eV}; \quad Z < 13 \\ 9.76 + 58.8Z^{-1.19} & \text{eV}; \quad Z \geq 13 \end{cases}$$

The energy loss per unit length is proportional to z^2; an α particle loses four times the energy that a proton loses in the same material. Thus measuring dE/dx allows to identify particles and measure the absolute value of their charge. For non-relativistic projectiles, the term $1/\beta^2$ dominates the velocity dependence. At the end of their range in matter, just before they are stopped, charged particles thus lose the most energy. This energy loss maximum is called the Bragg peak and used e.g. in irradiating cancerous tissue. All particles have a minimum of their energy loss at roughly the same momentum:

$$p/m = \beta\gamma \simeq 3 \text{ to } 3.5 \quad ; \quad -dE/dx|_{min} \simeq 1 \text{ to } 2\,\text{MeV cm}^2/\text{g}$$

At high projectile energies, the energy loss increases logarithmically $\propto \log(\beta\gamma)$. The ionisation process is of course statistical in nature. The actual energy loss per unit material thickness follows a very asymmetric Landau distribution [300] around its mean value given above.

Focus Box 8.2: Specific energy loss by excitation and ionisation

 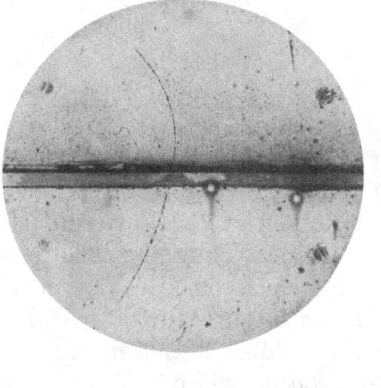

Figure 8.1 Left: Carl D. Anderson in front of his magnetic spectrometer for cosmic rays. Right: A cloud chamber photograph of a positron with kinetic energy 63 MeV, entering from below, traversing a 6mm thick lead plate losing energy, and continuing with a curvature corresponding to 23 MeV [252]. (Credit: Archives, California Institute of Technology)

of a gas filled metallic tube, serving as an ionisation chamber and cathode, with an axial anode wire strung in its centre and put under high voltage. When an ionising particle liberates electrons from the gas atoms, they are accelerated towards the anode. Since the field becomes very strong close to the thin anode wire, the drifting electrons gain momentum and can themselves ionise gas atoms. Thus an electron avalanche causes an appreciable current surge between anode and cathode. It can be made audible with a loudspeaker as a sharp snapping sound, and can be fed into an electronic circuit. The invention of the coincidence circuit for Geiger-Müller counters by Walther Bothe [234] and Bruno Rossi [392, p. 43f] enabled the systematic classification of particles in cosmic ray showers. In 1936 Carl Anderson and Seth Neddermeyer [269, 273] on the American west coast, Jabez Street and Edward Stevenson [277] on the east coast, simultaneously discovered the muon. It is a brother particle to the electron, but roughly 200 times heavier (see Section 9.1). Cecil Powell, César Lattes and Giuseppe Occhialini in 1947 discovered the pion, the lightest hadron[3]. The same year, Clifford Butler und George Rochester discovered the kaon, the first member of the second generation of hadrons, which were called "strange particles".

8.2 PARTICLE ACCELERATORS

Starting in the 1940s, accelerators[4] gradually took over as sources of energetic particles, to systematically produce new and unknown states and study

---

[3] A hadron is a bound state of quarks, see Section 9.1.

[4] A list of past and present accelerators for particle physics is offered by Wikipedia at https://en.wikipedia.org/wiki/List_of_accelerators_in_particle_physics

their properties. With that, one no longer depended on anecdotal observation of particles and their reactions, but could produce and study them under controlled conditions.

Accelerators were first developed in the early 1930s to push ahead nuclear research, which until then had been stuck with the use of α particles from nuclear decay, with fixed energies of a few MeV and low rates. Accelerators produce beams of charged particles using the Lorentz force discussed in Focus Box 2.5. Electric fields are used to accelerate particles by increasing their momentum. Magnetic fields serve to deflect them from their original direction. The simplest accelerator is the cathode ray tube (see Figure 2.5), where the accelerating structure is basically a capacitor. Static high voltages are generated from AC sources, like in the Cockcroft-Walton high voltage multiplier [245, 246], invented in Rutherford's Cavendish laboratory. Alternatively one can use mechanical transport of charges like in the Van de Graaf high voltage generator [254], invented at Princeton.

These inventions mark the beginning of the end of table-top experiments in nuclear and particle physics. The heavy and voluminous equipment hardly fitted into existing buildings. Careful engineering was required to make them work, in addition to good ideas. On the longer term this also led to a change in sociology of particle physics experiments. As detectors grew to cathedral size with increasing energies, research groups enlarged to several thousand members today. The second world war and the period immediate following it saw the creation of large research centres, able to create and maintain the required infrastructure, like the National Laboratories in the U.S. and CERN in Europe.

Both principles of high voltage generation for low currents are used until today to accelerate protons and ions; many demonstration experiments also use them. The reachable energy, given by the product of particle charge q and accelerating voltage V, $E = qV$, is limited by the stability of the high voltage insulation. Thus, it does not exceed about 25 MeV. To reach a higher energy requires that the projectile passes several times through an accelerating potential, like in a circular or linear accelerator.

The simplest circular accelerator is the cyclotron. The physics of this and other circular accelerators is explained in Focus Box 8.3. The cyclotron makes use of the fact that for non-relativistic velocities, the revolution frequency of a charged particle, the cyclotron frequency, is independent of energy and only depends on the magnetic field. Already in 1929, the 28-year-old professor Ernest O. Lawrence, who had just joined University of California's Berkeley campus, developed and built a first cyclotron. Lawrence patented this invention in 1934 as U.S. Patent 1,948,384 and obtained the Nobel prize in 1939. That same year, Berkeley disposed of the world's highest energy accelerator, the 60-inch cyclotron shown in Figure 8.2.

As the particle velocity approaches the speed of light, it depends less and less on the particle momentum or energy. This asymptotic saturation towards the speed of light is very useful, provided that the radius of curvature is

The simplest circular accelerator is the cyclotron. In a uniform and constant magnetic field B, a particle of charge e moves on a circle of radius $R = p/(eB)$ where p is its momentum (see Focus Box 2.1). The angular frequency of this movement is called cyclotron frequency $\omega_c = (e/m)B$.

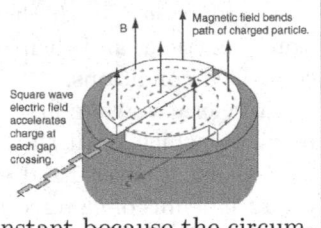

In the non-relativistic domain, this frequency is constant because the circumference of the orbit increases proportional to the velocity of the particle. The particle, typically a proton, is accelerated by the electric field present between the two D-shaped cavities. This field is provided by a radio frequency generator such that its frequency is an even multiple of the cyclotron frequency. The cyclotron has its limits in that the velocity of the particle does not remain proportional to the momentum, but asymptotically approaches the speed of light, $v/c = \sqrt{1 - m^2/E^2}$. In the relativistic limit, the velocity barely increases despite a steady increase in energy. Acceleration radio frequency and particle revolution frequency thus quickly fall out of phase in a cyclotron, at an energy of some tens of MeV for protons. One must then adjust the radio frequency to the relativistic velocity.

Alternatively, one can keep the field B proportional to the momentum, as in the *synchrotron*. In that case, one can work with a constant accelerating radio frequency. The RF field is transmitted to the beam by resonant cavities. The synchrotron principle requires a certain initial velocity sufficiently close to the speed of light. The beam is thus usually pre-accelerated before injection, typically by a linear accelerator. The acceleration process must stop when one reaches the maximum field of the dipole magnets. One then either extracts the beam or converts the accelerator into a storage ring. In this latter operational mode, radio frequency provides at each turn just the energy lost by the beam via bremsstrahlung.

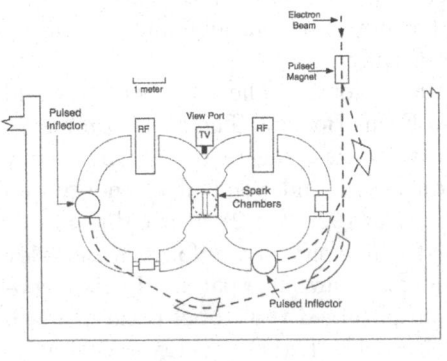

The separation of functions and the concentration of the components around the ring allows the combining of two rings to construct what is called a *collider*, originally invented by the Norwegian Rolf Widerøe working in war-time Germany [442]. A historical example is the Princeton-Stanford storage ring [385, 408]. It worked with electron beams in the two rings which intersect in the middle, where the two storage rings touch.

The detector, a set of spark chambers (see Section 8.3), determined the scattering angle. Obviously, one can also fold the two accelerators on top of each other, or even combine them to collide particles and antiparticles.

Focus Box 8.3: Circular particle accelerators

Figure 8.2 Left: The 60-inch cyclotron at Berkeley's Rad Lab in 1934. Ernest Lawrence is second from the left (Credit: Science Museum London/Science and Society Picture Library). Right: The 600-MeV Synchrocyclotron (SC), CERN's first accelerator, came into operation in 1957 (Credit: CERN).

held constant. In this case, the rotational frequency of the beam is again independent of energy. This is the principle of the synchrotron. By increasing the magnetic field proportional to the momentum of the particle, the radius of curvature remains constant. One does not have to fill a large volume with a magnetic field, but can concentrate it inside a vacuum chamber, whose shape approximates a circle. The synchrotron was invented by Vladimir Veksler [301] and independently by Edwin McMillan [304]. When particles are injected on the rising flank of the accelerating electric field, they gather automatically into synchronous bunches: late comers see more field, early ones less. The phase of the circulating particles with respect to the radio frequency is thus stabilised.

In 1954, University of California at Berkeley started exploiting the Bevatron [335], built again under the leadership of Ernest Lawrence. This proton accelerator, a synchrocyclotron shown in Figure 8.3, reached a maximum energy of 6.2 GeV, but with a wide beam of four square feet cross section. Its name comes from the energy "billion electron volts" (10^9 eV, today called Giga electron Volts, GeV). Its maximum energy was high enough to produce proton-antiproton pairs from a hydrogen target at rest. It was thus the first accelerator explicitly built to test a theoretical prediction, namely that each particle has an antiparticle with identical mass but opposite charge. For the electron, Anderson had already shown with cosmic rays that this was true, for baryons, i.e. nuclear building blocks, the proof required accelerators.

The transverse beam size can be dramatically reduced by magnetic lenses. Quadrupole and sextupole magnets are used for what is called strong focussing, to distinguish it from the by-product of weak focussing by dipole magnets. The first strongly focussing accelerator with much smaller beam size and thus higher luminosity (see Section 7.3) was the Alternating Gradient

Figure 8.3 Left: The Berkeley Bevatron in 1958 (Credit: The Regents of the University of California, Lawrence Berkeley National Laboratory, image used by permission). Right: A hydrogen bubble chamber built by the Berkeley group and NIST. The proton beam crosses the bathtub-shaped chamber body horizontally. The cameras are placed on the conical top. The whole chamber is immersed in a cryostat for cooling. Its cover is seen at the top. (Credit: National Institute of Standards and Technology).

Synchrotron (AGN)[5] at the Brookhaven National Laboratory, which reached a proton energy of 33 GeV in 1960. Next to energy, luminosity is the second important figure of merit for an accelerator. It is the proportionality factor relating the reaction rate to the cross section. The higher the luminosity, the higher the observable rate of a given reaction.

The first accelerator at the European centre for particle physics CERN, the Synchrocyclotron shown in Figure 8.2, started its regular service to experiments in 1957. A bit later followed the strong focussing Proton Synchrotron. Since then CERN has built and operates a whole series of accelerators[6], schematically shown in Figure 8.5, with steadily increasing energy and luminosity, such that today it is the leading particle physics laboratory in the world.

When an accelerator is used in what is called a fixed target experiment, a beam of particles is extracted and steered towards a stationary target where it interacts. The target is followed or surrounded by detectors to measure the final state. Alternatively, in a storage ring or collider, two particle beams are brought to collision, usually in the centre-of-mass of the two particles. The advantage is that no collision energy is lost in the recoil of the target, thus the whole centre-of-mass energy of projectile and target is available for the reaction. If both beams have the same energy, that centre-of-mass energy is simply twice the beam energy, whereas it is much smaller in a fixed target set-up, as shown in Focus Box 8.4. Recent examples of colliders are the Tevatron at the Fermi National Laboratory (FNAL) close to Chicago in the U.S., which

---

[5]See https://www.bnl.gov/about/history/accelerators.php
[6]See https://home.cern/science/accelerators/accelerator-complex

collided protons and antiprotons until 2011 and the Large Electron Positron collider, LEP, at CERN, which collided electrons and positrons. It was in operation between 1989 and 2000 in the tunnel which now houses the Large Hadron Collider, LHC, up to now the world's highest energy collider.

We will confront the kinematics of a fixed target and a collider experiment. In a fixed target experiment, a projectile of mass m and energy-momentum $(E, \vec{p})$ interacts with a target of mass M with $(M, 0, 0, 0)$. The total energy-momentum four-vector of the system is thus:

$$\begin{pmatrix} E + M \\ \vec{p} \end{pmatrix}$$

This four-vector is conserved in the collision, such that the total mass square of the final state, s, is:

$$s = (E + M)^2 - \vec{p}^2 = m^2 + M^2 + 2EM \simeq 2EM$$

The last approximation holds for $E \gg m, M$. To give a numerical example, if an electron of 100 GeV energy impacts on a proton, the total mass, $\sqrt{s}$, which can be created in the final state is about 14 GeV.

In a collision in the centre-of-mass system, the two incoming particles have four-momenta $(E, \vec{p})$ and $(E', -\vec{p})$. The total four-vector and the total mass are thus:

$$\begin{pmatrix} (E + E') \\ \vec{0} \end{pmatrix} \quad ; \quad s = (E + E')^2$$

For two particles of 100 GeV energy impacting on each other, the final state mass $\sqrt{s}$ is thus simply 200 GeV.

Focus Box 8.4: Centre-of-mass energy

Figure 8.4 shows that the centre-of-mass energy[7] reached by accelerators and storage rings increased roughly exponentially with time. The most powerful storage ring to date, the Large Hadron Collider (LHC)[8] of CERN, falls short of such a straight forward extrapolation from previous machines. However, its energy –currently 13 TeV and probably 14 TeV starting in 2021– is about 100,000 times higher than what the 60-inch cyclotron of Berkeley reached in proton-proton collisions 80 years earlier. In operation since 2008, the LHC proton-proton collider will dominate experimental particle physics for decades to come. Figure 8.5 shows the current complex of accelerators which CERN is operating. Planning for even larger machines is going on under the

---

[7] For better comparability, the diagram uses the mean quark energy and not the mean proton energy for hadron machines.

[8] See https://home.cern/science/accelerators/large-hadron-collider

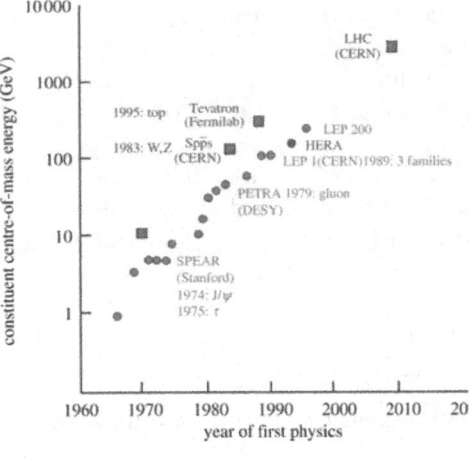

Figure 8.4 This Livingston diagram shows the maximum usable energy of accelerators and colliders as a function of time, electron accelerators (dots), proton machines (squares). HERA has so far been the only storage ring colliding electrons and protons. Next to the accelerator name, important discoveries are listed, which were made using it. (Credit: S. Myers, CERN)

auspices of the European Committee for Future Accelerators, ECFA[9]. CERN has become a world centre for accelerator based particle physics, with many contributions and contributors from outside Europe. A study is under way to assess the feasibility and physics of a giant circular collider at CERN, the Future Circular Collider[10], with collision energies ten times higher than the LHC.

Since light charged particles lose energy by bremsstrahlung when their trajectory is bent, linear accelerators are interesting alternatives to circular ones, especially for high energy electron beams. In fact, accelerator complexes all start with a linear accelerator right after the particle source, feeding beams into higher energy acceleration stages. The physics of linear accelerators is explained in Focus Box 8.5. A pioneer of this technology is the Stanford Linear Accelerator Center (SLAC), where major contributions to the discovery of quarks were made colliding high energy electron beams with protons and neutrons (see Section 9.1). While it is straight forward to direct the beam from a linear accelerator towards a stationary target, constructing a linear collider is rather complex. So far the only example, the Stanford Linear Collider, was also housed at SLAC. Plans also exist to build a major facility using colliding beams from linear accelerators[11].

Astrophysical phenomena can accelerate particles to energies which are many orders of magnitude higher than those man-made accelerators can ever hope to achieve. The highest energy particles observed so far reach beyond 10^{20} eV. They are, however, extremely rare. Their flux near Earth is as low as one particle per square kilometre per century. Cosmic rays, photons and neutrinos have become a subject of intense study again, in the emerging field of astroparticle physics. Because of their low fluxes at high energies, they are no longer

---

[9]See https://ecfa.web.cern.ch
[10]See https://home.cern/science/accelerators/future-circular-collider
[11]See https://ilchome.web.cern.ch and http://clic-study.web.cern.ch

In a circular accelerator of radius ρ, the power P lost at each turn is $P \propto \gamma^3/\rho$ with the relativistic factor $\gamma = E/m$, the particle energy E and mass m. The power lost is thus very much larger for electrons than for protons at the same radius and energy. One can eliminate this loss by increasing the curvature radius ρ, decreasing the bending field of the dipole magnets.

The limit $\rho \to \infty$ gives a linear accelerator sketched on the right. Obviously, in contrast to circular machines, the beam passes through each accelerating structure only once.

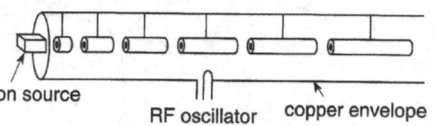

A large part of the vacuum tube forms itself an accelerating radiofrequency structure, in which an electromagnetic wave passes. Between the resonating cavities, there are field-free spacings realised by conducting tubes. The length l of these spacings must follow the velocity v of the accelerated beam, such that $l = vT/2$, where T is the period of the radiofrequency wave. The direction of the electric field in the cavities between tubes must obviously also be of the required strength and direction. Once the particle velocity approaches the speed of light, l becomes constant.

The particles are efficiently accelerated in this wave-guide structure, provided that:

- the electric field has a large longitudinal component along the beam direction and small transverse components;

- the phase velocity of the wave is roughly equal to that of the particles in the beam.

Since particles pass through the structure only once, one must aim for very high values of the electric field. This is usually done with superconducting cavities, which can reach electric fields of the order of 25 MV/m. A large accelerator nevertheless needs thousands of accelerating structures and cannot be too compact.

Focus Box 8.5: Linear accelerators

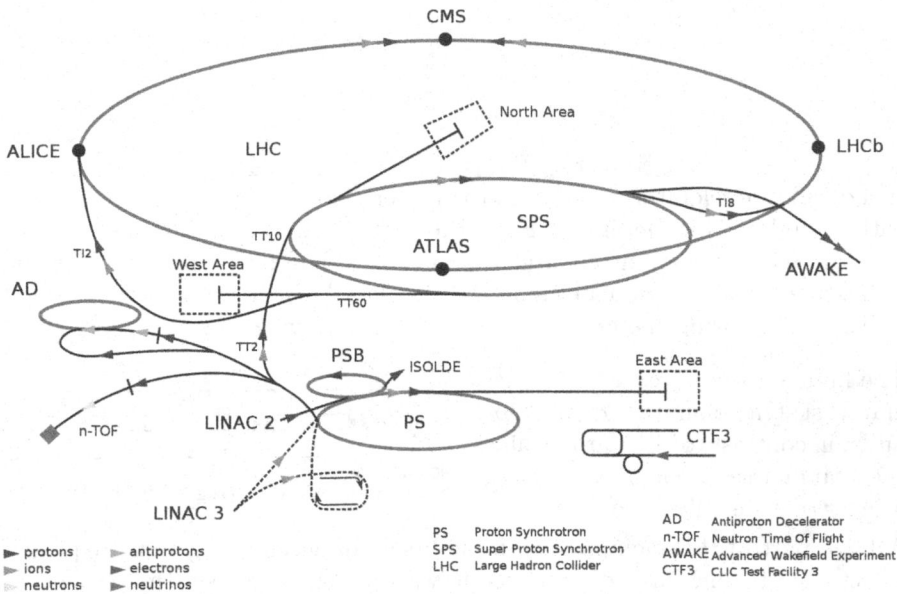

Figure 8.5 The CERN accelerator complex. Machine acronyms are indicated and explained in the legend. Acronyms of the four LHC experiments are also shown. Almost all machines built in over 60 years are still used, the smaller ones feeding pre-accelerated beams into the larger ones. (Credit: CERN)

used as a particle source. Instead, they carry important information about their astrophysical and cosmological sources and the accelerating mechanisms, which are able to reach these fantastic energies. More than a hundred years after their discovery, the production, acceleration and transport processes of cosmic rays are still mostly unknown and in the focus of experiments on Earth and in space (see Section 10.2).

8.3 PARTICLE DETECTORS

In parallel to the fast-paced development of accelerator technology, methods to detect and count them, and to measure their properties, have made impressive progress. In the beginning of the 20th century, particle detection was mostly about observing radioactive decay and cosmic rays, in addition to atomic spectroscopy. The counting rate of the Geiger and Marsden experiments, scattering α particles off nuclei never exceeded a few Hertz, and could be determined by observing a phosphorescent screen by eye. As far as the discovery of new particles was concerned, imaging techniques, like cloud chambers and photographic emulsions, played a crucial role. These devices not only visualise the impact of an ionising particle, as the phosphorescent screen does, but allow to follow its trajectory. Measuring the range and specific energy loss of charged particles (see Focus Box 8.2) together with their curvature in a magnetic field allows to estimate the mass and charge of particles. Yet the observations stayed anecdotal, a systematic measurement of particle properties and their reactions with significant statistics was not feasible with imaging techniques alone.

That changed with the invention of electronic methods of particle detection, pioneered by Hans Geiger and Walther Müller, and especially with the electronic registration of their signals via the coincidence method invented by Geiger, Bothe and Rossi (see Section 8.1 above). This is an electric circuit, using vacuum tubes at the time, which compares the time of arrival of two or more signals from detection devices, and triggers a counting or measuring process if they are coincident within a defined window. The method allows e.g. to do charged particle spectroscopy detecting them before and after an absorber, which defines an energy threshold. For imaging detectors, it allows triggering a camera whenever an interesting particle traverses the chamber.

An important progress in imaging detectors was the invention in the 1950s of the bubble chamber by Donald A. Glaser [320] at Ann Arbor, Michigan, working with his advisor Carl Anderson. Here a liquid is overheated under pressure. At the passage of particles, the pressure is released. Along the track of an ionising particle small bubbles form within microseconds, due to the local heating by ionisation. They are photographed using flash illumination. Pressure is then restored to be ready for the next passage. Since the thermodynamic process as well as pressurisation and release are not very rapid, this technique is limited to rates of the order of a Hertz. But it images particle reactions with rich detail. When the chamber is immersed in a magnetic field, it forms a spectrometer. The track curvature measures the particle momentum. The bubble density along the track measures the specific energy loss and allows to distinguish between light and heavy particles. Even photons, which do not themselves leave a track, can be observed if they convert into electron-positron pairs. Last but not least, bubble chamber pictures give an immediate visual impression of particle reactions. I for one was attracted to particle physics by precisely this feature.

In parallel to the accelerator development at Berkeley after World War II, the group of Luis Alvarez developed larger and larger bubble chambers filled with liquid hydrogen. One of their chambers, in the process of being inserted into its cooling dewar, is shown in Figure 8.3. Together with the Bevatron beam, the bubble chamber technique led to the discovery of a large number of hadrons and hadronic resonances. Both Glaser (1960) and Alvarez (1968) obtained the Nobel prize in physics for their important contributions to particle detection technology. The pictures of a bubble chamber had to be scanned by eye for interesting reactions. The particle tracks were digitised by hand and results typically stored on punched cards, later to be analysed by computers.

Louis Alvarez was a prolific inventor with no less than 17 U.S. patents. During the war he contributed to military efforts with the development of radar and an important contribution to Project Manhattan (see Chapter 6). His interests ranged from particle physics to dinosaur extinction, Egyptology and aviation.

Following the creation of intense proton beams at the CERN Proton Synchrotron and the rising interest in neutrino physics, the need arose for a massive detector capable to detect their very rare interactions [484]. This led to the proposal by André Lagarrigue of École Polytechnique in Orsay, France, to build a large bubble chamber filled with a heavy liquid like freon. The result was the heavy liquid bubble chamber Gargamelle, named after a character from the Renaissance fantasy novels by François Rabelais. It was almost 5m long and 2m in diameter, filled by liquid freon or propane, and immersed in a 2T magnetic field. The interior was viewed by eight cameras providing stereoscopic images of particle reactions. A photograph of the set-up is shown in Figure 8.6. It was operated with neutrino and antineutrino beams from 1970 to 1978. Until the 1980s, pictures from Gargamelle and its successor BEBC (1971 to 1984) were analysed by technical personnel and physicists using projection tables like the one shown in Figure 8.6. Again, particle tracks were digitised by hand in projection, then reconstructed in 3D using sophisticated computer programs. The results were stored on punched cards or magnetic tape for further analysis. The most important result from Gargamelle was the discovery of weak neutral current interactions, discussed in Section 9.2.

More target mass, but less spacial resolution and detail is provided by another imaging detector, the spark chamber invented by Shuji Fukui and Sigenori Miyamoto in the late 1950s [343, 350]. It consists of a series of parallel plate capacitors, filled with a gas mixture, typically on the basis of noble gases. A high voltage is applied to the metal plates with alternating polarity. When an ionising particle passes, a localised spark develops marking its position in every gap. The spark pattern can be photographed, it gives a coarse graphic representation of a reaction, which typically happens in one of the metal plates. The plates can be made of high density material like iron or lead, such that the target mass presented to the incoming beam is substantial. Therefore, spark chambers were preferentially used in experiments to observe small cross section reactions, like early neutrino experiments in Brookhaven and CERN. Spark chambers are still often used today to demonstrate the passage of cosmic rays.

The breakthrough towards digital technologies started with the invention of the multi-wire proportional counter by George Charpak (Nobel prize in 1992, Figure 8.7). The principles of operation of gas filled particle detectors are explained in Focus Box 8.6. In a proportional chamber, densely strung thin anode wires are set to high voltages, such that when an ionising particle passes, a cloud of electrons is created near the anode, but in contrast to a Geiger-Müller counter no break-down occurs. Under the influence of the strong electric field near the wires, electrons are accelerated towards the anode and

Figure 8.6 Left: The CERN bubble chamber Gargamelle. The large Helmholtz coils (white) generate a strong magnetic field orthogonal to the chamber axis, which is oriented along the direction of the incoming beam. The inside of the chamber body is lit by flash lights and viewed by eight cameras. Right: A projection table constructed by Saab in Sweden, projecting Gargamelle pictures on a table for scanning and digitisation. The tracks of final state particles are followed by hand and digitised. A computer program then reconstructs the reaction products in three dimensions. (Credit: CERN Courier, June 1973)

cause an avalanche. An electrical signal is thus induced which can be amplified and digitised. This allows to track the particle trajectory, with subsequent wire-by-wire measurements. The signal height is proportional to the specific energy loss of the particle, explaining the name proportional chamber, and allows to determine the absolute value of the particle charge. With this invention and the development of sensitive analog-to-digital converters, the step from analog to digital particle detection techniques was thus taken.

Even more precise localisation of particle tracks is achieved by semiconductor technology, which has largely taken over from gaseous detectors since the later part of the 20th century (see Focus Box 8.7). Thin silicon wafers, similar to computer chips, are doped to form diodes, with a potential barrier

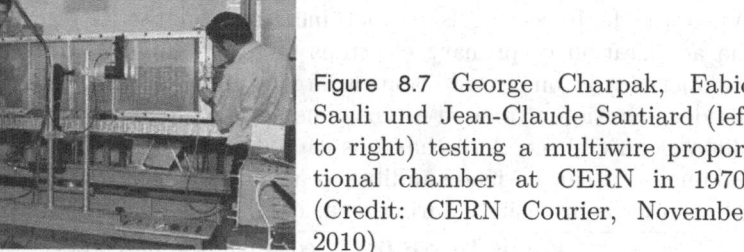

Figure 8.7 George Charpak, Fabio Sauli und Jean-Claude Santiard (left to right) testing a multiwire proportional chamber at CERN in 1970. (Credit: CERN Courier, November 2010)

Gaseous particle detectors register the passage of ionising particles by measuring the total charge of electrons and/or ions liberated in a gaseous medium.

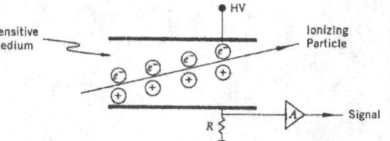

To recuperate electrons and ions before they recombine, there is an electric field collecting them on anode or cathode, respectively. The arriving charges cause a short current surge, detected after suitable amplification. The number of electron-ion pairs created is $N_i = -(dE/dx)\,d/W$, where dE/dx is given by the Bethe-Bloch formula (Focus Box 8.2), d is the thickness of gas traversed and W is the mean ionisation energy required for a single pair; in a gas $W \simeq 30$ eV.

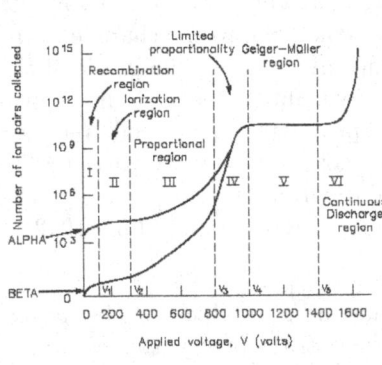

The main operational regions of gaseous detectors are shown on the left as a function of the applied voltage. In the *recombination region*, the field is too low to prevent recombination and the signal is very low. In the *ionisation region*, the charges drift towards the electrodes, but there is no avalanche. The collected charge is small, but directly measures dE/dx. This is the region where ionisation chambers like the ones used by Wolf or Hess work (see Focus Box 8.1).

In the *proportional region*, the signal is amplified by an ionisation avalanche close to the anode, still proportional to the number of ionisation pairs but amplified by a factor of 10^4 to 10^8. The signal is thus large, but subject to variations in amplification due to high voltage, temperature and other environmental factors.

This region is used in multiwire proportional chambers (MWPC) and drift chambers, examples of wire configurations are shown on the right.

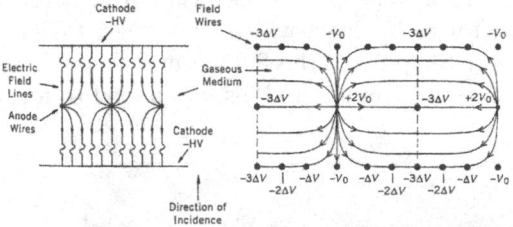

When the high voltage is further increased, in the *Geiger-Müller region*, the acceleration of primary electrons increases rapidly and they excite an avalanche early on. In addition, a large number of photons is produced in the de-excitation of ionised atoms. They also contribute to the avalanche by photo-electric effect. A discharge is caused giving a large electrical signal on the anode. It can be made audible as a sharp snap. Following the discharge, the counter is insensitive during what is called its dead time.

Focus Box 8.6: Gaseous particle detectors

in the junction zone. By applying a modest external voltage, this zone can be enlarged in a controlled way. When an ionising particle crosses, electrons are lifted from the valence to the conduction band and a short current surge passes the diode. The energy necessary to create an electron-hole pair in the semiconductor is ten times less than the ionisation energy of a typical gas atom. The signal is thus larger at comparable ionisation energy loss. It is also very localised, such that a structured metallisation of the surface allows spatial resolutions of the order of micrometers. Tracking devices together with a transverse magnetic field form a spectrometer, a detector to measure the particle momentum. Since the time integral of the signal current is proportional to the specific ionisation dE/dx, spectrometers can also contribute to particle identification.

A fundamentally different approach to the measurement of particle energy is the calorimeter. Its idea is to slow down and absorb particles completely, converting their energy loss into an electrical signal. Homogeneous materials which scintillate are an obvious choice. Scintillators are transparent inorganic or organic materials, which convert ionisation energy loss into light, with wavelengths often in the visible range. Examples are crystals like bismuth germanate or lead tungstate, lead glass and doped plastics, but also liquids like liquid noble gases. The light is collected and measured by a photodetector. Electrons liberated by photoelectric effect are multiplied by an avalanche in an electrical field; arrival time and signal height are measured. A layered construction, with passive metallic absorbers and active counting layers, allows absorbtion of high energy particles. In both homogeneous and layered calorimeters one uses the fact that high energy particles cause showers in matter[12], such that their energy is distributed over many particles and it suffices to count those to arrive at an energy measurement.

In addition to detectors which localise particles, determine their time of arrival and measure momentum or energy, there are many others which help identifying them. Examples are Cherenkov counters, transition radiation detectors and time-of-flight systems [503]. They are all sensitive to particle mass in one way or another. It is also common to all modern particle detectors that they generate electrical signals which can be digitised, analysed by computers and stored. In this way, synergy and redundancy between different detector types is used to optimise the reliability and accuracy of the measurement. This is of increasing importance, since modern particle physics experiments are often only repeatable in principle, due to the large effort necessary to construct accelerators and detectors. Prominent examples are the multipurpose detector systems ATLAS (`https://atlas.cern/`) (see Figure 9.14) and CMS (`https://cms.cern/`) at the Large Hadron Collider of CERN. The same is also true for modern particle experiments which do not use accelerators. Examples are the spectrometer for cosmic rays AMS (`http://www.ams02.org/`) (see Focus Box 10.3) on the International

---

[12]See e.g. `https://en.wikipedia.org/wiki/Particle_shower`

Modern ionisation detectors are often based on semiconductor sensors to localise particles and measure their specific energy loss dE/dx. A particle traversing a semiconducting solid, like e.g. a silicon diode, will create electron-hole pairs. Electrons can be lifted from the valence band into the conduction band by relatively little energy: only 3 eV are typically required, compared to typically 30 eV to ionise a gas molecule. This is due to the narrow gap between both bands and the fact that the Fermi energy, up to which the energy levels are completely filled, is situated right between the two. The probability density to find an electron at a given energy E at temperature T in a crystal is:

$$f(E) \;=\; \frac{1}{e^{(E-E_F)/kT}+1}$$

with the Boltzmann constant k (see Section 5.1). The Fermi level E_F is defined as the energy where $f(E_F) = 1/2$.

The sketch shows filled (hatched) and unfilled (blank) electron bands in a typical conductor, semiconductor and insulator. The Fermi level is also indicated.

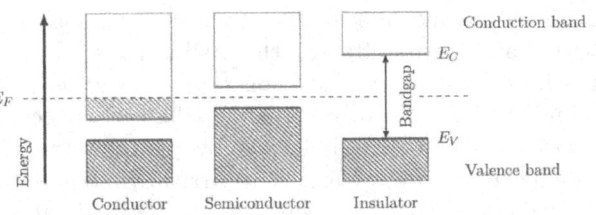

The gap between valence and conduction band is typically 1.12 eV wide for silicon at 300 K. The basis of a semiconductor diode are two layers, one of which is doped with positive (holes p), one with negative charges (electrons n).

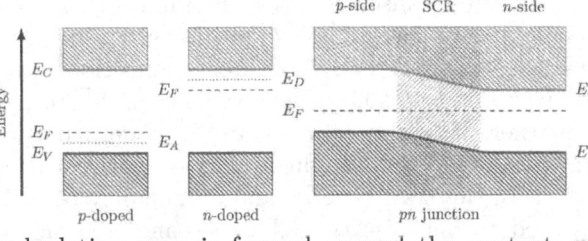

Where the two layers are in contact, a $p - n$ junction forms and charges migrate until an equilibrium is reached.

A depletion zone is formed around the contact, which works like a diode. When one applies an external voltage, a so-called reverse bias, the depletion zone can be enlarged. Electrons and holes liberated in this zone migrate to the two faces, within nanoseconds and very locally. Because of the low energy required to create mobile charges, the efficiency of such a detector is high, its response is linear in dE/dx and its energy resolution is very good. Only for very large ionisation densities, like those of heavy ions, saturation effects set in. To localise particles, the surfaces of semiconductor sensors are metallised with fine structures where charges are collected and fed into further electronic stages. Besides particle physics, semiconductor detectors are used in light detection, e.g. by CCD cameras.

Focus Box 8.7: Solid state particle detectors

Space Station, and the neutrino observatory Ice Cube at the South Pole (`https://icecube.wisc.edu/`).

The electronic signals from modern detectors are digitised and stored. They are first used online to decide whether an interesting reaction has occurred. If so, the data are kept for further processing. Data from different parts of the detector, but relevant to the same final state particle, are then coordinated and used to reconstruct the final state of the reaction with as much detail and accuracy as possible. These data can then be compared to theory predictions. For this purpose, reactions and their final states are generated in a computer program following the prescriptions of a given theory. The response of the detector to the final state particles is then simulated. This numeric process, called Monte-Carlo simulation, was pioneered during the Manhattan project (see Section 6.4). Reaction cross sections and distributions of final state properties can thus be statistically compared between experiment and theory.

The high luminosities of modern colliders lead to substantial reaction rates even for processes with tiny cross sections. Systems of particle detectors deliver extremely detailed information about every single event. Particle physics experiments thus need to use every step up in computing, data storage and analysis technology. From Rossi's coincidence circuit to neural networks in modern experiments, experimental particle physics has accompanied and often enough driven this progress. An excellent example is the invention of the World Wide Web by Tim Berners-Lee at CERN in 1989[13]. Originally meant to organise and access data and documentation of physics experiments, it has caused a revolution in everybody's way of communicating and access to knowledge.

8.4 FAILED PROJECTS

From the above you may get the impression that the history of particle physics technology consists of an uninterrupted sequence of breakthroughs and shining successes. While this is true in general, history is also marked by important set-backs, which have taught us how to live with failures and learn from them.

There are major accelerator projects, which have never been put to a successful completion. A first prominent example is the ISABELLE collider project at the Brookhaven National Laboratory. It was supposed to be an intersecting storage ring, colliding 400 GeV on 400 GeV protons, succeeding the pioneering and successful CERN ISR [582], which had 31+31 GeV centre-of-mass energy and ran from 1971 to 1984.

The idea of high energy proton colliders was based on original work by Gerard K. O'Neill in 1956 [337], who proposed two separate storage rings side-by-side with a single interaction point in common, like the Princeton-Stanford electron collider [385, 408] (see Focus Box 8.3). What was proposed instead at

---

[13] For a copy of his original proposal see `http://info.cern.ch/Proposal.html`

Brookhaven in 1973 were two storage rings on top of each other with multiple intersections, in the spirit of the CERN design, but with superconducting magnets [414].

A year later, the U.S. High Energy Physics Advisory Panel recommended that ISABELLE should be built. Construction began in 1978, with a prototype magnet successfully tested in parallel. Series production of these magnets, however, faced serious quenching problems, and caused considerable delays. Meanwhile, the construction of the Sp$\bar{p}$S collider at CERN and the subsequent discovery of the W and Z bosons (see Section 9.2) preempted the original physics justification of the collider.

In addition, starting in 1976 a proposal for a much larger machine, eventually called the Superconducting Super Collider, a proton-proton system to collide 20 TeV protons on 20 TeV protons, was discussed. On July 13, 1983, the U.S. Department of Energy cancelled the ISABELLE project after spending more than 200 million U.S. dollars on it. However, not all was lost. Parts of the tunnel, experimental hall and magnet infrastructure built for ISABELLE were finally reused to construct the Relativistic Heavy Ion Collider (RHIC)[14], which began operation in 2000.

A far bigger blow to U.S. particle physics was the failure of the subsequent Superconducting Super Collider (SSC) project. In the beginning of the 1980s a plan was put forward, with prominent supporters like Fermi National Laboratory director Leon Lederman and theorist Steven Weinberg as well as a majority of the U.S. particle physics community. In the first half of 1984 an ad hoc group prepared the Reference Designs Study [463], in which three possible designs for the SSC were developed and compared. At that same time a Central Design Group was created and Maurice Tigner of Cornell was appointed as its director. In October 1984 the CDG officially began its work of coordination and guidance for a national research and development effort at Brookhaven National Laboratory, Fermi National Laboratory, Lawrence Berkeley Laboratory, the Texas Accelerator Center, and participating universities. The group delivered a Conceptual Design Report in March 1986, which was reviewed by the DoE. The project gained presidential approval by the Reagan administration in January 1987, in "a deliberate attempt to reestablish U.S. leadership in a scientific discipline the nation had long dominated. If other nations were to become involved, they would have had to do so as junior partners in a multibillion-dollar enterprise led by U.S. physicists" [638]. Subsequently, candidate sites were invited to make proposals. The DoE received 43 proposals by September 1987, of which 36 were initially retained as qualified[15]. A committee narrowed down the choice to a short list of seven sites. Among them was the established Fermi National Laboratory close to Chicago, but also sites without major research infrastructure. One of the latter was selected. The site –2600 hectares of fields and pastures outside

---

[14]See https://en.wikipedia.org/wiki/Relativistic_Heavy_Ion_Collider
[15]See https://lss.fnal.gov/archive/other/ssc/ssc-doe-eis-0138-22.pdf

of Waxahatchee in Ellis County, Texas– was announced a day after the Texan George Bush Sr. was elected president in 1988. The announcement had been delayed to "avoid the appearance of backroom favouritism" [465]. Congress in 1989 voted 331 to 92 to proceed with the construction.

During the design and the first construction stage, a heated debate ensued about the high cost of the project. In 1987, Congress had been told the project could be completed for 4.4 billion U.S. dollars, an adventurous estimate considering that there was no existing infrastructure nor a strong national laboratory existing on site. In 1993, the NGO Project on Government Oversight, in an open letter to president Clinton, released a draft audit report by the Department of Energy's Inspector General heavily criticising the project for its high costs and poor management [472]. By then, the cost estimate had evolved to almost three times the initial estimate.

As far as politicians were concerned, cost overruns and poor management were the official reasons for cancelling the project. But in hindsight, many reasons have been identified for its failure. The analysis which I –as a European particle physicist– find most convincing has been summarised by Michael Riordan [624, 638]. He argues that the daring decision by a single nation, however powerful, to build a collider –twenty times as large and twenty times as costly as the existing Tevatron at the Fermi National Laboratory– in the Texan wilderness was just "a bridge too far." He argues, that a more incremental approach, as the one chosen by CERN in Europe (see Figure 8.5), and relying on an existing national centre like Fermilab with existing infrastructure and expert manpower would have had more chance of success. According to Riordan, a multilateral approach, inviting other nations to make major contributions and giving them a say in shaping the project, would have protected the project better from U.S. partisan politics. However, recent unilateral withdrawal of the U.S. from major international initiatives casts some doubt on this assumption. In addition to a slowing economy approaching a recession, Riordan identifies a clash of culture between physicists and managers from the military-industrial complex as a major problem, as well as the reluctance of high level physicists and engineers to venture a long time engagement in an unattractive area.

The termination of the SSC left a deep shock in the U.S. particle physics community. As an immediate reaction, Steven Weinberg [474, p. 277ff] feared that with this cancellation, the U.S. seemed "permanently to be saying goodbye to any hope of ever having a responsible program of research in elementary particle physics." Fortunately, this harsh judgement turned out to be overly pessimistic. In the aftermath, U.S. particle physicists defined a vigorous program of neutrino, heavy flavour and heavy ion physics at accelerators, as well a rich non-accelerator particle physics agenda. Moreover, U.S. physicists have an important participation in experiments at the European LHC.

The cancellation of the SSC indeed gave a push to CERN's less ambitious LHC project, which had been selected unanimously in 1991 by its governing body, the CERN Council, in which the 23 member states have equal votes.

The full scientific, technical and financial proposal was presented by the designated CERN director general Chris Llewellyn-Smith and approved in December 1993. In addition to its less ambitious goal of 14+14 TeV centre-of-mass energy, LHC's advantage in terms of cost was the use of the pre-existing engineering infrastructure and the 27 km long underground cavern of the Large Electron-Positron Collider. The construction was completed in 2008 under the leadership of its Welsh project director, Lyn Evans [577]. The LHC costed 4.6 billion Swiss Francs to build, equivalent to about the same sum in U.S. dollars. It produced first beams in August 2008. A month later, a faulty connection in one of its superconducting magnets caused a damaging quench, explosively releasing helium from the magnet cold mass into the tunnel. Nobody was hurt, but the real start-up was delayed by about a year to prevent similar incidents. The LHC has been running according to schedule ever since, including regular upgrades.

In contrast to abandoned accelerator projects, failures of particle physics experiments are less visible. To my knowledge, none associated with large accelerator programs have failed because of technical or financial problems. This may well be because when an experiment fails, the participating scientists are basically the only victims. This is a fabulous incentive to overcome technological or financial crisis. Also, regular peer review prevents major problems to pass below the radar.

Non-accelerator projects are more vulnerable. An example of a major incident is the 2001 accident at the Super-Kamiokande neutrino observatory in Japan. After maintenance work, for which the 50,000 ton water tank of this giant Cherenkov detector had been emptied, one of the 11,200 photomultiplier tubes imploded, sending out a shock wave that destroyed half of the tubes[16]. Almost immediately, the director of the Kamioka laboratory, where the detector is housed, reassured the scientific community [526]: "We will rebuild the detector. There is no question." Indeed it restarted physics a year later with half the density of tubes, and in 2006 with a fully equipped detector.

Large technological and operational risks are of course encountered by balloon and space experiments [612]. During launch, mechanical stress applies, which sensitive particle detectors do not always withstand. In the atmosphere, weather is unpredictable and so are balloon trajectories. In space, temperature excursions are very large and the radiation environment is harsh. Thus numerous probes and satellites have been lost or failed to deliver the expected data[17]. Repairs in orbit, as e.g. done for the Hubble Space Telescope (https://hubblesite.org) and the AMS cosmic ray observatory on the ISS (see Focus Box 10.3), are a very rare exception.

What is far more interesting is the case of experiments or whole programs which technically worked, but did not have the expected outcome. It is

---

[16]For a complete account of the accident, see http://www-sk.icrr.u-tokyo.ac.jp/cause-
-committee/1st/report-nov22e.pdf

[17]For a list see e.g. https://en.wikipedia.org/wiki/Timeline_of_artificial_satellites-
_and_space_probes

Discoveries in Physics

Facility	Original purpose, Expert Opinion	Discovery with Precision Instrument
P.S. CERN	π N interactions	Neutral Currents -> Z, W
Brookhaven	π N interactions	ν_e, ν_μ CP violation, J
FNAL	Neutrino physics	b, t quarks
SLAC Spear	ep, QED	Scaling, Ψ, τ
PETRA	t quark	Gluon
Super Kamiokande	Proton decay	Neutrino oscillations

Figure 8.8 Mismatch between original purpose and actual discoveries made with major particle physics infrastructure, as seen by Nobel laureate Samuel C.C. Ting. (Credit: S.C.C. Ting)

after all the purpose of scientific activity to probe the unknown and discover what nobody had thought of before. Nobel laureate Samuel C.C. Ting always makes this point, as shown in Figure 8.8. Thus rarely the actual outcome of experiments figures in the original proposal. These unexpected outcomes of experimental programs are of course not failures but the very definition of successful experimentation.

FURTHER READING

Luis W. Alvarez, *Adventures of a Physicist*, Basic Books, 1987.

Wiliam R. Leo, *Techniques for Nuclear and Particle Physics Experiments*, Springer, 1994.

Edmund J.N. Wilson, *An Introduction to Particle Accelerators*, Oxford University Press, 2001.

Michael Riordan, Lilian Hoddeson and Adrienne W. Kolb, *Tunnel Visions: The Rise and Fall of the Superconducting Super Collider*, University of Chicago Press, 2015.

CHAPTER 9

The Standard Model of matter and forces

The Standard Model is expressed in equations governing the various fields, but cannot be deduced from mathematics alone. Nor does it follow straightforwardly from observation of nature... Nor can the Standard Model be deduced from philosophical preconceptions. Rather, the Standard Model is a product of guesswork, guided by aesthetic judgment, and validated by the success of many of its predictions. Though the Standard Model has many unexplained aspects, we expect that at least some of these features will be explained by whatever deeper theory succeeds it.

S. Weinberg, *To Explain the World*, 2015 [628]

I N THIS CHAPTER I describe the Standard Model of matter, electroweak and strong interactions and its vacuum. This is the *pièce de résistance* of the book.

9.1 MATTER

In the beginning of the 1930s, electron and positron were the only known leptons. In the years 1936/37, Anderson and Neddermeyer installed their spectrometer on Pike's Peak, with an altitude of 4300m one of the highest summits in the Rocky Mountains. There they identified a charged particle with a mass between that of the electron and that of the proton and suggestively called it "mesotron" [269, 273]. Shortly after, Street and Stevenson [277] confirmed their discovery of the new particle. It was wrongly conjectured at the time to be the force particle that should transmit the nuclear force according to Hideki Yukawa's trial field theory [268]. Consequently it was wrongly categorised as a meson [446]. In the group of Marcello Conversi in Rome, as well as by Evan J. Williams and George E. Roberts at University College of Wales, the decay of the new particle was first observed [490]. Under the most difficult conditions of World War II, Conversi, Oreste Piccioni and Ettore Pancini determined its lifetime of about 2 μs. However, their findings were not generally known until after the war [303, 309]. After Enrico Fermi's emigration to the U.S., Franco Rasetti, a close collaborator of his, went to Canada in 1938. He measured, in parallel to Conversi's group, the lifetime of "mesotrons" stopped in matter [294, 592]. After the war, Conversi and his colleagues showed that the muon had nothing to do with nuclear forces [309]. It was thus recognised as a heavy "brother" of the electron, the name mesotron disappeared in favour of the less suggestive muon (μ). However, its is not at all an excited state of the electron, as the absence of the radiative decay $\mu^\pm \to e^\pm \gamma$ shows.

The neutral leptons, i.e. neutrinos (ν), are much harder to detect, since they only interact via weak interactions (see Section 9.2 below). The existence of such a rarely interacting particle had been boldly conjectured by Wolfgang Pauli in 1930, based on the requirement of energy and angular momentum conservation in beta-decays of nuclei with atomic number A and charge Z, $(A, Z) \to (A, Z + 1)e^- \bar{\nu}_e$. He had called it neutron, the real neutron had not been discovered yet (see Section 5.5). In his famous letter to the "dear radioactive ladies and gentlemen"[1] he had also predicted that the particle would be hard to find. Indeed the cross sections of neutrinos with matter are the smallest of all elementary particles, such that a neutrino can easily traverse Earth without interacting. It is thus no wonder that their discovery had to wait for high neutrino fluxes and massive detectors to become available. The first idea came from nuclear bomb research at Los Alamos [492] (see Section 6.4). Frederick Reines and Clyde Cowan had realised that nuclear beta decay can be reversed into the reaction $\bar{\nu}_e(A, Z) \to (A, Z - 1)e^+$, leading to a detectable positron. When the reaction takes place on a hydrogen nucleus, $\bar{\nu}_e p \to n e^+$, also the final state neutron can be used. Their first plan sounds crazy to us today: they considered using neutrinos released in the test of a nuclear bomb. In reality and more realistically, the nuclear reactors at Hanford and Savannah River were used as neutrino sources. The experiment required massive detectors, capable of detecting electron-positron annihilation into photons, $e^+ e^- \to \gamma\gamma$, as well as neutron capture in a nucleus with subsequent photon emission. This became possible with the development of liquid scintillators, organic materials in solution which react to passing ionising particles by emitting short pulses of light. They have a high hydrogen content and were originally also developed for military purposes. Counting neutrino reactions against the intense background coming from the nuclear reactor and cosmic rays was difficult. It succeeded because the signal from neutron capture is delayed by a few microseconds from the positron signal. In addition, random coincidences with an even longer delay give a direct measure of the background. This procedure, called side-band subtraction, is still routinely used today to identify small signals in a large background. In June 1956, Reines and Cowan communicated

---

[1] A copy of the letter in German can be found at cds.cern.ch/record/83282/files/-meitner_0393.pdf

the evidence for neutrino reactions in a telegram to Wolfgang Pauli. In July a short publication followed [332].

Muon neutrinos and antineutrinos, neutral leptons of the second generation, are produced in the decay chain of charged pions: $\pi^- \to \mu^- \bar{\nu}_\mu$; $\mu^- \to e^- \bar{\nu}_e \nu_\mu$ (resp. $\pi^+ \to \mu^+ \nu_\mu$; $\mu^+ \to e^+ \nu_e \bar{\nu}_\mu$). Similar processes happen in the decays of charged kaons. The search for their interactions required powerful proton accelerators with the necessary intensity to create massive amounts of pions and kaons. The first available was the Brookhaven AGN, where a focalised neutrino beam was first constructed in the early 1960s. Since neutrinos are electrically neutral, they cannot themselves be focalised to increase the flux. Instead, following ideas by Bruno Pontecorvo [344] and Mel Schwartz [348], one can steer the proton beam towards a beryllium target and focalise the pions and kaons, which are produced in large numbers. One then lets them decay. A thick metal absorber stops all leftover particles but the barely interacting neutrinos. In Brookhaven, steel plates from a scrapped military vessel with a total thickness of 13.5m were used. A massive detector –in Brookhaven a sandwich of aluminium plates, spark chambers and scintillators with a total mass of 100 tons– allowed detecting interactions of muon neutrinos with nucleons, $\nu_\mu \mathcal{N} \to \mu^- X$, where $\mathcal{N}$ denotes a proton or neutron and X an arbitrary hadronic final state. Details of the hadronic final state in fact don't matter, detecting the rather penetrating muon suffices to show that the reaction is not due to electron neutrinos or hadrons. In summer of 1962, Leon Lederman, Schwartz and Jack Steinberger published first results on the discovery of muon neutrino interactions [357]. The three were awarded the Nobel prize in physics of 1988.

Bruno Pontecorvo, one of the founding fathers of neutrino physics, was a fascinating physicists and a somewhat mysterious person [616]. Having worked for the British Atomic Energy Research Establishment in Oxfordshire, he defected to the Soviet Union in 1950 and disappeared from public view for over five months. There have been speculations ever since about him having been a Soviet spy (see Section 6.5), an accusation he always denied. He did most of his best work while in Dubna, at the Joint Institute for Nuclear Research. He died of Parkinson's disease in 1993, shortly after the end of the Cold War.

The high mass resolution of bubble chambers and the increasing energy reach of accelerators led to an explosion in the number of newly detected hadrons and resonances[2] in the 1950s and 1960s. Putting order into the richer and richer spectrum of particles was badly needed. Starting in 1957, Arthur H. Rosenfeld[3], Walter H. Barkas, Murray Gell-Mann [339, 371] and Matts

---

[2] A resonance is a short lived hadronic state, seen as a broad peak in the mass distribution of its decay products. Other methods to detect resonances are the Dalitz plot and the Argand diagram.

[3] A first edition of the Rosenfeld table was a report from the University of California Radiation Laboratory, W.H. Barkas and A.H. Rosenfeld, *Data for Elementary Particle Physics*, UCRL-8030, which was updated several time and published in Review of Modern Physics from 1964 onward.

Roos [368, 382] did pioneering work in this field editing the so-called Rosenfeld tables. They listed the measured properties of all known particles, with experimental details and systematic averaging wherever more than one measurement was available. Rosenfeld was Enrico Fermi's last PhD student. Roos, a theoretician mastering statistical methods, played an important role in developing well founded methods of kinematic and statistical analysis[4] in particle physics. The Rosenfeld tables developed into the Review of Particle Properties (RPP), which was published for the first time in 1968. Brought together by an international network, the Particle Data Group (http://pdg.lbl.gov/), the Review has grown today to a volume of more than 2000 pages [658]. It is not only the generally accepted reference for carefully and critically edited results concerning particle properties, but also a comprehensive reference for the methodology of particle physics. Since 1995, it is thus rightly called the Review of Particle Physics. Its pocket edition, which started as a wallet card in 1957 and is now the Particle Physics Booklet of about 300 pages, has been the daily companion of several generations of particle physicists.

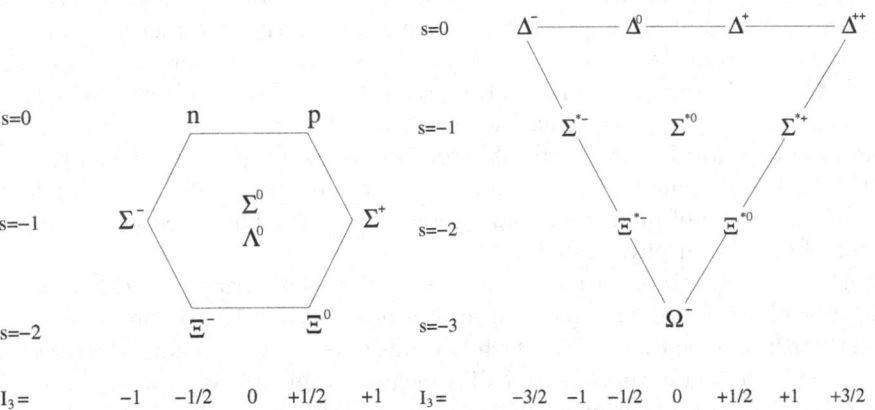

Figure 9.1   Left: Octet of baryons with spin $1/2$ (left) and decuplet of baryons with spin $3/2$ (right). The particles are ordered according to their strong isospin (horizontal axis) and their strangeness (vertical axis). The discovery of the Ω^- was a triumph of the Eightfold Way.

An organising pattern in the properties of mesons and baryons was found in group theory. Groups are formed by transformations, if their sequential application always leads to another member of the group. Lie groups are those where transformations in an infinitesimal neighbourhood to the unity transformation –the one which changes nothing– define the properties of the whole group. Translations and rotations in three-dimensional space are com-

---

[4]See the standard reference book by W.T. Eadie, D. Drijard, F.E. James, M. Roos and B. Sadoulet, *Statistical Methods in Experimental Physics*, North Holland, Amsterdam, 1971, current edition by Frederick James [563].

mon examples of Lie groups, for which all transformations can be decomposed into a series of infinitesimal ones. Hadrons with the same spin and similar masses can be classified according to their quantum numbers strangeness and strong isospin[5] [354, 358, 365, 366]. The former distinguishes strange hadrons, like Λ or K, from their non-strange siblings, like n and π. The latter has mathematical properties analogous to spin, it is a two-component vector defined by its length I and its third component I_3. It distinguishes e.g. the proton with $I_3 = +1/2$ from the neutron with $I_3 = -1/2$. Transformations which form the group SU(3), relate members of a so-called representation of the group; their properties are similar. The lightest baryons with spin 1/2 and with or without strangeness belong to an octet or singlet representation of SU(3), those with spin 3/2 to a decuplet, as shown in Figure 9.1. Inspired by buddhist ethics, Murray Gell-Mann baptised this order scheme the "Eightfold Way" [351], based on the irreducible representations of SU(3)[6].

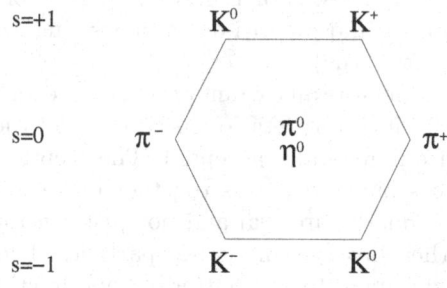

Figure 9.2 Octet of light mesons.

The triply strange baryon Ω^- was still missing when the order scheme was proposed; it was discovered in 1964 at the 30 GeV proton synchrotron of the Brookhaven National Laboratory, again with a hydrogen-filled bubble chamber [372]. Herwig Schopper, who became the CERN Director General from 1981 to 1988, wrote about this discovery [376]: "This time it is not just any new particle, but the discovery of the Ω^- is an *experimentum crucis* for a new theory, which could become for the systematics of elementary particles what the Balmer formula was for the structure of atoms." There is of course also an analogy to the prediction and discovery of chemical elements based on Mendeleev's periodic system.

Strong interactions, which hold together hadrons, are insensitive to isospin and strangeness, such that masses in a multiplet are similar. They would even be equal, if the strong SU(3) symmetry were exact. This suggests the

---

[5]This strong isospin is not to be confused with the weak isospin we will introduce in Section 9.2.

[6]Irreducible representations are those which cannot be decomposed into less populated ones. Examples are $3 \times 3 \times 3 = 10 + 8 + 8 + 1$ (Fig. 9.1) and $3 \times \bar{3} = 8 + 1$ (Fig. 9.2).

hypothesis that the observed symmetry properties are due to a substructure of light mesons and baryons. This is the basis of the quark model, simultaneously invented by Murray Gell-Mann [373] and the CERN theorist George Zweig [384] in 1964. The name quark is borrowed from the powerful creative language of James Joyce's "Finnegans Wake"[7]. In its original form, the model is based on a triplet of quarks, called *up*, u = $(1, 0, 0)$, *down*, d = $(0, 1, 0)$ and *strange*, s = $(0, 0, 1)$. The coordinates denote the position in an abstract space of strong isospin and strangeness, which today count among the more numerous flavour quantum numbers. The antiquarks $\bar{u}$, $\bar{d}$ and $\bar{s}$ have the same, but negative quantum numbers. Rotations in this abstract space, which do not change the number of quarks, belong to the group SU(3).

The electric charge is $+\frac{2}{3}e$ for the up quark and $-\frac{1}{3}e$ for the two others; all three are fermions with spin $\frac{1}{2}$. With these settings, all baryons of Figure 9.1 can be composed by three quarks. As an example, the proton is a state consisting of two up and one down quark, (uud), the neutron has one up and two down quarks, (udd). The mesons of Figure 9.2 consist of a quark-antiquark pair, like the π^+ of an up and an anti-down quark, (u$\bar{d}$), or the K$^+$ of an up and an anti-strange quark, (u$\bar{s}$).

While the quark model was an elegant order scheme for the hadrons known at the end of the 1960s, its interpretation as a model for the internal structure of hadrons was far from generally accepted. The scepticism was due to the fact that the dynamics binding quarks together into hadrons was unknown. The direct proof that quarks are real and not just a fiction of group theory, was assumed to be their observation as free particles. The signature is clear: they are distinguished from ordinary matter by their fractional electric charge. However, all attempts to find free quarks were fruitless; a concise overview of the methods and results of these searches is e.g. given in [550].

Instead the proof that hadrons really have a material substructure was left to scattering experiments à la Rutherford, but with electrons instead of α particles as probes. The reason is that the interactions of high energy electrons with charged particles were well understood and calculable thanks to Quantum Electrodynamics. The ones of nuclei were not. In addition, point-like particles like electrons can probe deeply into protons and neutrons, without being themselves subject to strong interactions.

Experiments at the Stanford Linear Accelerator Center (SLAC) in California were of fundamental importance for this research into the structure of nucleons. In the early 1950s, Robert Hofstadter, then a young associate professor at Stanford University, initiated a program on electron scattering off heavy and light nuclei. He used electron beams of a few hundred MeV energy and a one-arm spectrometer rotatable around a target to identify and measure the outgoing electron. In 1954, first experiments with a hydrogen gas target were conducted [353].

---

[7]The fourth chapter begins with a poem: "Three quarks for Muster Mark! / Sure he hasn't got much of a bark / And sure any he has it's all beside the mark." Thus quarks owe their name to an alliteration.

We consider the elastic scattering of electrons off protons, $e^- p \rightarrow e^- p$. Let us start with the simplest case, the elastic scattering of a relativistic electron off a scalar, point-like target with charge Z. In QED, the cross section for this relativistic process is described by the so-called Mott cross section [232, 251]:

$$\left(\frac{d\sigma}{d\Omega}\right)_{\text{Mott}} = \frac{\alpha^2}{4E^2 \sin^4 \frac{\theta}{2}} \frac{E'}{E} \left[1 - \sin^2 \frac{\theta}{2}\right]$$

Here E and E' are the energies of the incoming and outgoing electron, M the target mass and θ the scattering angle. The fine structure constant $\alpha = e^2/(4\pi)$ determines the order of magnitude of the cross section. The first term reminds us of the Rutherford cross section of Focus Box 4.5. Its denominator results from the photon propagator, inversely proportional to the square of the exchanged four-momentum q:

$$\frac{1}{q^2} = \frac{1}{(k' - k)^2} \simeq \frac{1}{2(EE' - \vec{k}\,\vec{k}')} \simeq \frac{1}{4E^2 \sin^2 \frac{\theta}{2}}$$

where k and k' are the four-momenta of incoming and outgoing electron. The second term, $E'/E = 1/(1 + 2E/M \sin^2 \theta/2)$, takes into account the recoil of the target. The third term takes into account the electron spin, i.e. the conservation of its helicity. Helicity is the projection of spin on the momentum vector of a particle. It is positive for spin oriented in the direction of motion, negative otherwise. Helicity conservation requires that the cross section vanishes for backscattering, $\theta = \pi$. If the target is a fermion, with magnetic moment $\mu = e/2M$, a magnetic term is added [316], which only becomes important in the relativistic limit:

$$\left(\frac{d\sigma}{d\Omega}\right)_{\text{Mott}} = \left[\frac{\alpha^2}{4E^2 \sin^4 \frac{\theta}{2}}\right] \left[\frac{E'}{E}\right] \left[\cos^2 \frac{\theta}{2} + \frac{q^2}{2M^2} \sin^2 \frac{\theta}{2}\right]$$

The three terms thus describe static and point-like target; target recoil; and spin effects, respectively. The last term is especially important. It has two concurrent angular dependences, one proportional to $\cos^2 \theta/2$, the other to $\sin^2 \theta/2$. We will find back these two terms when the target has an internal structure.

Focus Box 9.1: Electron scattering off a structureless target

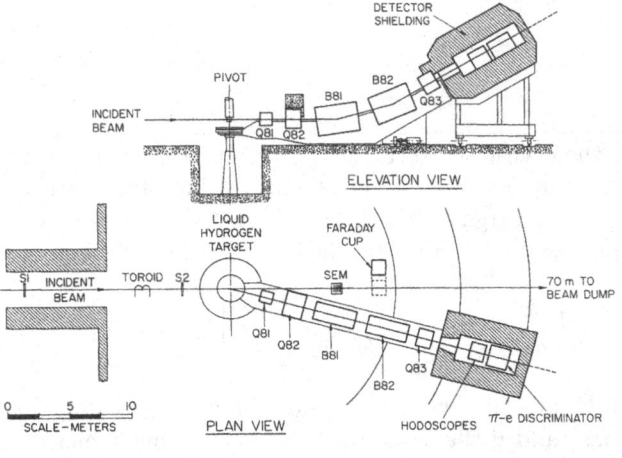

Figure 9.3 The SLAC 8GeV electron spectrometer [469]. The electron beam enters from the left and hits a target of liquid hydrogen. The beam intensity is measured by an upstream toroid coil. The outgoing electron is identified and its momentum measured in the downstream spectrometer arm. This arm can be rotated around the target to cover different scattering angles.

At small electron energies, less than about 1 GeV, and small momentum transfers from electron to nucleon, the target particle stays intact and the scattering is elastic. The scattering cross section then resembles the one for Rutherford scattering, except that the recoil of the target and the spins of both reaction partners have to be taken into account, as detailed in Focus Box 9.1. The interaction takes place with the charge distribution of the target as a whole. If the target is not point-like but extended, the elastic cross section is diminished in a characteristic way, by what is called a form factor. As explained in Focus Box 9.2, the form factor can be measured from the angular distribution, and converted into a measure of the size of the charge distribution. When the electron is relativistic, also the magnetic moment of the target influences the cross section; there is then a second form factor related to the distribution of the magnetic moment inside the nucleon. For the proton, both correspond to a size of about 0.7×10^{-15}m, close to 1 fm [331, 336]. Modern precision measurements give a larger value of 0.831×10^{-15}m [671]. For the neutron, the magnetic form factor indicates a similar size for the distribution of its magnetic moment. For this first look into the structure of the nucleon, Hofstadter shared the 1961 Nobel prize in physics with Ludwig Mössbauer.

Alternatively, the proton radius can be determined from the atomic spectra of hydrogen, since ground state electrons traverse the nucleus all the time. For normal hydrogen, values different from those quoted above are obtained [652, 669]. When the atomic electron is replaced by a muon, which supposedly gives more precise values, spectroscopy results agree with the scattering result [580]. New spectroscopy experiments are under way at Paul Scherrer Institute in Switzerland to resolve this discrepancy.

Here we extend the discussion of Focus Box 9.1 to targets which are not point-like, and to nucleons in general. If the target has a finite size, its charge is distributed over space, such that the cross section is diminished by a so-called form factor $|F(q^2)|^2$, a function of the four-momentum transfer q between projectile and target. In non-relativistic approximation, the form factor is the Fourrier transform of the spatial charge distribution $\rho(\vec{x})$:

$$F(\vec{q}) = \int e^{i\vec{q}\vec{x}} \rho(\vec{x}) d^3x = 1 - \frac{1}{6}|\vec{q}|^2 \langle r^2 \rangle + \dots$$

On the right is an approximation for small momentum transfers to a spherically symmetric charge distribution with a mean quadratic extension $\langle r^2 \rangle$. If the charge distribution is exponential, $\rho(r) \sim \exp(-\Lambda r)$, one finds a form factor $F(|\vec{q}|) \sim (1 + |\vec{q}|^2/\Lambda^2)^{-2}$. For any shape, the form factor diminishes rapidly with increasing momentum transfer.

Things are a bit more complex if we consider spin and a charge distribution which is not static. Interactions are not limited to Coulomb scattering, but magnetic effects come in. Of course, the role of "electric" and "magnetic" terms depends on the reference frame. In the laboratory system, one finds [316]:

$$\left.\frac{d\sigma}{d\Omega}\right|_{\text{Lab}} = \left(\frac{\alpha^2}{4E^2 \sin^4 \frac{\theta}{2}}\right) \frac{E'}{E} \left(\frac{G_E^2 + \tau G_M^2}{1 + \tau} \cos^2 \frac{\theta}{2} + 2\tau G_M^2 \sin^2 \frac{\theta}{2}\right)$$

with the parameter $\tau \equiv -q^2/(4M^2)$, depending on the ratio of the four-momentum transfer q and the target mass M. One somewhat sloppily calls G_E and G_M the electric and magnetic form factors of the target. Experimentally, the form $G \sim (1 + |\vec{q}|^2/\Lambda^2)^{-2}$ describes both form factors of the proton rather well, with a common parameter $\Lambda \simeq 0.84$ GeV. For the neutron, the electric form factor is compatible with zero, the magnetic one is roughly compatible with the one of the proton, multiplied by the neutron magnetic moment.

Focus Box 9.2: Electron scattering off an extended target

While elastic scattering leaves the target intact, its cross section falls off quickly with increasing momentum transfer from the electron to the target. At larger momentum transfer, inelastic scattering sets in. If the nucleon has internal structure, a short lived excited state can be produced. This happens when the photon probing the nucleus has a wavelength comparable to the nucleon size, i.e. at momentum transfers of the order of the nucleon mass, roughly 1 GeV. Excited states of the nucleon are called resonances, in analogy to other resonance phenomena in physics (see Focus Box 9.12). They occur indeed when the photon resonates with an internal structure of the nucleon. An example is the reaction $e^-p \to \Delta^+ \to p\pi^0$ or $\to n\pi^+$. The resonance Δ^+ itself is not observable due to its short lifetime, of the order of 10^{-24}s, but it manifests itself in the invariant mass distribution of the final state particles. It has a broad distribution around the mass of the resonance, its width is the inverse lifetime, as explained in Focus Box 7.7.

When one increases the electron energy and the momentum transfer further, beyond a few GeV, the photon exchanged between projectile and target has a wavelength much smaller than the size of a nucleon. It can thus resolve its internal structure. If quarks really exist inside the nucleon, the photon can probe them. The nucleon is broken up and many additional mesons or even baryon-antibaryon pairs are produced. The invariant mass of the final state is no longer a fixed quantity –neither the proton mass as in elastic scattering, nor the less precisely fixed resonance mass. It becomes an additional kinematic variable only limited by energy-momentum conservation. A useful choice of kinematic variables is the energy transfer from the electron to the nucleon, thus the photon energy, and the square of the four-momentum transfer, the squared photon mass. The latter is of course not zero since the photon is virtual. The cross section can then be expressed in a very general way as a function of two so-called structure functions, which are a function of both variables, as explained in Focus Box 9.3.

Starting in the late 1960s, the SLAC linear accelerator provided the intense and high energy electron beams required for experiments at energies of a few GeV, and opened the path to reaching the required resolution. A series of experiments to measure elastic and inelastic scattering off hydrogen and deuterium nuclei were conducted between 1968 and 1973 [469]. The experimental approach was in opposition to the doctrine of S-matrix theory, which stated that the dominant processes at low momentum transfer between projectile and target contained all the physics of the process. Since the S-matrix approach denies any substructure of hadrons, no probe was supposed to penetrate them anyway. In contradiction to this ideology, the experiments at SLAC clearly demonstrated that protons and neutrons indeed have an inner structure. Jerome Friedman, Henry Kendall und Richard Taylor were awarded the Nobel prize in physics for these seminal experiments in 1990. The experiments used a so-called one-arm spectrometer which only identified and measured the final state electron, as shown in Figure 9.3. The momentum vector of the initial state electron being known from the accelerator beam line, the final momentum vector suffices to calculate the kinematic properties of the hadronic final state, without actually measuring them.

The structure function measurements at SLAC established the following picture of inelastic electron-nucleon scattering [468, 469]:

- At moderate momentum transfer the cross section has a series of broad peaks, which correspond to short-lived resonances, excited states of the nucleon.

The kinematic region where the nucleon is broken up by a high energy photon is also called deep inelastic scattering. Not knowing the internal structure of the nucleon, one can parametrise the differential cross section with two terms, as for elastic scattering (see Focus Box 9.2):

$$\frac{d^2\sigma}{dE'd\Omega}\bigg|_{\text{lab}} = \left(\frac{\alpha^2}{4e^2\sin^4\frac{\theta}{2}}\right)\left\{W_2(\nu, q^2)\cos^2\frac{\theta}{2} + 2W_1(\nu, q^2)\sin^2\frac{\theta}{2}\right\}$$

The two structure functions, W_1 and W_2, which take the role of the form factors, parametrise the unknown distribution and dynamics of the nucleon constituents, if it has any. They depend on the photon energy, $\nu = E - E'$, and its squared invariant mass, q^2. The description requires two kinematic variables, since the hadronic final state no longer has a fixed mass; its total mass M_{had} is related to these variables and the nucleon mass M by $M_{had}^2 = M^2 + 2M\nu + q^2$. The structure functions are measured by decomposing the observed rate as a function of scattering angle θ and electron energy E'.

The great discovery at SLAC, in the late 1960s, was that starting at q^2 of a few GeV^2, the structure functions do not depend on ν and q^2 separately, but on their ratio $x_{\text{BJ}} = q^2/(2M\nu)$, a dimensionless number between 0 and 1. The phenomenon was named scaling at that time, based on the very general observation that the variable x_{BJ} does not depend on any scale, neither of mass nor of energy. The structure functions show this behaviour, if the nucleon has a substructure of charged particles inside, i.e. if quarks are really its constituents. The photon can then interact elastically with those as sketched here:

$$E, p \qquad = \Sigma \int dx\, e_i^2 \qquad E, p \qquad xE, xp$$

This interpretation is valid when the binding force between quarks becomes negligible with respect to the momentum transfer by the photon, and the photons interacts with quarks incoherently. To obey kinematical constraints, the photon must hit a quark which carries a fraction $p_q/p_N = x_{\text{BJ}}$ of the energy-momentum of the nucleon. Consequently, the cross section must be proportional to the probability to find such a quark. In the limit $q^2 \to \infty$, the structure functions then tend to be directly proportional to the quark momentum distributions, $f_i(x) = dp_i/dx$, for quarks of type i with charge e_i:

$$\nu W_2(\nu, Q^2) \to F_2(x) = \sum_i e_i^2 f_i(x)x \quad ; \quad MW_1(\nu, Q^2) \to F_1(x) = \frac{1}{2x}F_2(x)$$

Structure function thus measure quark charges and momentum distributions.

Focus Box 9.3: Deep inelastic electron-nucleon scattering

- If one increases the momentum transfer further, thus reducing the photon wavelength, the cross section not only becomes a smooth function, it also only depends on a single kinematic variable instead of two. This variable is called Bjorken-x, x_{BJ}, after one of its inventors [401], James D. Bjorken, whom his friends and colleagues call BJ. It is a dimensionless real number which can take values between 0 and 1.

A single kinematic variable immediately means that the scattering is elastic: the photon hits a constituent instead of the nucleon as a whole. The Bjorken variable has two roles. On the one hand, it describes the ratio between energy and mass of the exchanged photon. On the other, it denotes the fraction of the nucleon's energy-momentum which the struck quark carries. This is due to the fact that energy-momentum can only be conserved if the two ratios agree. For Bjorken-x close to 1, the photon interacts with a quark which carries almost the full nucleon momentum. For values around $\frac{1}{3}$, it interacts with one out of three quarks sharing it equally.

The hypothesis behind this interpretation of structure functions is that quarks behave like free particles when they are probed at high momentum transfer. One calls this hypothesis asymptotic freedom. It is a general property of a theory, if the coupling becomes small at small distances. We come back to this important property of strong interactions in Section 9.2. In summary, the experimental findings based on nucleon structure functions are:

- The three so-called valence quarks, (uud) for the proton, (udd) for the neutron, together carry about half of the energy-momentum of the nucleon. The rest must be carried by the particles which bind quarks together. They are called gluons.

- The mean fraction of the proton energy-momentum carried by u quarks is $\bar{x}_u \simeq 0.36$, for d quarks it is $\bar{x}_d \simeq 0.18$. In the neutron, the two values are interchanged. This agrees with expectations from the quark model, that quarks share their half of the nucleon energy-momentum roughly equally. The x-distributions are very asymmetric with long tails towards large values.

- Large separation between quarks, which would be necessary to observe free quarks, cannot be realised, not even at extreme energy transfers much larger than the nucleon mass. Quarks are always bound in hadrons.

This latter fact is complementary to asymptotic freedom, which says that at large momentum transfers, i.e. small distances, even bound quarks behave like free particles. The quark model does not explain the dynamics of quark binding inside hadrons. This requires a dynamical theory of strong interactions, provided by Quantum Chromodynamics (see Section 9.2). The momentum distributions of quarks and gluons in the proton and neutron have been measured with impressive precision [633] from 1992 to 2007 by the experiments H1 and ZEUS at the electron-proton collider HERA in Hamburg, Germany.

The results of the SLAC experiments in the late 1960s and similar findings with neutrino and muon scattering off nucleons convinced most physicists of the existence of substructure inside hadrons, if not necessarily of the existence of quarks as real particles. However, well into the 1970s there remained an influential sceptical minority under the leadership of Werner Heisenberg. In a plenary talk at the annual meeting of the German Physical Society in 1975 [427] –incidentally the first physics conference I attended–, entitled "What is an elementary particle?", Heisenberg made clear that he considered the quest for hadron constituents to be stillborn: "Here the question has obviously been asked: what do protons consist of? But it has been forgotten, that the term 'consist of' makes sense only if one succeeds to dismantle the particle, using a small amount of energy, into components with a mass much larger than this energy; otherwise, the word 'consist of' has lost its meaning." This verdict corresponds to his deep conviction that only observables should enter into a theory of the physical world. And structure functions in his view did not qualify. Until the dynamics of the binding forces were not understood, one was not supposed to discuss constituents [427]: "One can exaggerate somewhat and say that good physics has been spoiled by bad philosophy." Heisenberg's opinion was that hadrons were to be understood as representations of a symmetry group; respecting certain conservation laws, any one could be transformed into any other, like the S-matrix has it. Quantum Chromodynamics, the field theory which precisely describes the dynamics of quarks and gluons but excludes the existence of free quarks, has completely refuted this ansatz.

Admitting this exception, for most physicists at the end of the 1960s matter consisted of the three quarks up, down and strange, as well as four leptons, the charged electron and muon and the corresponding neutrinos, ν_e and ν_μ. Yet in 1964, James Bjorken and Sheldon Glashow [369] speculated that there ought to be a fourth quark reestablishing the symmetry between leptons and quarks. The explicit prediction in 1970 for this fourth quark, named charm, and its properties is due to Glashow, John Iliopoulos and Luciano Maiani [406]. In a big leap ahead, their aim was to show under which circumstance weak neutral interactions (which were not yet discovered!) could avoid changing the flavour quantum numbers of quarks. Such violations of flavour were already ruled out by experiments on weak decays of hadrons.

Consequently, their prediction at first attracted little attention. That changed promptly, when in November of 1974 the groups of Samuel C.C. Ting at the Brookhaven AGN [419] and Burton Richter at SLAC [418] simultaneously announced the discovery of a new heavy meson. Their two papers appeared next to each other in Physical Review Letters. Ting called the new particle J, Richter named it Ψ. Since none of them can claim precedence, the meson is called J/Ψ, the only particle with a double name. It is the ground state of a bound pair of charm-anticharm quarks. The second generation of quarks was thus completed. With this so-called November revolution, the Eightfold Way arrived at its end, strong isospin and strangeness merged with charm into the flavour quantum numbers. Richter and Ting were awarded the 1976 Nobel prize in physics for their discovery.

As if that were not enough, a little later Martin L. Perl and his group at the SPEAR electron-positron collider discovered a third charged lepton, the tau [424]. Perl, the thesis advisor of Sam Ting, shared the 1995 Nobel prize with Frederick Reines. The evidence was a significant number of e^+e^- annihilations, where the observed final state contained nothing but an electron and a muon of the opposite charge, in an apparent violation of the conserved lepton number. In addition, the final state showed an imbalance of energy momentum, indicating additional unobserved particles in the final state. It was thus clear that the reaction was indeed $e^+e^- \to \tau^+\tau^- \to (e^+\nu_e\bar{\nu}_\tau)(\mu^-\bar{\nu}_\mu\nu_\tau)$ such that the neutrinos, which escape detection, assure conservation of lepton numbers and energy-momentum. The demonstration of tau neutrino interactions, however, had to wait another 25 years [529, 659] because of their rareness and minute cross section with matter. Anyway, nobody had really doubted the existence of a third neutrino since the experiments at the LEP electron-positron collider had determined the number of light neutrino species to be exactly three[8]. To reestablish the symmetry between quarks and leptons, a third generation of quarks also had to exist [413, 425]. The properties of these quarks were known, but not their masses, so their discovery was a matter of accelerator technology. The periodic system of quarks and leptons was finally completed –as far as we know today– with the discovery of the third generation bottom quark in 1977 [431] and the top quark in 1995 [477, 478], both at the Fermi National Laboratory, after an intensive experimental search guided by theory.

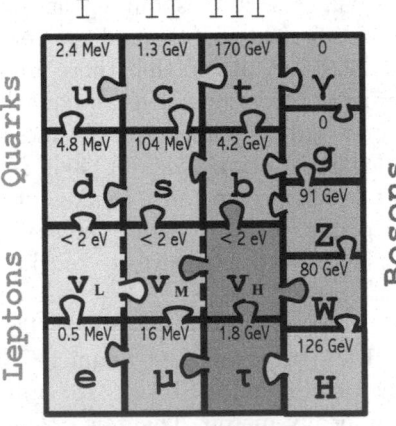

Figure 9.4 The periodic table of quarks and leptons as it is known today. Rough values of the masses are given above the particle symbol. Roman numbers denote the generations. The twelve matter particles on the left are fermions, the four force carriers on the right are bosons.

The periodic table of the quarks and leptons known today is shown in Figure 9.4. The left side shows the twelve constituents of matter, which we are entitled to take as elementary until further notice, i.e. point-like and without substructure. They are classified into three generations with increasing masses,

---

[8]For a review of these results see [625].

and four families. The family members behave in a similar way as far as interactions are concerned. The families are:

- quarks with electric charge $+\frac{2}{3}e$, also called up-type quarks;

- quarks with electric charge $-\frac{1}{3}$, also called down-type quarks;

- neutrinos with electric charge 0; and

- charged leptons with electric charge $-e$.

For each particle there is an antiparticle with equal mass and opposite charges. The term charge not only denotes electric charge, but charges of all types, which belong to the other forces of the Standard Model, i.e. weak and strong charge. We will discuss these in the next Section.

In many such representations of the periodic system of quarks and leptons, even those of the Particle Data Group, you will find the mixed states, ν_e, ν_μ and ν_τ, which interact with the W-boson, instead of the "real" neutrinos of the three generations, ν_L, ν_M and ν_H. In the above table I list only eigenstates of the mass operator, i.e. particles which evolve in space-time with a velocity given by their energy-momentum[9]. Please excuse me for this somewhat pedantic detail. That weak interactions interact with mixtures of particles –instead of pure states– is a special feature of theirs which we will also discuss in the next section.

9.2 FORCES

Weak interactions are the only known forces which can transform quarks and leptons into different flavours, even across generations. It is thus understandable that the periodic system of Figure 9.4 for elementary matter constituents emerged from the study of weak interactions.

The first product of this force was observed by Henri Becquerel in 1896 (see Section 2.2), when he discovered the penetrating, negatively charged radiation which results from the beta decay of uranium salts. He read no less than seven papers to the French academy of sciences in 1896 and many more in the following years [107]. His work on radioactivity was recognised together with that of Marie and Pierre Curie with the Nobel prize in physics of 1903.

Rutherford and his group developed the hypothesis that this beta radiation comes from the decay of the atomic nucleus. Measurements of the spectrum were made by Otto von Baeyer, Lise Meitner and Otto Hahn [137] in 1911 and more precise ones by James Chadwick three years later in Hans Geiger's laboratory [153] at Physikalisch Technische Reichsanstalt in Berlin, using Geiger's newly developed counter. They demonstrated convincingly that the spectrum

---

[9]The upper limit on neutrino masses has recently been lowered to 1.1eV, with a precise measurement of the endpoint of the electron spectrum from tritium β decay [682]. This determines an upper limit on $m_\nu^2 = |U_{eL}|^2 m_L^2 + |U_{eM}|^2 m_M^2 + |U_{eH}|^2 m_H^2$. See Focus Box 9.14.

of this type of radiation did not have a sharp peak, as one would expect for a two-body decay. Rather it had a continuous distribution between zero and a maximum energy. We know today that this maximum energy is the mass difference between neutron and proton, but the neutron was only discovered in 1932 by the same James Chadwick [244]. The continuous spectrum led Wolfgang Pauli in 1930 to the daring hypothesis, that in nuclear beta decay an additional, undetected particle was emitted. Edoardo Amaldi and Enrico Fermi named it neutrino because of its electric neutrality and small mass [452]. The negative decay product was identified by Maurice and Gertrude Goldhaber as identical to the atomic electron [312]. The beta decay of the neutron is thus the reaction $n \to pe^- \bar{\nu}_e$. The cross sections of neutrinos are so tiny[10] that the discovery of their interactions with matter had to wait for the availability of nuclear reactors as intense sources of this elusive particle (see Section 9.1).

In 1934, Enrico Fermi formulated a theory of nuclear beta decay [260], closely following the example of early predecessors of quantum electrodynamics. In this theory, beta decay proceeds via a direct coupling of a nucleon current, which transforms the neutron into a proton, with a lepton current, which connects electron and electron antineutrino. No force particle for weak interactions was postulated. Thus no propagator enters the amplitude calculation, rather a global constant represents the coupling, which is called the Fermi constant, G_F. As long as momentum transfers between nucleon and lepton current are small compared to the mass of the force quantum, this effective field theory makes rather precise predictions. However, it does not allow to go beyond first order in perturbation theory.

Hideki Yukawa [268] extended the theory by including the exchange of a force quantum, with the corresponding propagator. To allow Fermi's approximation to work well, this particle had to be very heavy. At the time, "heavy" meant a mass of several GeV, in reality weak bosons weigh almost 100 GeV. Consequently, weak interactions theory stayed an effective one for a long time. What was astonishing –and still is– is the fact that weak interactions not only violate flavour symmetry, but even so fundamental ones as parity, also called mirror symmetry. The discrete parity transformation transforms three-vectors into their opposite, e.g. the position vector $(x, y, z) \to (-x, -y, -z)$. Instead, it keeps pseudo-vectors like spin untouched. The potential violation of parity symmetry was first discussed theoretically by Tsung-Dao Lee und Chen-Ning Yang [334] in 1956. A year later, parity violation in β-decay was experimentally demonstrated by Chien-Chung Wu [338], as explained in Focus Box 9.4. Lee and Yang obtained the Nobel prize in physics of 1957, Wu's experimental results were mentioned in the acceptance speech, but she was not among the awardees. As if parity violation was not enough of a disrespect for fundamental symmetries, it turned out that even the symmetry between matter and antimatter, a combination of charge and space reversal called CP, is not

---

[10]Already in 1934 Bethe and Peierls calculated that neutrinos can easily cross the Earth without interacting [257].

For their beta decay experiments, Chien-Chung Wu and collaborators used a heavy nucleus with strong nuclear spin, ^{60}Co with $J = 5$. If one cools the sample adiabatically in a strong magnetic field down to cryogenic temperatures, all spins end up aligned in the direction of the magnetic field. Beta decay of one of the nuclear neutrons can –respecting angular momentum conservation– proceed in the two ways sketched below: a right-handed electron can be emitted in the direction of the nuclear spin, or a left-handed electron is emitted opposite to it. The corresponding angular distributions are:

$$\frac{d\Gamma}{d\Omega}\bigg|_a = \Gamma_0 \left(1 + \frac{\vec{\sigma}\vec{p}}{E}\right) \quad ; \quad \frac{d\Gamma}{d\Omega}\bigg|_b = \Gamma_0 \left(1 - \frac{\vec{\sigma}\vec{p}}{E}\right)$$

with the spin vector $\vec{\sigma}$ and the electron momentum vector $\vec{p}$. Only the latter distribution is observed, the electron is always emitted opposite to the direction of the nuclear spin.

Conservation of parity would require that both configurations are realised with the same probability. Any deviation from equality would indicate a violation of parity. That only configuration b) is observed means that parity is violated in a maximum way. Only left-handed fermions and right-handed antifermions participate in weak interaction. This does, however, not mean that Noether's theorem is violated. Since parity is a discrete transformation, Noether's theorem does not protect the corresponding quantum number.

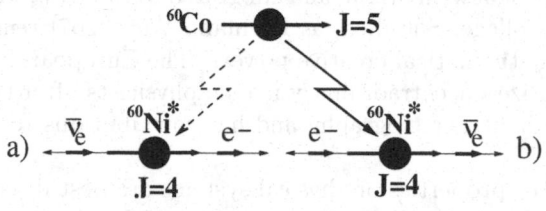

Focus Box 9.4: Wu experiment on parity violation

respected by weak interactions [380]. This violation of CP symmetry won the 1980 Nobel prize for its discoverers, James Cronin and Val Fitch. In summary, weak interactions only couple to left-handed fermions, i.e. those with spin opposite to the direction of motion, and right-handed antifermions, with spin parallel to momentum, but not exactly in the same way.

A field theory for weak interactions exists since the work of Steven Weinberg [396], Sheldon Glashow [352] and Abdus Salam [398]. In the introduction to his publication, Weinberg names his motivation:"Leptons interact only with photons, and with the intermediate bosons that presumably mediate weak interactions. What could be more natural than to unite these spin-one bosons into a multiplet of gauge fields?" The term "gauge" shows up again here, like it did in our discussion of electrodynamics. In order to understand why the freedom to gauge potentials in a range of different ways is of fundamental importance, we must discuss Emmy Noether's theorem [176, 407]. The theorem says that for each invariance of a physical system under a group of continuous symmetry transformations, there is a conserved quantum number. Examples from classical physics are the connection between energy-momentum conservation on one hand, and invariance under translations in time and space on the other hand. And of course the gauge freedom of the electromagnetic field, which leads to conservation of the electric charge. These examples are detailed in Focus Box 9.6.

The proof of this theorem is due to Amalie Emmy Noether, a genius mathematician from the city of Erlangen in Germany and one of the first female students and lecturers at its university. Even though women could not be enrolled at the time, she was given permission in 1900 to attend lectures, including those by eminent mathematicians like Schwarzschild, Minkowski, Klein and Hilbert. In 1904 women were finally officially admitted. Noether was promoted in 1908 and habilitated in 1919, and was given an unpaid position as extraordinary professor. In 1933, she had to leave university like all other researchers of Jewish origin, and emigrated to the U.S., where she taught at Bryn Mawr College. She died, as Hermann Weyl [267] remarked, "at the summit of her mathematical creative power." The European Physical Society has named its prize for extraordinary female physicists after this unique personality. More about her biography and her contributions to physics can be found in [482].

The symmetry properties of physical systems are best discussed using the Lagrange formulation as explained in Focus Box 9.5 for classical mechanics and in Focus Box 9.7 for quantum fields. At the end of the 18th century, Joseph-Louis Lagrange invented a single scalar function [5], or operator in quantum physics, which contains all necessary information to predict the evolution of a system. The so-called Euler-Lagrange equations transform the function into equations of motion for the system. The interesting feature of the Lagrange function or operator is that its invariance under a given set of transformations is transmitted to the equations of motion. As shown in Focus Box 9.7, the Lagrange operators for free bosons and fermions are all scalar under Lorentz

A classical system can be characterised by a single scalar function, the Lagrange function, $L(q, \dot{q}, t)$, which was introduced by the Italian-French mathematician and astronomer Joseph-Louis Lagrange in 1788 [5]. It can depend on the generalised coordinates q, which do not need to be Cartesian, velocities $\dot{q}$ and also explicitly on time t. The classical action S results from the Lagrange function by time integration:

$$S = \int_{t_1}^{t_2} dt L(q, \dot{q}, t)$$

The equations of motion result from Hamilton's principle of minimal action, which postulates that S is stationary for the physical trajectory which the system will choose for its evolution between t_1 and t_2:

$$\delta S = \delta \int_{t_1}^{t_2} dt L(q, \dot{q}) = 0$$

The variation symbol δ denotes the difference between two possible trajectories, such that the variation of coordinates is zero at the start and end point, $\delta q_i(t_1) = \delta q_i(t_2) = 0$. If L does not explicitly depend on time, one obtains from Hamilton's principle the Euler-Lagrange equations:

$$\frac{d}{dt}\left(\frac{\partial L}{\partial \dot{q}_i}\right) - \frac{\partial L}{\partial q_i} = 0$$

For conservative systems (see Focus Box 2.2), L is the difference between kinetic energy T and potential energy V, $L = T - V$. It thus has the dimension of an energy. The Euler-Lagrange equations connect the Lagrange function to unique equations of motion. It follows, that a transformation which leaves L invariant does not change the evolution of the system; it represents one of its symmetries.

Let us take as an example a particle of mass m in a unidimensional system. Assuming a potential $V(x)$, its Lagrange function is $L = \frac{1}{2}m\dot{x}^2 - V(x)$. The Euler-Lagrange equations then lead to the well-known Newtonian equations of motion for a conservative force F (see Focus Box 2.1), $m\ddot{x} = -\partial V/\partial x = F(x)$.

Focus Box 9.5: Lagrange mechanics

Noether's theorem says that continuous symmetries of a system result in a conservation law for one of its properties. Energy-momentum conservation follows from the invariance of physical laws under translations in space and time. This fact is well known for classical systems and is best discussed in the framework of Lagrangian mechanics, introduced in Focus Box 9.5. Let us consider a conservative system, where the potential V only depends on the position vector $q_i (i = 1, 2, 3)$ of its ingredients, and is independent of their velocity vector $\dot{q}_i$. For those, the Lagrange function is $L = T(\dot{q}_i) - V(q_i)$. A shift in coordinate position along an axis i does not change the kinetic energy T. The Euler-Lagrange equations thus simplify to:

$$\frac{d}{dt}\left(\frac{\partial T}{\partial \dot{q}_i}\right) = -\frac{\partial V}{\partial q_i}$$

On the left side of the equation you see the time derivative of the momentum, on the right side the gradient of the potential, i.e. the force. If the potential does not explicitly depend on a given coordinate q_i, the Lagrange function is invariant under shifts in this direction and the right hand side of this equation is zero. The corresponding component of the momentum is then a constant of motion. Translational symmetry of the system has led to momentum conservation. In an analogous way, invariance against time shifts leads to energy conservation, invariance against rotations to conservation of angular momentum.

The gauge freedom of the electromagnetic potential leads to the conservation of charge. One can see that using a simple Gedankenexperiment. Two observers look at an electron at rest in an isolated box. One observer defines the electric potential at the position of the electron as zero volts, the other as 5V, gauge freedom entitles them to do so. If one could thus make the electron disappear –without introducing any additional particle– the second observer would gain 5eV of energy, the first would gain nothing. So if energy is conserved, gauge freedom leads to the conservation of electric charge. Of course one can make the electron disappear by introducing a positron into the box, without violating energy conservation: for the first observer, this would require no energy (if done adiabatically), the second one would have to work against the potential.

Focus Box 9.6: Noether's theorem in classical physics

transformations. Consequently, the derived equations of motion are covariant and can describe relativistic particles. In the light of Noether's theorem we can say that the invariance of the Lagrange function under Lorentz transformations causes energy-momentum conservation in relativistic terms.

Lagrange operators can be used to describe the interaction of particles. And again, the properties of the interaction follow from a symmetry principle. The symmetries we have thus far considered concern the coordinates with which the fields are described, thus transformations of the type $\psi(x) \to \psi(x')$. We find it natural that the choice of the coordinate system does not change the physics one describes. We will now consider changes to the fields themselves, thus transformations of the type $\psi(x) \to \psi'(x)$. A well-known example from classical physics are the gauge transformations of the electromagnetic potential, discussed in Focus Box 2.3 and already used when discussing electromagnetic waves in Focus Box 2.4. It is shown there that there is a class of global transformations of the electromagnetic potential which leave the electric and magnetic fields unchanged. They are called global since they concern all potentials alike, everywhere and for all times. Take the electric potential as an example: only potential differences matter, not the absolute values. Thus the 5V marked on your battery denote the difference in potential between the two poles; there is no absolute zero of the potential other than by convention. Focus Box 9.6 explains that this symmetry is responsible for the conservation of electric charge.

However, the experiments à la Aharonov and Bohm, which we introduced in Focus Box 7.9, convinced us that electromagnetic potentials are not arbitrary. They change the phase angle of matter fields. But since only phase *differences* can be made observable by interference experiments, we are not worried about that. Noether's theorem says that there ought to be a conserved quantity caused by this global symmetry. Focus Box 9.8 shows that it is the electromagnetic four-vector current density which is conserved. The local charge density can only change when a divergent or convergent current is present; global gauge invariance leads to local charge conservation. Remember that the equations of motion of the fields, the Klein-Gordon and the Dirac equations, already followed from the requirement of covariance under Lorentz transformations, which leads to the conservation of energy-momentum. Clearly, symmetries have momentous consequences for a field theory.

Modern field theories of fundamental interactions take this argument a step further. Until now, we have considered only global gauge changes to the potentials, leading to global phase changes of all matter fields. These stay unobservable, since only phase differences can be made apparent in interference experiments. The quadrature of the wave function makes the absolute phase disappear when the probability for finding the particle is calculated. Modern gauge theories make the more radical requirement that this is also true when the gauge of the potential is changed locally, leading to arbitrary phase changes of the matter fields in space and time. These are benign only, if we add a well defined interaction term to the Lagrange density. As shown in

The Lagrangian formulation of a quantum field theory allows summarising the properties of a system of particles by a single scalar operator, the Lagrange operator. The principle difference to Lagrange mechanics (see Focus Box 9.5) is that one no longer uses discrete coordinates $q_i(t)$ but continuous fields $\psi(t, \vec{x})$. The field can be seen as a generalised coordinate at each value of its argument, the space-time coordinates. This then represents a system with an infinite number of degrees of freedom. The description of the system is based on the Lagrange density operator, $\mathcal{L} = \mathcal{L}(\psi, \partial_\mu \psi)$, which depends on the field $\psi(x)$ and its four-gradient $\partial_\mu \psi(x)$. The action integral thus becomes:

$$S = \int_{t_1}^{t_2} dt \int d^3\vec{x} \, \mathcal{L}(\psi, \partial_\mu \psi)$$

The inner integral changes the Lagrange density into the Lagrange operator, $L = \int d^3\vec{x} \, \mathcal{L}(\psi, \partial_\mu \psi)$. Hamilton's principle of minimal action determines the evolution of the system:

$$\delta \int_{x_1}^{x_2} d^4x \, \mathcal{L}(\psi, \partial_\mu \psi) = 0$$

with initial and final conditions $\delta\psi(x_1) = \delta\psi(x_2) = 0$. By the same reasoning as for classical systems, one obtains the Euler-Lagrange equations for a field theory:

$$\frac{\partial \mathcal{L}}{\partial \psi} - \partial_\mu \frac{\partial \mathcal{L}}{\partial(\partial_\mu \psi)} = 0$$

They lead to the equations of motion for the fields, which inherit invariance properties from symmetries of the Lagrange operator. When $\mathcal{L}$ is a scalar operator under Lorentz transformations, thus relativistically invariant, the equations of motion will be covariant. Examples:

- The Lagrange density which leads to the Klein-Gordon equation (see Focus Box 7.2) for a free scalar boson is $\mathcal{L}_{s=0} = (\partial_\mu \psi)^* (\partial^\mu \psi) - m^2 \psi^* \psi$.

- For the spinor field of a free fermion, the Lagrange density $\mathcal{L}_{s=1/2} = \bar{\psi} (i\gamma^\mu \partial_\mu - m) \psi$, with $\bar{\psi}(x) = \psi^\dagger(x)\gamma^0$, leads to the Dirac equation (see Focus Box 7.4).

- A free vector boson like the photon A_μ has a Lagrange density $\mathcal{L}_{s=1} = -\frac{1}{4} F_{\mu\nu} F^{\mu\nu}$ with the field tensor $F_{\mu\nu} = \partial_\mu A_\nu - \partial_\nu A_\mu$ (see Focus Box 3.7).

All terms above are products of four-vectors or scalars, and thus fulfil the invariance condition for Lorentz transformations.

Focus Box 9.7: Lagrange formalism for free quantum fields

A global gauge transformation changes the phase of matter fields ψ:

$$\psi(x) \to \psi'(x) = e^{i\alpha Q} \psi(x)$$

with a constant angle $\alpha \neq \alpha(x)$ and the charge operator Q with $Q\psi(x) = q\psi(x)$. The same angle applies to all fields and all of space-time. We will show invariance for the example of a complex scalar field $\psi(x)$. Its equation of motion follows from a Lagrange density (see Focus Box 9.7):

$$\mathcal{L} = (\partial_\mu \psi)^* (\partial^\mu \psi) - m^2 \psi^* \psi$$

The global gauge transformations form the Lie group U(1) –rotations around a single axis– defined by infinitesimal transformations:

$$\psi \to (1 + i\alpha Q)\psi$$

Invariance of the Lagrange density means that the change in $\mathcal{L}$ is zero. For an electron:

$$
\begin{aligned}
\delta\mathcal{L} &= \frac{\partial\mathcal{L}}{\partial\psi}\delta\psi + \frac{\partial\mathcal{L}}{\partial(\partial_\mu\psi)}\delta(\partial_\mu\psi) + \frac{\partial\mathcal{L}}{\partial\psi^*}\delta\psi^* + \frac{\partial\mathcal{L}}{\partial(\partial_\mu\psi^*)}\delta(\partial_\mu\psi^*) \\
&= i\alpha e \left[\frac{\partial\mathcal{L}}{\partial\psi} - \partial_\mu\left(\frac{\partial\mathcal{L}}{\partial(\partial_\mu\psi)}\right)\right]\psi + i\alpha e\partial_\mu\left(\frac{\partial\mathcal{L}}{\partial(\partial_\mu\psi)}\psi\right) + \ldots
\end{aligned}
$$

where the dots indicate the complex conjugates of the first two terms. The first term and its complex conjugate disappear because of the Euler-Lagrange equation. Invariance thus requires that the second term and its complex conjugate add up to zero for all values of α:

$$\partial_\mu\left(e\frac{\partial\mathcal{L}}{\partial(\partial_\mu\psi)}\psi - e\psi^*\frac{\partial\mathcal{L}}{\partial(\partial_\mu\psi^*)}\right) = 0$$

The term in parentheses is the electromagnetic current density (see Focus Box 3.7):

$$j^\mu = -e\left(\psi^*\partial^\mu\psi - \psi\partial^\mu\psi^*\right)$$

It must thus be conserved, $\partial_\mu j^\mu = 0$. The time-like component, $j^0 = -e\frac{\partial}{\partial t}(\psi^*\psi)$ is the change in the local charge density $e\rho$, with ρ itself the particle number density. The spatial components, $\vec{j} = -e\vec{\nabla}(\psi^*\psi)$, form the vector of electromagnetic current density. The continuity equation $\partial_\mu j^\mu = 0$ thus means that the local charge density can only change when a divergent or convergent current of charges is present. Integrating over all space one sees that the total charge of the Universe is conserved, $\frac{\partial q}{\partial t} = \frac{\partial}{\partial t}\int e\rho\, d^3x = 0$.

Focus Box 9.8: Global U(1) gauge symmetry and conserved current

...

Focus Box 9.9 it must have exactly the form corresponding to Maxwell's equations. To preserve local gauge invariance, photons must thus always accompany charged particles wherever they are, to react appropriately to local phase changes and keep them from becoming observable. One thus calls the photon a gauge boson. The Feynman graphs of elementary interactions between photons and matter, which result from this requirement, are shown on the bottom of Figure 9.5.

The Lagrange density of a free fermion (see Focus Box 9.7) is not invariant against local gauge transformations, $\psi(x) \to \psi'(x) = e^{i\alpha(x)\mathcal{Q}}\psi(x)$, because of its term $\partial_\mu \psi$. It becomes invariant if and only if one adds an interaction term:

$$\mathcal{L} = \bar{\psi}\left(i\gamma^\mu \partial_\mu - m\right)\psi + e\bar{\psi}\gamma^\mu \psi A_\mu$$

The four-vector A_μ must transform like $A_\mu \to A_\mu + (1/e)\partial_\mu \alpha(x)$ to preserve invariance. We can thus identify A_μ as the electromagnetic four-vector potential, which has exactly this gauge freedom: the four-vector gradient of an arbitrary function can be added to it. Invariance is thus guaranteed if an electromagnetic potential is present wherever there is a charged particle. And the interaction between the two is not arbitrary. The electromagnetic current density must couple to the photon field just as Maxwell's equations have it. Without spoiling local gauge invariance we can add the kinetic term for the free photon and obtain the Lagrange density of Quantum Electrodynamics:

$$\mathcal{L}_{QED} = \bar{\psi}\left(i\gamma^\mu \partial_\mu - m\right)\psi + e\bar{\psi}\gamma^\mu \psi A_\mu - \frac{1}{4}F_{\mu\nu}F^{\mu\nu}$$

Thus all of QED can be derived from invariance requirements.

Focus Box 9.9: Local U(1) gauge symmetry and electron-photon interaction

One could discount this fact purely as an aesthetically pleasing feature, if it did not have far reaching consequences. In the beginning of the 1970s, Gerard 't Hooft and Martinus "Tini" Veltman found a connection between gauge invariance and the renormalisability of a field theory. Renormalisation and regularisation are part of the systematic treatment of infinities which show up in field theories when higher order processes are included. They should in principle disappear when one sums over all orders of perturbation theory. In practical calculation they are absorbed into measured parameters like masses and coupling constants, which are known and finite. Veltman and 't Hooft demonstrated that gauge theories with massless gauge bosons are renormalisable. It was thus shown that gauge invariance leads to theories which are predictive, in principle up to arbitrary order and thus high precision, once their parameters are precisely measured. Veltman and his former graduate student 't Hooft received the physics Nobel prize in 1999 for this important proof. The two men are of very different character. Veltman, outspoken and never shy of controversy, refused to tell their individual contributions apart

[500, p. 168]: "I once said to 't Hooft that once you collaborate you forget about who invented what." 'T Hooft, soft spoken but equally forceful, took offence and the two drifted apart.

With this discovery, the interest of the theory community in field theories of weak and strong interactions was revived. A class of such theories had been proposed twenty years earlier by Chen-Ning Yang and Robert L. Mills from Brookhaven in a paper [328] which represents a highlight in the field, really a course on gauge theories in a nutshell. It would be superficial to think that renormalisability is just a practical property, allowing to increase precision in a perturbative approach. Rather, effective theories, like the one by Fermi for weak interactions, already show divergences in second order terms, justifying criticism against field theory on principle grounds. These arguments were now removed and the end of S-matrix theory was near.

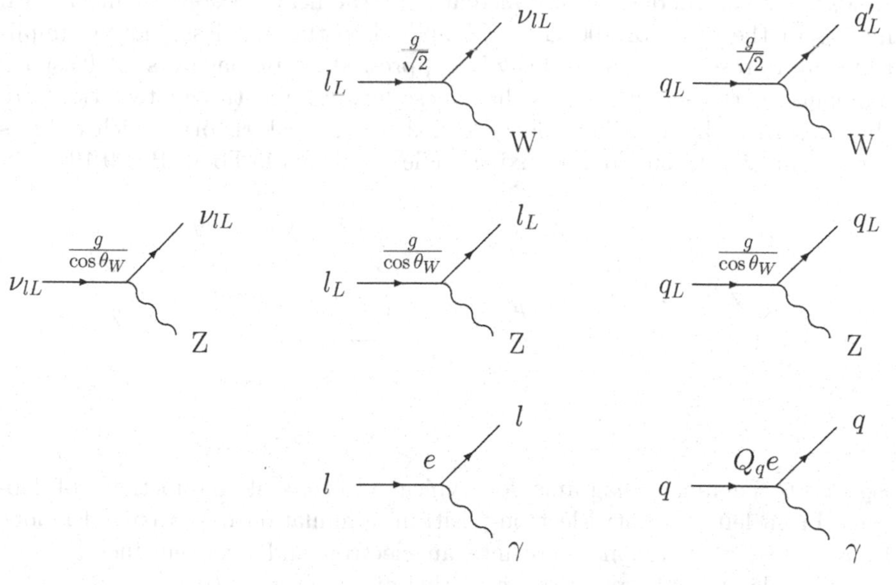

Figure 9.5 Feynman graphs for the vertices of electromagnetic and weak interactions of leptons l and quarks q. Downwards from the top: charged weak interactions, neutral weak interactions and electromagnetic interactions. Because of their maximal parity violation, weak interactions only act on the left-handed components of the fields, indicated by the index L. Above the vertices the coupling constants are noted.

Weinberg had already conjectured in 1967 [396], that his unified gauge theory of electromagnetic and weak interactions might be renormalisable. He classified matter and force particles in representations of the joint symmetry group $SU(2) \times U(1)$ which combines rotations around two and one axes, respectively. Here the transformations of SU(2) again change the phase of matter fields, but apply only to their left-handed components. The consequence is a conserved weak charge, called weak isospin (T, T_3) in analogy to spin, which is explained in Focus Box 9.10. The U(1) transformations, which belong to electromagnetism, of course apply to the whole fields.

Figure 9.5 shows the Feynman graphs of elementary weak interaction vertices together with those of electromagnetic interactions. They all conserve weak isospin as well as electric charge. The neutral interactions, mediated by the Z boson or photon γ and also called neutral currents, do not change the charge or the flavour quantum number of matter particles. The charged weak interactions, interactions with the $W^\pm$ bosons, also called charged currents, change a charged lepton into its neutrino, e.g. $\mu^- \leftrightarrow \nu_\mu$, or a quark of type up into a quark of type down, like $c \leftrightarrow s$, not necessarily of the same generation. The coupling constants of all three interaction types are of similar size. However, weak interactions are mediated by the heavy bosons W and Z with masses of the order of 100 GeV. Compared to the massless photon, amplitudes and cross sections are heavily suppressed by propagators, at least for momentum transfers much less than these masses. The three interactions are thus described by a unified theory, called electroweak theory, which reduces three couplings to one and a mixing angle, as shown in Focus Box 9.10.

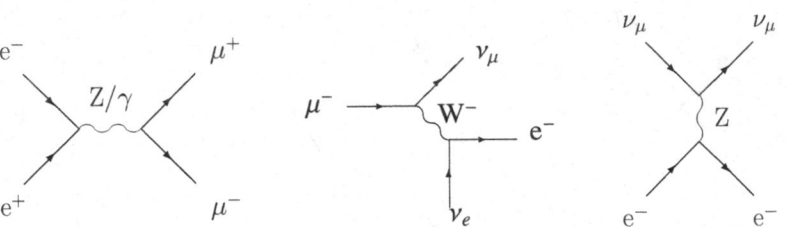

Figure 9.6 Feynman diagrams for typical electroweak interactions of leptons. From left to right: Electron-positron annihilation into a pair of muons $(e^+e^- \rightarrow \mu^+\mu^-)$; muon decay into an electron and two neutrinos $(\mu^- \rightarrow e^- \bar{\nu}_e \nu_\mu)$; elastic muon neutrino scattering off an electron $(\nu_\mu e^- \rightarrow \nu_\mu e^-)$.

Figure 9.6 shows Feynman graphs for typical leptonic electroweak interactions. Wherever initial and final state allow, like in the electron-positron annihilation of the left graph, photon as well as Z exchange contribute to the invariant amplitude. The cross section then has three terms, corresponding to the probabilities for the two mechanisms plus an interference term. Muon decay, shown in the middle graph, is the mother of all charged weak interactions; it serves today to determine the Fermi constant with high precision[11]. The rightmost process, neutrino-electron scattering, only proceeds

---

[11]See the discussion by the Particle Data Group, http://pdg.lbl.gov/2019/reviews/ rpp2019-rev-standard-model.pdf, p. 3.

The Lagrange density of free fermions is invariant against global SU(2) transformations, as it was for U(1). The corresponding conserved quantity is the weak charge with two components, called weak isospin. In analogy to spin, weak isospin is specified by its absolute value T and its third component T_3. The weak charges of leptons then are:

$$\begin{pmatrix} \nu_e \\ e^- \end{pmatrix}_L \quad T = 1/2 \; ; \; T_3 = \begin{cases} +1/2 \\ -1/2 \end{cases} \qquad e_R \quad T = 0 \; ; \; T_3 = 0$$

$$\begin{pmatrix} \nu_\mu \\ \mu^- \end{pmatrix}_L \quad T = 1/2 \; ; \; T_3 = \begin{cases} +1/2 \\ -1/2 \end{cases} \qquad \mu_R \quad T = 0 \; ; \; T_3 = 0$$

$$\begin{pmatrix} \nu_\tau \\ \tau^- \end{pmatrix}_L \quad T = 1/2 \; ; \; T_3 = \begin{cases} +1/2 \\ -1/2 \end{cases} \qquad \tau_R \quad T = 0 \; ; \; T_3 = 0$$

This way, only the doublets of the left-handed field components have a weak charge, the singlets of right-handed components don't, in agreement with maximum parity violation. In the same way, the quark charges are:

$$\begin{pmatrix} u \\ d \end{pmatrix}_L \quad T = 1/2 \; ; \; T_3 = \begin{cases} +1/2 \\ -1/2 \end{cases} \qquad u_R, d_R \quad T = 0 \; ; \; T_3 = 0$$

$$\begin{pmatrix} c \\ s \end{pmatrix}_L \quad T = 1/2 \; ; \; T_3 = \begin{cases} +1/2 \\ -1/2 \end{cases} \qquad c_R, s_R \quad T = 0 \; ; \; T_3 = 0$$

$$\begin{pmatrix} t \\ b \end{pmatrix}_L \quad T = 1/2 \; ; \; T_3 = \begin{cases} +1/2 \\ -1/2 \end{cases} \qquad t_R, b_R \quad T = 0 \; ; \; T_3 = 0$$

Local gauge invariance is ensured if the Lagrange density contains interaction terms between matter fields and three gauge bosons, $(Z, W^\pm)$, which form a triplet of weak isospin, (W^+, Z, W^-) with $T = 1$, $T_3 = (+1, 0, -1)$. The coupling constants of weak charged (g), weak neutral (g') and electromagnetic interactions (e) are quite similar and connected by the unification relation $e = g \sin\theta_W = g' \cos\theta_W$. It uses the so-called weak or Weinberg angle $\cos\theta_W = M_W/M_Z$, $\theta_W \simeq 30°$. At small momentum transfer q^2, like in weak decays, amplitudes are suppressed by propagator effects. Instead of $1/q^2$ for the massless photon, the weak propagator is $1/(q^2 - M^2 + iM\Gamma)$, with masses $M_Z \simeq 90\text{GeV}$, $M_W \simeq 80\text{GeV}$ and widths $\Gamma_Z \simeq 2.5$ GeV and $\Gamma_W \simeq 2.0$ GeV. Weak cross sections are thus small for $q^2 \ll M^2$, with an effective coupling like in the Fermi theory:

$$\frac{G_F}{\sqrt{2}} = \frac{g^2}{8M_W^2}$$

At momentum transfers comparable to the boson masses, $q^2 \simeq M^2$, weak cross sections have a resonance pole and become much larger than electromagnetic ones. However there is an inherent problem. Local gauge invariance is only preserved if all gauge bosons are massless like the photon. But Z and $W^\pm$ clearly aren't.

Focus Box 9.10: SU(2) gauge symmetry and weak isospin

via Z exchange, since the neutrino has no electric charge and cannot emit a photon. It has led to the discovery of neutral weak interactions.

The electroweak gauge theory of Glashow, Weinberg and Salam thus converted Fermi's effective theory of weak interactions into a renormalisable field theory. It predicted that parallel to electromagnetic interactions, there ought to be a neutral weak interaction which alters neither fermion charge Q, nor T_3. It is called a weak neutral current, to distinguish it from the charged current interaction of neutron or muon decay. Excellent candidate experiments to check this prediction are neutrino interactions. Neutral currents were discovered using the heavy liquid bubble chamber Gargamelle in the intense neutrino beam at the CERN PS (see Section 8.3). Since neutrinos interact only weakly, the scattering of neutrinos off atomic electrons is essentially free of background. Ever since Simon Van der Meer and his collaborators had invented focussing by a magnetic horn, there was a muon-neutrino beam of sufficient intensity. Gargamelle with its heavy freon filling provided enough target mass to search for these rare processes. It was thus only a question of time and patience to prove or disprove electroweak theory by observing (or not!) the elastic scattering process $\nu_\mu e^- \to \nu_\mu e^-$ on a bubble chamber picture[12]. I remember that in the room at RWTH Aachen, where the scanning team analysed Gargamelle pictures, there was a check list, counting the weeks left until electroweak theory was proven wrong. Instead in 1973, the event reproduced in Figure 9.7 was found [411, 428]. The final state electron is clearly identified by the minimum ionisation and logarithmic curling of its track; it also emits bremsstrahlung which converts into electron-positron pairs in the heavy liquid. Shortly afterwards, also neutral weak interactions of the neutrino with quarks were demonstrated [410, 421]. This hadronic part of weak interactions, however, was experimentally controversial, since there is a background process, neutron scattering off nuclei, which is unavoidably present and must be carefully controlled. This was first done successfully by Dieter Haidt using reactions, where neutron source and interaction are both visible on the same bubble chamber picture. He received the physics prize of the German Physical Society of 1975 for his method. The Gargamelle collaboration received the High Energy and Particle Physics Prize of the European Physical Society in 2009 for their discovery.

The view on the history of the discovery of hadronic weak neutral currents is quite different on both sides of the Atlantic. While U.S. authors, like David B. Cline [471, 493] emphasise that the publications from the Gargamelle experiment at CERN and the concurrent HPWF experiment at Fermilab [415, 416, 417] were almost simultaneous, European authors like Haidt and Donald Perkins [473, 497] take pride in claiming precedence for neutral weak interactions with both leptons and quarks. In any case, the discovery was promptly

---

[12]To filter such rare events out of the many thousand (mostly empty!) bubble chamber pictures is an impressive achievement. The RWTH Aachen technical staff at the time consisted of: Sylvia Jakobs, Heidi Klamann, Christa Malms, Karin Pohlmann, Marlene Stommel, Hannelore von Hoegen and Gisela Weerts.

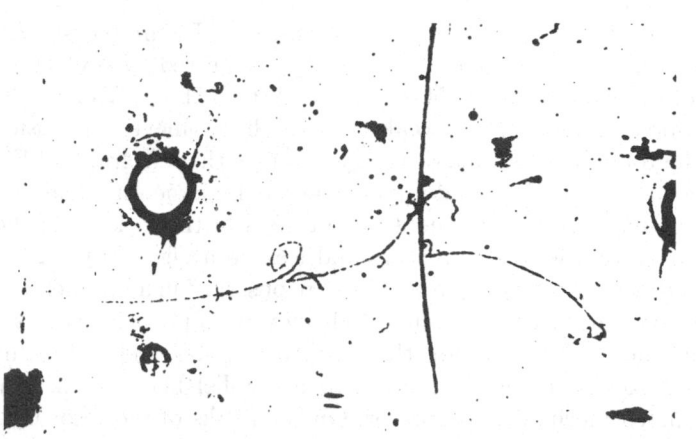

Figure 9.7 Bubble chamber picture of the first observed elastic muon neutrino scattering off an electron ($\nu_\mu e^- \to \nu_\mu e^-$). The neutrino enters from the left and hits an atomic electron projecting it to the right. The electron is seen as a minimum ionising particle. It emits a bremsstrahlung photon which converts into an electron-positron pair. The energy loss in the heavy liquid makes the electron then curl up logarithmically.

confirmed by two independent experiments. However, the conflict may have influenced the decision of the Nobel committee to award the 1979 prize in physics to the three theorists Glashow, Weinberg and Salam and ignore experimentalists [436].

The discovery of Z and W bosons as real particles required the construction of a new collider for protons and antiprotons at CERN. For this purpose, the Super-Proton-Synchrotron was converted into a proton-antiproton collider. The revolutionary method of stochastic cooling, again invented by Simon Van der Meer [437], allowed to obtain beams of smaller size and spread in energy. The experiments UA1 [444, 445] and UA2 [451, 447] reported the production and leptonic decay of both weak bosons in 1983. Van der Meer and Carlo Rubbia obtained the Nobel prize in 1984 for the concept and construction of the collider. To further investigate the properties of the weak bosons, CERN decided even before their discovery to construct the Large Electron-Positron collider LEP with four experiments. From 1989 to 2000, these experiments have investigated weak interactions with unprecedented precision and found the predictions of electroweak theory brilliantly confirmed. In parallel, the SLAC Linear Collider SLC, providing data with polarised electron and positron beams, and the Tevatron storage ring at Fermilab, providing measurements of the mass of W boson and t quark, have contributed to these precision measurements. Since there was no controversy, encouraged by CERN, SLAC and Fermilab management, the experimental collaborations got together to combine their results [561, 599]. As an example, Figure 9.8 shows the total cross section for the reaction $e^+e^- \to \mu^+\mu^-$ collected by several generations of electron-positron colliders [506], including the LEP experiment L3, together with the prediction of electroweak theory. The momentum transfer relevant here is the total centre-of-mass energy, $\sqrt{s}$ (see the left graph of Figure 9.6). The electromagnetic process dominates at small momentum transfer, with its propagator term $1/s$. At momentum transfers in the vicinity of the Z mass, neutral weak interactions dominate and lead to a gigantic peak in the cross section. Resonance phenomena like this are observed in all branches of physics, from acoustics and electrodynamics all the way to particle physics. They are thus a wonderful example for the unity of physics and nature in general. In Focus Boxes 9.11 and 9.12 we draw a parallel between the Z resonance and a damped mechanical oscillator. On the flanks of the resonance, there is notable interference between Z and photon interactions, destructive below and constructive above the resonance peak, which also has a mechanical analog. Interference shows up in particular in the asymmetry between scattering into the forward and the backward hemisphere, which is also shown in Figure 9.6. Experiments thus not only demonstrate the workings of the Z boson as the field quantum of neutral weak interactions, but also the validity of the superposition principle in quantum physics far below the atomic scale, and the wave nature of particle interactions.

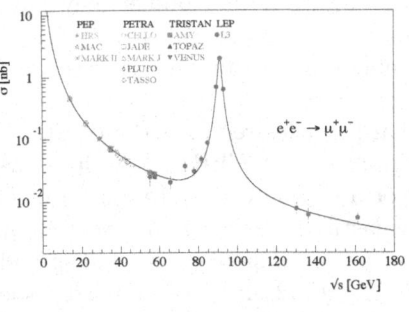
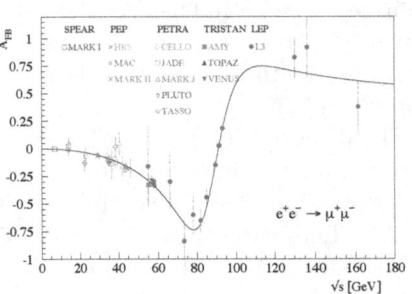

Figure 9.8 Cross section (left) and forward-backward asymmetry (right) for the reaction $e^+e^- \to \mu^+\mu^-$ measured at LEP and at lower energies [506]. The asymmetry and the overlaid theory curves are explained in Focus Box 9.11.

With these measurements, the gauge theory of electroweak interactions is confirmed down to per mille level accuracy, if not better. But there is an inherent contradiction. Local gauge invariance can only be ensured for massless gauge bosons like the photon. Mass terms for gauge or matter particles break gauge symmetry. The Z and W bosons are not only massive, but really heavy; the top quarks weighs as much as a Hafnium nucleus. So why is electroweak theory still renormalisable? Is gauge theory a wrong track or are all

The cross section for $e^+e^- \rightarrow \mu^+\mu^-$ according to electroweak theory has three terms, $\sigma_{tot}(e^+e^- \rightarrow \mu^+\mu^-) = \sigma_\gamma + \sigma_Z + \sigma_{int}$. The first term contains the square of the amplitude for photon exchange:

$$\frac{d\sigma_\gamma}{d\Omega} = \frac{\alpha^2}{4s}\left(1 + \cos^2\theta\right) \quad ; \quad \sigma_\gamma = \frac{4\pi\alpha^2}{3s}$$

The second comes from Z exchange, the third describes the interference between the two. The angular distribution taking into account all three terms is:

$$\frac{d\sigma}{d\Omega} = \frac{\alpha^2}{4s}\left[A_0\left(1 + \cos^2\theta\right) + A_1\cos\theta\right]$$

with the scattering angle $\theta = \angle(e^-, \mu^-)$. The first term is symmetric under $\cos\theta \leftrightarrow -\cos\theta$, the second is asymmetric. Their coefficients are $A_0 = 1 + 2\text{Re}(r)g_V^2 + |r|^2\left(g_v^2 + g_A^2\right)^2$ and $A_1 = 2\,\text{Re}(r)g_A^2 + 8|r|^2 g_V^2 g_A^2$. The pole term r contains the Breit-Wigner function, characteristic for the decay of heavy particles:

$$r = \frac{\sqrt{2}G_F}{e^2}\frac{sM_Z^2}{s - M_Z^2 + iM_Z\Gamma_Z}$$

The couplings g_V et g_A come from the electroweak charges, $g_V = T_3 - 2Q\sin^2\theta_W$ and $g_A = T_3$, with the third component of weak isospin T_3 and the electric charge Q. The index of these constants indicates that they refer to the vector and axialvector components of the weak neutral current. Pure electromagnetic interactions correspond to $A_0 = 1$ and $A_1 = 0$. The purely weak terms are proportional to $|r|^2 \propto g_{V/A}^4$, the interference terms to the real part $\text{Re}(r) \propto g_{V/A}^2$. Major observables are the total cross section σ, where the asymmetric term cancels, and the forward-backward asymmetry A_{FB}, which isolates the interference effect:

$$\sigma = \frac{4\pi\alpha^2}{3s}A_0 \quad ; \quad A_{FB} = \frac{\sigma_F - \sigma_B}{\sigma_F + \sigma_B} = \frac{3A_1}{8A_0}$$

with the partial cross sections σ_F for forward ($\cos\theta > 0$) and σ_B for backward ($\cos\theta < 0$) scattering. The asymmetry is negative for energies below the resonance, goes through zero at $s = M_Z^2$ and becomes positive above. Experimental data for the cross section and the asymmetry as a function of the centre-of-mass energy are shown in Figure 9.8.

Focus Box 9.11: The Z resonance in $e^+e^- \rightarrow \mu^+\mu^-$

A simple one-dimensional mechanical oscillator system is shown in the figure below on the left. A periodic force of amplitude F_0 and angular frequency ω excites a harmonic oscillator characerised by the spring constant k and the mass m. A damping device is seen on the bottom, with a friction coefficient b. The amplitude of the movement follows from Newton's equation:

$$-kx - b\frac{dx}{dt} + F_0 \cos \omega t = m\frac{d^2 x}{dt^2}$$

The solution is:

$$x(t) = \frac{F_0 \sin (\omega t - \phi)}{\sqrt{m^2(\omega^2 - \omega_o^2) + b^2 \omega^2}}$$

Here ϕ is the phase shift between the exciting force and the oscillation. The eigenfrequency $\omega_0 = \sqrt{k/m}$ depends on the ratio of spring constant and mass. The two figures on the right show how the maximum elongation x_{max} and the phase shift ϕ depend on the ratio between exciting frequency and eigenfrequency, ω/ω_0. Comparison with the equations ruling the Z resonance and Figure 9.8 reveal the close analogy between the two processes. The resonance term has the same form. The cross section corresponds to the square of the mechanical amplitude, the sine of the phase shift to the angular asymmetry. The centre-of-mass energy corresponds to the exciting frequency, k to the coupling constant and b to the resonance width. All in all a nice example of the unity of physical laws in very different branches.

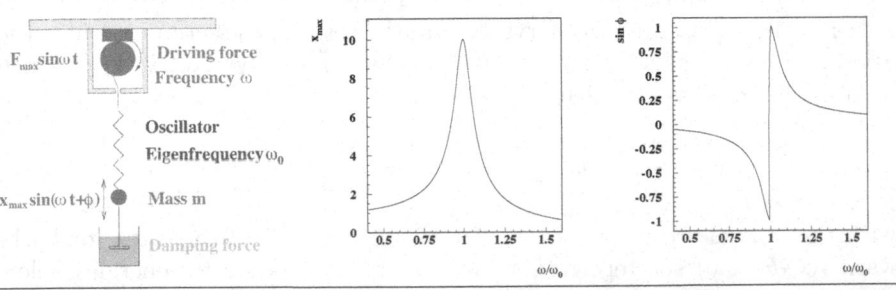

Focus Box 9.12: The damped mechanical oscillator

particles really massless? We come back to the resolution of this mystery in Section 9.3.

Weak interactions break many symmetries which the other interactions preserve:

- Charged weak interactions change the flavour quantum number. The W does not interact with the quarks and leptons of Figure 9.4 but with mixtures specified in Focus Boxes 9.13 and 9.14. These are generated by rotations in flavour space, described by two unitary matrices, named after Nicola Cabibbo [356], Makoto Kobayashi and Toshihide Maskawa [413] for quarks, Bruno Pontecorvo [340], Ziro Maki, Masami Nakagawa and Shoichi Sakata [367] for leptons. These matrices rotate the eigenstates of the mass operator of Figure 9.4 into the mixtures which interact with the W bosons. Both are dominated by their diagonal elements, but have non-negligible entries off the diagonal. The latter lead to interactions among members of different generations, such as s $\to$ uW$^-$ or $\tau^- \to \nu_M$W$^-$. Even rarer are processes which jump two generations. Mixtures are generated by weak interactions, but what propagates through space-time are the "real" particles, and with different velocities since they do not have the same mass. When their flavour is probed again, the mixture thus may have changed and flavour oscillations occur.

- Weak interactions also do not respect parity, but violate its symmetry in a maximum way, as we have seen in Focus Box 9.4. They only interact with the left-handed part of matter fields.

- Moreover, they also violate the symmetry between matter and antimatter, the so-called CP symmetry, but only a little bit this time[13]. This violation manifests itself in the decays of heavy quarks, in that particles and antiparticles have slightly different decay properties. The electroweak theory describes this phenomenon by a complex phase in the mixing matrix (see Focus Box 9.13). For quarks it has been measured, not yet for leptons.

In contrast to the photon as a carrier of electromagnetic interactions, W and Z carry themselves the charges responsible for electroweak interactions. The neutral boson has $Q = T_3 = 0$, but $T = 1$. The charged bosons have $Q = T_3 = \pm 1$. For this reason, the electroweak bosons interact with themselves as the Feynman graphs of Figure 9.9 show. At the same time, these interactions fix the electroweak couplings of matter particles, and make them identical for all generations. The underlying reason for boson self-coupling is the non-Abelian nature of the group SU(2), i.e. the fact that rotations around two axes do not commute.

This property of boson self-coupling is shared by the third force of the Standard Model, the strong force that binds quarks together into hadrons.

---

[13]See http://pdg.lbl.gov/2019/reviews/rpp2019-rev-cp-violation.pdf

Weak transitions between quarks (and leptons) happen predominantly but not exclusively among members of the same generation. Charged pion decay $\pi^+ \to \mu^+ \nu_\mu$ is an example of the dominant type. Kaon decay $K^+ \to \mu^+ \nu_\mu$ is a process involving quarks of the first and second generation. Otherwise they are very similar:

Due to the much larger phase space for the kaon decay, one would expected a ratio of the partial widths $\Gamma(K^+ \to \mu^+ \nu_\mu)/\Gamma(\pi^+ \to \mu^+ \nu_\mu) = \mathcal{O}(100)$. The observed ratio is of order 1 instead. One concludes that cross-generation couplings are very much disfavoured compared to intra-generation couplings, by at least a factor 10 in the amplitude. In 1962, Nicola Cabibbo proposed a way to describe this fact [356]. States which interact with the $W^\pm$ are not the quarks which are listed in Figure 9.4, but mixtures, $d' = d\cos\theta_C + s\sin\theta_C$ and $s' = -d\sin\theta_C + s\cos\theta_C$, using the Cabibbo angle $\cos\theta_C$. The states d', s' and b' are not eigenstates of the mass operator. Weak interactions mix a little bit of a strange quark into the wave function of a down quark and vice versa. Consequently, a factor of $\cos\theta_C$ is included in the intra-generation amplitude. For inter-generation transitions, a factor of $\sin\theta_C$ appears. In the ratio of decay widths, we get a factor of $\tan^2\theta_C$, from the measured rates $\theta_C \simeq 10°$. Kobayashi and Maskawa enlarged the mechanism to three generations [413], introducing the Cabibbo-Kobayashi-Maskawa matrix:

$$\begin{pmatrix} d' \\ s' \\ b' \end{pmatrix} = \begin{pmatrix} V_{ud} & V_{us} & V_{ub} \\ V_{cd} & V_{cs} & V_{cb} \\ V_{td} & V_{ts} & V_{tb} \end{pmatrix} \begin{pmatrix} d \\ s \\ b \end{pmatrix}$$

There is a hierarchy of elements: intra-generation elements dominate, those which jump a single generation are still appreciable, those which jump two are small. Methodology of the measurement and current values are found in http://pdg.lbl.gov/2019/reviews/rpp2019-rev-ckm-matrix.pdf. The mixing mechanism leads to oscillations between particle and antiparticle for neutral hadrons, like $K^0 \leftrightarrow \bar{K}^0$, $D^0 \leftrightarrow \bar{D}^0$ and $B^0 \leftrightarrow \bar{B}^0$. Note that the matrix elements can be complex. There can thus be a complex phase which cannot be eliminated by redefining the flavour base. Such a phase violates the symmetry between matter and antimatter in the amplitudes where it appears, it is the way in which the Standard Model describes CP violation.

Focus Box 9.13: Weakly interacting quark states

In the leptonic sector, weak interactions also introduce a mixing mechanism. The weak bosons do not interact with the mass eigenstates of Figure 9.4, but with superpositions:

$$
\begin{pmatrix} \nu_e \\ \nu_\mu \\ \nu_\tau \end{pmatrix} = \begin{pmatrix} U_{eL} & U_{eM} & U_{eH} \\ U_{\mu L} & U_{\mu M} & U_{\mu H} \\ U_{\tau L} & U_{\tau M} & U_{\tau H} \end{pmatrix} \begin{pmatrix} \nu_L \\ \nu_M \\ \nu_H \end{pmatrix}
$$

The matrix is called Pontecorvo-Maki-Nakagawa-Sakata matrix after its inventors. For an up-to-date review of the related methodologies, please see http://pdg.lbl.gov/2019/reviews/rpp2019-rev-neutrino-mixing.pdf. The superposition leads to oscillations between weak neutrino flavours if the masses of ν_L, ν_M and ν_H are not the same. When a pure ν_μ beam is produced, e.g. by weak pion or kaon decay, different mass eigenstates propagate with different velocity. When the beam composition is then probed, again by weak interactions, at a distance from the production point, the beam composition will have changed. Thus one observes e.g. ν_e or ν_τ interactions in a ν_μ beam.

Focus Box 9.14: Weakly interacting lepton states

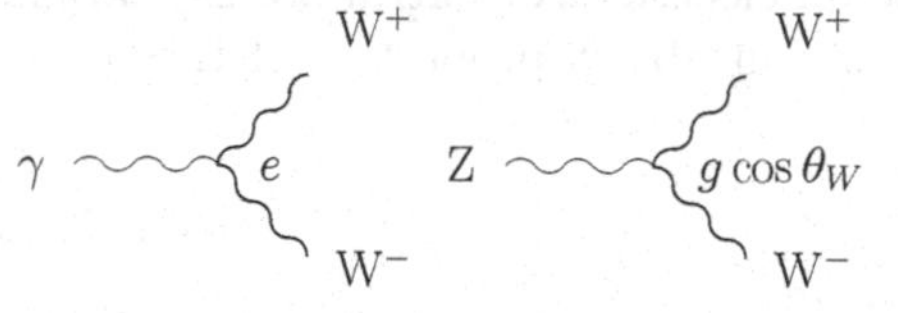

Figure 9.9 Feynman graphs for the self coupling of electroweak bosons.

Its unknown dynamics had kept Heisenberg from accepting the quark model. It was already clear from spectroscopy that quarks needed an additional new quantum number. A state made of three identical quarks without relative angular momentum and with total spin 3/2, like Δ^{++}, is not possible without an additional quantum number which makes the total wave function asymmetric, as the Pauli principle requires.

The gauge theory of strong interactions, quantum chromodynamics (QCD) takes this into account. It is based on the symmetry group SU(3) of rotations around three axes, again a non-Abelian group where transformations do not commute, as we explain in Focus Box 9.15. QCD has many fathers, a short account is found in [590], a more detailed one in [609]; an early reference is [412]. An interesting introduction very much centred on aesthetic arguments has been given by Frank Wilczek [629]. The global version of SU(3) symmetry causes the conservation of a three-component strong charge, called colour. One chooses the basic colours red (R), green (G) and blue (B) like in colour TV. A red quark, q_R, has the red colour quantum number equal to 1, the two others equal to zero, $(1, 0, 0)$. An antiquark of the same kind carries an anticolour, $\bar{q}_{\bar{R}}$ has a colour of $(-1, 0, 0)$. Electroweak interactions are colour blind, leptons and electroweak bosons do not carry a colour charge. We go into technical details in Focus Box 9.15. Bound states of coloured quarks have no net colour, in a symmetric way for mesons, an antisymmetric way for baryons. We can symbolically decompose their wave functions as follows:

$$(q\bar{q}) = (q_R\bar{q}_{\bar{R}} + q_G\bar{q}_{\bar{G}} + q_B\bar{q}_{\bar{B}})$$

$$(qqq) = (q_Rq_Gq_B - q_Gq_Rq_B + q_Bq_Rq_G - q_Rq_Bq_G + q_Gq_Bq_R - q_Bq_Gq_R)$$

The field quanta of strong interactions are called gluons since their role is to bind quarks into hadrons. Like the electroweak bosons, they carry the required charge to interact among themselves. In fact they carry a colour and and an anticolour; a red-antigreen gluon thus has colour charge $(1, -1, 0)$. When a quark emits or absorbs a gluon, its own colour charge changes as sketched in Figure 9.11. The total colour of the hadron always stays zero in this process. In particular proton and neutron thus have no net colour and cannot as a whole emit or absorb gluons. The nuclear force that binds them into nuclei is just a residual of the strong force, analogous to the tiny Van der Waals force that acts between electrically neutral atoms and molecules in gases and liquids. More information about properties of the nuclear force is found in Section 5.5.

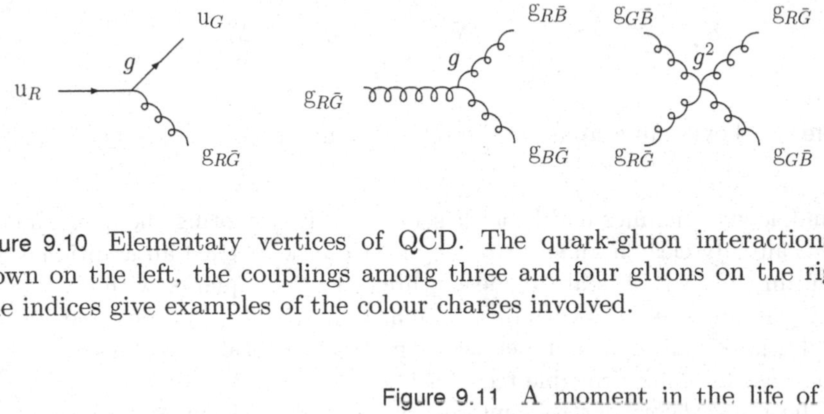

Figure 9.10 Elementary vertices of QCD. The quark-gluon interactions is shown on the left, the couplings among three and four gluons on the right. The indices give examples of the colour charges involved.

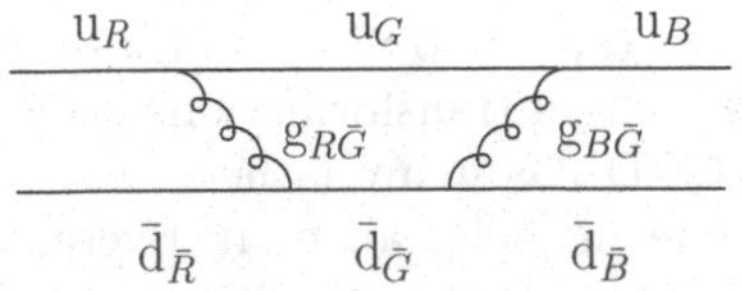

Figure 9.11 A moment in the life of the positive pion. Its main ingredients are an up quark u and an anti-down quark $\bar{\text{d}}$. They bind by constantly exchanging gluons, their colour charge changes but the net colour of the hadron always stays zero.

Compared to electroweak couplings, the strong coupling constant g is rather large. It enters the scattering amplitude as $\alpha_s = g^2/(4\pi)$. This is the analog of the fine structure constant of QED, but at least one order of magnitude larger than its electromagnetic counterpart. Strong interactions thus

The phase transformations of the group SU(3):

$$\psi_q(x) \to \psi_q'(x) = e^{i\alpha_k \frac{\lambda_k}{2}} \psi_q(x)$$

are generated by a linear combination of the eight Gell-Mann matrices $\lambda_k, k = 1\ldots8$. The Dirac equation is invariant under global transformations with constant coefficients $\alpha_k \neq \alpha_k(x)$. According to Noether's theorem, there is a corresponding charge with three components, which are individually conserved. It is called colour, its basis can be chosen as red, green and blue:

$$C_R = \begin{pmatrix} 1 \\ 0 \\ 0 \end{pmatrix} \quad ; \quad C_G = \begin{pmatrix} 0 \\ 1 \\ 0 \end{pmatrix} \quad ; \quad C_B = \begin{pmatrix} 0 \\ 0 \\ 1 \end{pmatrix}$$

A quark carries a single colour, the up-quark has thus three states, u_R, u_G et u_B. Antiquarks carry an anticolour, like $\bar{u}_{\bar{R}}$, $\bar{u}_{\bar{G}}$ and $\bar{u}_{\bar{B}}$. Interactions between quarks involve a colour change by emitting or absorbing one of eight gluons. Gluons carry one colour and one anticolour each, with e.g. the combinations:

$$R\bar{B}, R\bar{G}, B\bar{G}, G\bar{R}, G\bar{B}, B\bar{R}, \frac{1}{\sqrt{2}}(R\bar{R} - B\bar{B}), \frac{1}{\sqrt{6}}(R\bar{R} + B\bar{B} - 2G\bar{G})$$

With three colours and anticolours one could have thought that there were nine gluons; however, the symmetric combination $(R\bar{R} + B\bar{B} + G\bar{G})/\sqrt{3}$ has no net colour and cannot interact. The remaining eight are spin-one bosons, neutral with respect to electroweak interactions and massless. If one imposes invariance under local SU(3) phase transformations, by a local angle $\alpha_k(x)$, the following Lagrange density results:

$$\mathcal{L}_{QCD} = \bar{\psi}_q \left(i\gamma^\mu \partial_\mu - m\right) \psi_q - g \left(\bar{\psi}_q \gamma^\mu \frac{\lambda_k}{2} \psi_q\right) G_\mu^k - \frac{1}{4} G_{\mu\nu}^k G_k^{\mu\nu}$$

The first term describes the free quark field, the second its interaction with the eight gluon fields G_μ^k, as shown on the left of Figure 9.10. The last term involves the eight gluon field tensors $G_{\mu\nu}^k$:

$$G_{\mu\nu}^k = \left(\partial_\mu G_\nu^k - \partial_\nu G_\mu^k\right) - g f_{kjl} G_\mu^j G_\nu^l$$

with the SU(3) structure constants f_{kjl}. The terms in brackets remind us of the electromagnetic analog $F_{\mu\nu}$, they describe the motion of the gluon field. The third term, quadratic in the gluon fields, leads to the self-interactions among gluons, shown on the right of Figure 9.10:

$$g \left(\partial_\mu G_\nu^k\right) \left(G_j^\mu G_l^\nu\right) \sim G^3 \quad ; \quad g^2 \left(G_\mu^j G_\nu^l\right) \left(G_j^\mu G_l^\nu\right) \sim G^4$$

They have important consequences for the large distance behaviour of strong interactions.

Focus Box 9.15: SU(3) symmetry and the strong force

really deserve their name. However, their range is extremely limited despite the fact that their force quanta are massless. This is due to the fact that the strong coupling is anything but constant, but depends on momentum transfer more strongly than all other couplings.

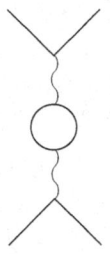 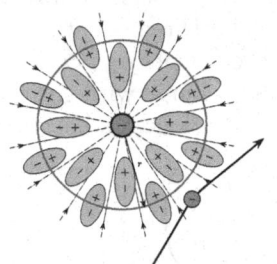

Figure 9.12 Vacuum polarisation in QED. The Feynman graph on the left shows the contribution of fermion loops to the photon propagator. It leads to fermion pairs of opposite charge close to the source as sketched on the right.

The fact that couplings are not constant is due to higher order contribution to the propagators of gauge bosons. When we measure the charge of a matter particle, we of course do that experimentally. The measurement automatically contains all orders of perturbation theory and is made at a given momentum transfer. However, we insert this coupling constant into an amplitude of fixed order and eventually at a different momentum transfer. This is taken into account by letting the coupling constant "run". Contributions to the photon and gluon propagator are shown in Figures 9.12 and 9.13, respectively. In both cases, the contributions of fermion loops, called vacuum polarisation, lead to a screening of the electric and colour charge. And the more so, the larger the distance between projectile and target, i.e. the smaller the momentum transfer. The electric charge is thus a little bit smaller at large distances, a little larger at small distances.

In contrast to that, the strong vacuum polarisation receives an additional contribution by gluon loops, and these strengthen the charge since they are made of bosons[14]. They are also more numerous than fermion loops, such that the strong coupling grows rapidly with increasing distance. Colour charge is thus large at large distances, bound states cannot be separated.

Strong vacuum polarisation, on the one hand, leads to a steep decrease of the colour charge at small distance, i.e. high momentum transfer. The quarks bound inside hadrons then behave much like free particles, asymptotic freedom is an intrinsic property of the theory. H. David Politzer, David J. Gross and Frank A. Wilczek shared the 2004 Nobel prize in physics for their discovery

---

[14]Due to the electroweak self-couplings, also the photon and Z propagator receive strengthening contributions from weak boson loops. However, these do not play a significant role because of their high mass.

When bound quarks are struck by a gauge boson, or created as quark-antiquark pairs *in vacuo* by a photon or Z, they move apart with high momentum. As they do that, a colour field builds up between them –as an electric field does when electric charges are pulled apart. However, the electric field becomes weaker with increasing distance, the colour field quickly becomes much stronger.

When the distance between the quarks is about 1fm= 10^{-15}m, the potential energy stored in the field, which comes at the expense of the kinetic energy of the quarks, is sufficient to form an additional quark-antiquark pair out of the vacuum, as sketched on the right.

This process continues until the kinetic energy is exhausted. It creates a multitude of hadrons moving along essentially in the direction of the initial quarks. The result is a hadron jet, examples are shown in Figure 9.14. One can model this process of jet formation, also called hadronisation or fragmentation, in a less chromostatic and more dynamic way as follows.

One of the high momentum quark can emit a gluon. This gluon can split into two according to gluon self-coupling, or into a quark-antiquark pair. In fact, the two-gluon spilt is more likely, since there are more allowed colour combinations. A cascade of quarks and gluons is thus created.

When one of the quark-antiquark pairs comes close enough in phase space and the colours match, a hadronic bound state is formed. The end result is the same as in the chromostatic model: the quark transforms itself into a jet of hadrons around its initial direction.

Focus Box 9.16: Hadron jets

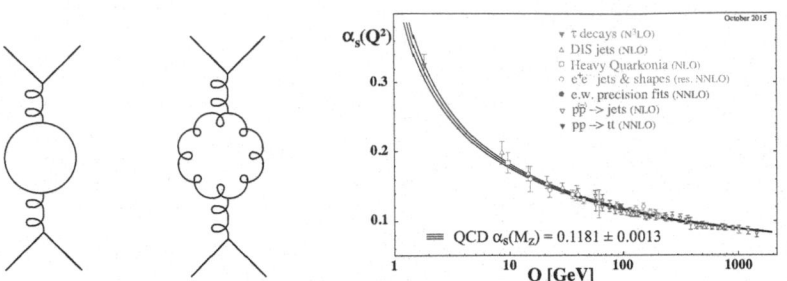

Figure 9.13 Vacuum polarisation in QCD. The Feynman graphs on the left show the contributions of fermion and gluon loops to the gluon propagator. The gluon loop contribution is larger and leads to an amplification of the colour charge at large distances, i.e. small momentum transfer. This is experimentally verified by the dependence of α_s on momentum transfer Q, shown on the right [658].

of this feature of quantum chromodynamics. On the contrary, the coupling increases rapidly at large distances, such that quarks are forever trapped inside hadrons. At a distance of about a femtometer, the energy stored in the gluon field between two quarks is so large, that quark-antiquark pairs are spontaneously created. When the quarks are torn further apart, this process repeats until the kinetic energy of the coloured pair is exhausted. What remains are mesons, colourless bound states of quarks and antiquarks, but no free colour charges. This leads to the formation of hadron jets, which accompany the original quark direction with a small opening angle. The mechanism is explained further in Focus Box 9.16. It happens with probability one, so does not modify elementary cross sections. Figure 9.14 shows examples of hadronic jets in high energy reactions. Events with multiple jets, as shown on the right, come from gluon bremsstrahlung, here in the reaction $e^+e^- \rightarrow q\bar{q}g$. The cross section is quoted in Focus Box 9.17. Such events were first observed by experiments at the PETRA collider of DESY, Hamburg. The gluon as a force carrier of strong interactions was thus discovered. There were four experiments running simultaneously when the collider reached an energy where the phenomenon became detectable. There is no consensus on which one can claim precedence in convincingly demonstrating the existence of gluons [481, 501, 607]. Once again, the experimental proof thus did not receive recognition by a Nobel prize, but the existence for gluons is at the same level as that for quarks, and so are asymptotic freedom and gluon self-coupling [568]. QCD thus exhibits exactly the properties that electron-nucleon scattering experiments require (see Section 9.1).

Calculations based on QCD are difficult at small momentum transfers, due to the large coupling strength which limits the application of perturbation theory. Instead one can choose to build models based on specific properties of

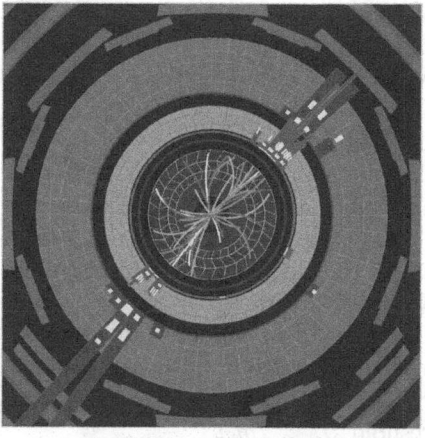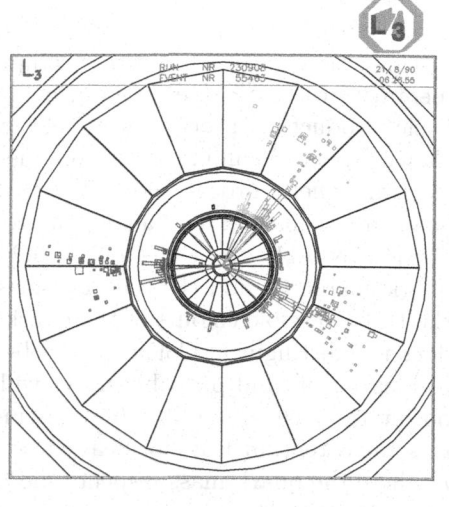

Figure 9.14 Left: Reaction with two back-to-back hadronic jets registered by the ATLAS experiment at the LHC collider. (Credit: ATLAS © CERN) Right: Electron-positron annihilation into a hadronic final state with three jets, registered by the L3 experiment at the LEP collider. (Credit: L3 © CERN) The tracks of charged particles are shown as curved lines in the inner detectors, energy deposits in the calorimeters as histograms.

QCD, which describe the phenomenology of strong interactions rather well. An example is the formation of jets [433, 434]. Bound states cannot be described by perturbative methods at all. Implementations of QCD on a lattice, i.e. with a minimal distance between quarks, which is then shrunk towards zero, deliver numerical results for relative bound state masses with minimal input. The reproduction of the spectrum of light baryons and mesons [573] shows that QCD is the right theory for strong interactions also in the non-perturbative regime. Heisenberg's requirement of a dynamic theory for quarks in hadrons is thus brilliantly fulfilled. Moreover, implementations of QCD on a lattice even reproduce successful models of the nuclear binding potential (see Focus Box 5.12), a residual force transmitted by colourless objects.

9.3 SPACE-TIME

Emmy Noether's theorem dictates that continuous symmetries lead to the conservation of all elementary charges. The three interactions described by the Standard Model, electromagnetic, weak and strong, have observed properties in agreement with a field theory based on the requirement of local gauge invariance under the joint group U(1)×SU(2)×SU(3). Gauge invariance has been uncovered by 't Hooft and Veltman as a necessary condition for the renormalisability of a field theory. Without it, it makes no predictions which can be confronted with precision experiments.

The reaction $e^+e^- \to$ hadrons comprises many elementary processes: $e^+e^- \to q\bar{q}$, $q\bar{q}g$, $q\bar{q}gg$ etc. The basic production process is an electroweak one, as shown in the Feynman graph on the right. For the electromagnetic process alone, the cross section for a

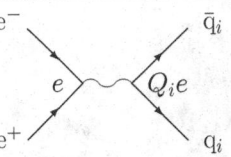

given quark flavour i is analogous to the muon pair production quoted in Focus Box 9.11, $\sigma(e^+e^- \to q\bar{q}) = 3Q_i^2(4\pi\alpha^2/3s)$, with the fractional electric quark charge Q_i and the center-of-mass energy $\sqrt{s}$. The factor 3 comes from the three possible colours of the quark, which the electromagnetic interaction does not distinguish. Colour is an observable, although we do not know how to measure it, and probably never will. Thus the cross sections add up, with no interference. If one does not distinguish quark flavour (although one can to some extent, at least for heavy flavours), one can also sum over all flavours i, where the quark mass respects the threshold $\sqrt{s} > 2m_i$:

$$\sigma(e^+e^- \to \text{hadrons}) = 3\frac{4\pi\alpha^2}{3s}\sum_i Q_i^2$$

The colour and electric charge factor $3\sum Q_i^2$ makes a jump at every threshold:

$3[(\frac{2}{3})^2 + 2(\frac{1}{3})^2] = 2$	u, d, s	$\sqrt{s} < 2m_c \simeq 3.7$ GeV
$2 + 3(\frac{2}{3})^2 = \frac{10}{3}$	u, d, s, c	$3.7 < \sqrt{s} < 2m_b \simeq 10$ GeV
$\frac{10}{3} + 3(\frac{1}{3})^2 = \frac{11}{3}$	u, d, s, c, b	$10 < \sqrt{s} < 2m_t \simeq 350$ GeV
$\frac{13}{3} + 3(\frac{2}{3})^2 = 5$	u, d, s, c, b, t	$350 < \sqrt{s}$

Gluon bremsstrahlung in the final state is sketched in the Feynman graphs on the left. The gluon emission happens after the elementary pair creation by the photon, as does hadronisation. So this process cannot be distinguished from the one without gluon bremsstrahlung. There is thus an interference term of order $\alpha^2\alpha_s$ between the two (in addition to the smaller $\alpha^2\alpha_s^2$ probability for the process itself), adding to the cross section:

$$\sigma(e^+e^- \to \text{hadrons}) = 3\frac{4\pi\alpha^2}{3s}\sum_i Q_i^2\left(1 + \frac{\alpha_s}{\pi}\right)$$

Focus Box 9.17: Electron-positron annihilation into hadrons

Yet non-vanishing masses for matter and force particles are incompatible with gauge invariance, and consequently lead to divergences in the amplitudes and cross sections. One can see that e.g. by the Feynman graph of Figure 9.15. In this so-called box graph, two virtual force particles are exchanged between a pair of fermions. Momentum conservation does not limit the individual momentum transfers; any q exchanged in the first can be compensated by $-q$ in the second. One must thus integrate the amplitude over the momentum transfer up to infinity. If the propagators of the virtual particles diminish like $1/q^2$, like for photons or gluons, there is no problem, the integral converges. But if the propagator has a pole at $q^2 = M^2 \neq 0$, i.e. if the force particle is massive, the integral diverges and cannot be calculated without artificially limiting the momentum transfer. Such a cut-off can be introduced, with the hope that it corresponds to a new scale in the process. But new cut-offs would be necessary at every higher order, the theory makes no prediction beyond lowest order. Similar divergences occur when masses are taken into account for the virtual fermions in our example.

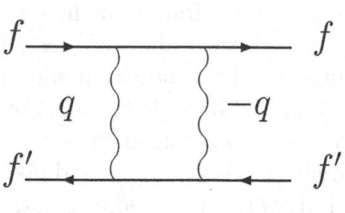

Figure 9.15 A so-called box graph, a higher order contribution to the scattering between two fermions. A momentum transfer q between f and f' is compensated by an opposite one in the second boson exchange. The energy-momentum circulating in the box is thus not limited by the kinematics of the incoming and outgoing fermions.

There is thus a serious problem. Either no particle has an intrinsic mass, or the Standard Model is only valid as an effective theory. Matter particles and gauge bosons do have a mass, and some not even a small one. One way to solve this contradiction is to view mass not as an intrinsic property of particles, but as a dynamically generated one. A mechanism must be found by which massless particles are prevented from propagating at the speed of light. We then would have misinterpreted this mechanism hindering free motion as an intrinsic mass. The dynamical generation of mass is not at all an unfamiliar phenomenon as seen in Figure 9.16. While the masses of crystals, atoms and even atomic nuclei are dominated by their constituents, this is no longer the case for hadrons. The contribution of the up- and down-quark masses to the mass of protons and neutrons is small. Nucleon mass is dominated by the binding energy between quarks, provided by massless gluons.

So the remaining question is: Can mass also be dynamically produced for point-like particles without an internal structure? Their mass spectrum is indeed impressively wide, as Figure 9.4 shows. It ranges from the neutrino mass, probably of the order of 10^{-3}eV, to the top quark with 1.7×10^{14}eV,

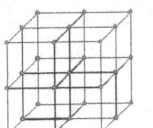   

$$M \simeq \sum M_{\text{Atom}} \qquad M \simeq \sum m_{e^-} + M_{\text{Kern}} \qquad M \simeq \sum M_{\text{p}} + M_{\text{n}} \qquad M_N \gg \sum m_{\text{q}}$$

$$1\,\text{eV} \;<\; E_{\text{coh}}/A < 10\,\text{eV} \qquad 5\,\text{eV} \;<\; E_I/e^- < 25\,\text{eV} \qquad E_B/N \;\simeq\; 8\,\text{MeV} \qquad m_{\text{u}} \simeq m_{\text{d}} \simeq O(\text{MeV})$$

Figure 9.16 Contributions of constituent mass and binding energy to the masses of crystals, atoms, nuclei and nucleons. From left to right, binding energy makes larger and larger contributions, for nucleons it dominates.

spanning 17 orders of magnitude. While photons and gluons are massless, weak bosons also weigh of the order of 10^{14} eV.

The quest for dynamically generated particle masses has found an answer in solid state physics in the 1960s, long before massive gauge bosons were discovered [355]. The Scotsman Peter Higgs and others later found out how similar dynamics can be applied to the gauge bosons of the Standard Model [370, 375, 377, 378]. If massive fields break gauge symmetry, the opposite should also be true. A spontaneous breakdown of symmetry should be able to turn massless fields into massive ones. That breaking a symmetry can make massless systems massive can be made plausible with a simple mechanical Gedankenexperiment shown in Figure 9.17. Imagine an infinitely thin needle, balanced vertically on a flat surface. It has no moment of inertia around its axis. Now we stress the needle along its axis by putting a weight on top, large and beyond its elastic limit. The equilibrium is evidently unstable, a small fluctuation in the metal lattice will bend the needle spontaneously in an unpredictable direction. The system thus falls into its ground state of minimal potential energy. The axial symmetry of the system is broken, simultaneously it has developed a moment of inertia around the vertical axis. The loss of symmetry has generated a kind of mass.

The mechanism of spontaneous symmetry breaking in particle physics, named after Peter Higgs, is also based on a scenario where a system in its ground state no longer exhibits a symmetry, which it does have in general. We discuss a simple example in Focus Box 9.18 to show how it works. If a ubiquitous scalar field, the Higgs field, results from such a spontaneous symmetry breaking, it can confer an apparent mass to all particles to which it couples, including itself. The quantum of this field is a spin zero particle, the Higgs boson. Its special feature is that it is present everywhere and all the time, as a property of the vacuum. It does not have to be produced, but interacts with particles anyway, if they have the required coupling. This way

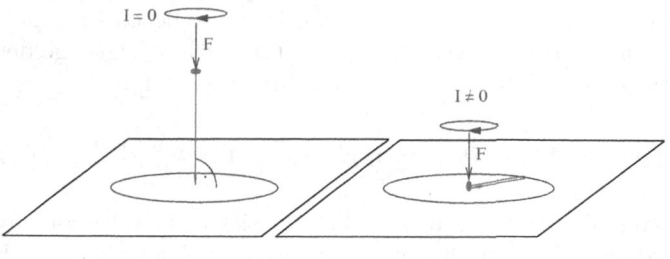

Figure 9.17 A mechanical example demonstrating that the breaking of symmetry can lead to mass-like phenomena, in this case a moment of inertia I. A symmetric unstable equilibrium with $I = 0$ is shown in the left, the asymmetric ground state with $I \neq 0$ on the right.

the Higgs field keeps them from evolving at the speed of light; we call this artefact their mass. One can picture this like the motion through a viscous liquid, except that particles do not lose energy by bouncing off the Higgs boson. Massless particles can thus indeed move at less than the speed of light. The apparent mass they have is determined by their coupling strength to the Higgs field. Of course, a single Higgs field is just the minimal scenario. More than one may be present.

With the exception of its mass, which after all it generates itself, all properties of the Higgs boson were known since the 1960s. But of course not, if it really existed as a solution to the mass problem. Its profile is given by the role it is supposed to fill: it is a scalar boson without charge, coupling preferentially to heavy matter particles and gauge bosons. It also decays preferentially into the heaviest particles kinematically accessible, but all final states compatible with conservation laws occur. Its wanted poster hung on the wall of several generations of accelerators and colliders. It has finally been identified in 2012 by the experiments at the CERN Large Hadron Collider in a spectacular way, at a mass of about 125 GeV. An overview of the history of this search and discovery has been given e.g. by Karl Jakobs and Chris Seez [621]. Figure 9.18 shows an example reaction recorded by the ATLAS experiment in 2016. The Higgs boson decays here into two Z bosons, one of which in turn decays into a pair of muons, the other one into an electron-positron pair. Since its discovery, many of the properties of the particle have been measured and found to be in agreement with the ones a Higgs boson should have[15]. Nonetheless it remains uncertain if it is the only such particle; the search will continue. What is already clear is that the vacuum is not an empty theatre, where particles evolve and interact with each other. It is a rather bustling environment, which actively influences what happens.

---

[15]See e.g. Heather Gray and Bruno Marsoulié, *The Higgs boson: the hunt, the discovery, the study and some future perspectives,*
https://atlas.cern/updates/atlas-feature/higgs-boson

We construct a toy model with a photon-like gauge boson to demonstrate the functioning of the Higgs mechanism. We start with a Lagrange density for a complex scalar field Φ interacting with a boson field A_μ:

$$\mathcal{L} = (\partial^\mu + ieA^\mu)\Phi^*(\partial_\mu - ieA_\mu)\Phi - \mu^2\Phi^*\Phi - \lambda(\Phi^*\Phi)^2 - \frac{1}{4}F_{\mu\nu}F^{\mu\nu}$$

It is invariant under U(1) gauge transformations, since we already included the required term $j_\mu A^\mu$ with the scalar current of Focus Box 7.2. For $\mu^2 > 0$ this is the Lagrange density of QED from Focus Box 9.9, for a scalar field with mass μ, extended by a self-interaction term between four Φ fields with coupling λ. The potential energy density, $V = \mu^2\Phi^2 + \lambda\Phi^4$, has a minimum for $\Phi = 0$, easy to verify by calculating its first and second derivative with respect to Φ. This is normal: when there is no field, the energy density should be minimum. The interpretation of the first term in the potential changes drastically when we choose $\mu^2 < 0$. The term proportional to Φ^2 is no longer a mass term, since μ is imaginary, Φ is a massless field. Moreover, the potential has a local maximum at $\Phi = 0$.

A minimum is found instead for a finite value of the field, $|\Phi| = v = \sqrt{-\mu^2/\lambda}$. The potential as a function of the real and imaginary part of Φ has the shape shown on the right. It still is symmetric under U(1) phase transformation of the Φ field. However, if we want to perturbatively expand our toy theory, we must choose one of the (infinitely many) minima of its potential energy density. Only then the system is stable agains small perturbations. Our perturbative "vacuum" thus cannot be empty space-time, but contains a non-zero field $\Phi(x)$ everywhere. Let us choose to expand our field around the specific minimum $\Phi = (v/\sqrt{2})e^{i\theta(x)}$. Any θ is allowed by gauge freedom, but we have to chose one at each x. We can then write small perturbations as:

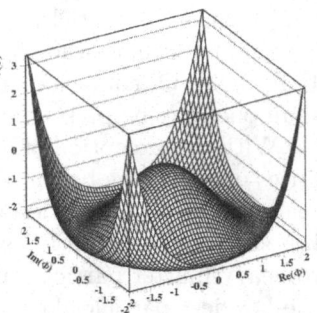

$$\Phi(x) = \frac{1}{\sqrt{2}}[v + h(x)]e^{i\theta(x)}$$

The Lagrange density in terms of the real field h is:

$$\mathcal{L}' = \frac{1}{2}(\partial_\mu h)^2 - v^2\lambda h^2 + \frac{1}{2}e^2v^2 A_\mu A^\mu - \frac{1}{4}F_{\mu\nu}F^{\mu\nu} + \text{Int}(h, A)$$

The last term is a short for interactions and self-interactions of the fields. The angle θ does not appear, A_μ absorbs it. But we now have mass terms for h with $m_h = \sqrt{2\lambda v^2}$ and the gauge field A_μ, with $m_A = ev$, while there were no mass terms in the original Lagrange density. The U(1) symmetry of the original Lagrangian has disappeared, mass terms appear [361]. The field h is called a Higgs field, its quantum is a Higgs boson.

Focus Box 9.18: A toy example of the Higgs mechanism

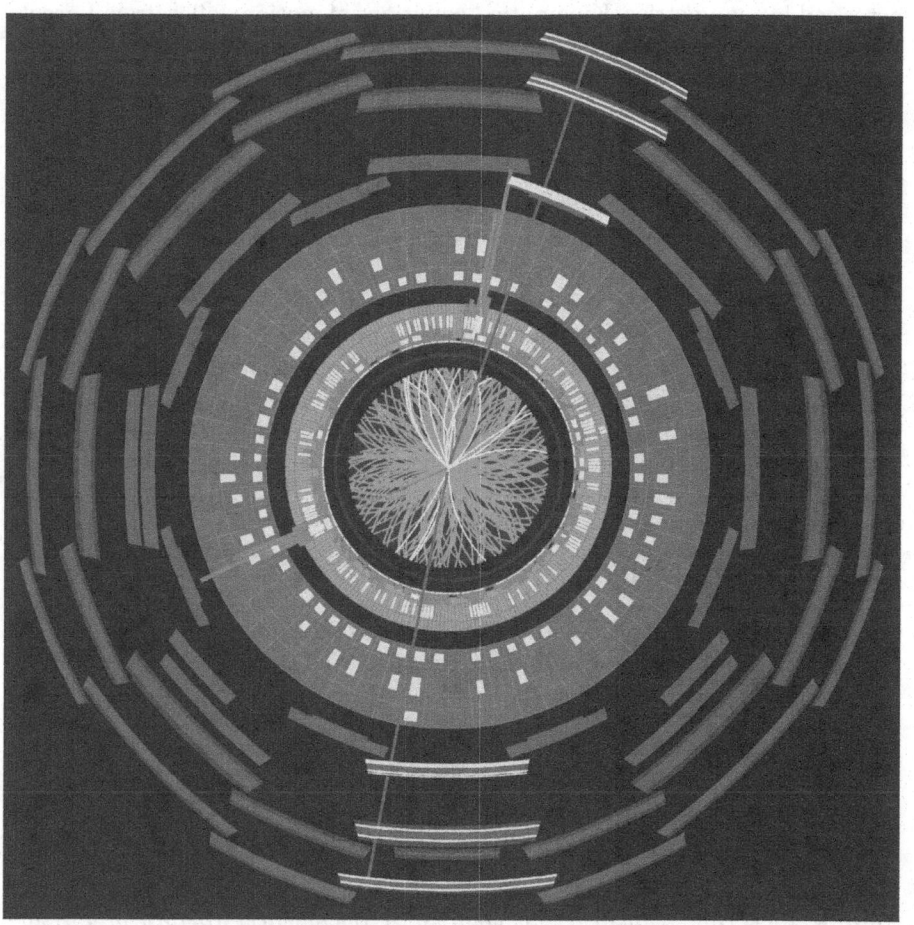

Figure 9.18 A display of a candidate Higgs boson event from proton-proton collisions recorded by the ATLAS experiment with LHC at a collision energy of 13 TeV. The candidate event is reconstructed in the $\mu^+\mu^-e^+e^-$ final state. The long grey lines show the path of the two muons including the hits in the muon spectrometer, the short grey lines show the paths of the two electrons together with the energy deposit in the electromagnetic calorimeter. The lighter tracks in the inner detector are the remaining charged particles from the Higgs boson candidate vertex. The other tracks come from additional proton-proton reactions recorded simultaneously because of the high luminosity of the collider. (Credit: ATLAS © CERN)

The scenario of the Standard Model is thus completed. The systematics of matter, the motion of particles and also the forces which act on them are characterised by symmetries. Kinematics, equations of motion, follow from Lorentz invariance, dynamics from gauge invariance. The charges of electroweak and strong interactions are thus conserved, as the Noether theorem requires, just like energy and momentum. Obviously all these statements must constantly be challenged by experiment, to the accuracy achievable at any moment. On the other hand, quantum numbers, which are not protected by a symmetry, are at maximum approximately conserved. Examples are the flavour quantum numbers and particle properties under discrete symmetries, like the one between matter and antimatter.

Thus symmetries are the key to microcosm. They allow the reduction of an enormous variety of physical phenomena to a few basic laws, as sketched schematically in Figure 1. One is allowed to hope that this reduction may continue, but one doesn't have to. If a further step in reduction were possible, it would at least have to include a few phenomena outside the Standard Model, which we know about today and which we will discuss briefly in the final chapter.

FURTHER READING

Carsten Jensen, *Controversy and Consensus: Nuclear Beta Decay 1911-1934*, Finn Aaserud, Helge Kragh, Erik Rüdinger and Roger H. Stuewer (Edts.), Birkhäuser Verlag, 2000.

Frank Close, *The Infinity Puzzle: Quantum Field Theory and the Hunt for an Orderly Universe*, Basic Books, 2011.

Francis Halzen and Allan D. Martin, *Quarks and Leptons*, John Wiley and Sons, 1984.

Andrew Pickering, *Constructing Quarks: A Sociological History of Particle Physics*, University of Chicago Press, 1984.

Peter L. Galison, *How Experiments End*, University of Chicago Press, 1987.

Martinus Veltman, *Facts and Mysteries in Elementary Particle Physics*, World Scientific, 2018.

CHAPTER 10

Pushing the boundaries

> In moving from a theory to the theory that supersedes it, we do not save only the verified empirical content of the old theory, but more. This "more" is a central concern for good physics. It is the source, I think, of the spectacular and undeniable predictive power of theoretical physics. I think that by playing it down one risks misleading theoretical research into a less effective methodology.
>
> ... figuring out where the true insights are and finding a way of making them work together is the work of fundamental physics. This work is grounded on *confidence* in the old theories, not on a random search for new ones.
>
> Carlo Rovelli, *Quantum Gravity*, 2004 [551][1]

IN THIS CHAPTER I describe known deficiencies of the Standard Model, including dark matter, dark energy and quantum gravity. I sketch some current directions of research trying to include them in a way compatible with the introductory quote. But without being sure that I include the right directions. Much of the physics later in this chapter is really beyond me. I am walking on rather thin ice here and what I write may be out of date tomorrow. So please take care when you decide to follow me onward.

10.1 DARK MATTER

With the discovery of the Higgs boson, the Standard Model as described in Chapter 9 is internally consistent. It might well be, that there is no fundamentally new physics concerning particles and their interactions [581] until the Planck mass is reached, about 10^{19} GeV, or a million billion times the LHC energy. There might well be no supersymmetry, no multiple Higgs bosons, no structure to quarks and leptons. But even in this case, there is compelling evidence that the matter and energy contents of the Standard Model is incomplete.

---

[1]Reproduced with permission of the Licensor through PLSclear.

The best know example is the astrophysical evidence for the existence of dark matter, a form of matter that emits no light, does not reflect or absorb light. Speculation about elusive forms of matter is as old as philosophical thinking [649]. The first to use gravitational observation to find an object invisible to astronomical telescopes was the mathematician Friedrich Bessel. In the 1840s he used accurate measurements of the stellar parallax[2] to conclude that Sirius had a dark companion, today known as Sirius B. His work also contributed to the discovery of Neptune a few years later. Thus even if they only manifest themselves through gravity, objects do not go undetected.

The American astronomer of Swiss origin Fritz Zwicky is often cited as the discoverer of dark matter. This is arguably also because he had a very strong personality[3]; Freeman Dyson characterises him as "intensely Swiss" [690]. In 1933 he studied the velocities of galaxies in the Coma cluster [256, 279]. He applied the virial theorem to link the mass of the cluster to the width of the velocity distribution measured by Edwin P. Hubble [240]. The virial theorem applied to astrophysics and Zwicky's argument are explained in Focus Box 10.1. From estimates of the number of galaxies in the cluster and their average mass he arrived at a total mass corresponding to an expected velocity variance of about 80 km/s. The observed velocity spread is much larger, 1000 km/s along the line of light, estimated via Doppler shift measurements [240]. The visible mass is thus much too small to explain the variance of velocities. Zwicky concluded [256]: "If this should prove true, one would get the surprising result that dark matter is present in much larger density than luminous matter." In 1936, Sinclair Smith published a similar estimate [272] for the Virgo cluster. He also found a much larger mass than the value estimated from luminosity measurements by Hubble [261].

One of the most convincing proofs for the existence of dark matter are the rotation velocities of objects around their galaxies, thus one step in scale down from the cluster argument of Zwicky and Smith. They are also measured by the electromagnetic Doppler effect, which shifts the line spectra of abundant elements in young stars, like hydrogen and nitrogen, towards the red when the source recedes, towards the blue when it approaches. Spectroscopy for astronomy [601] works according to the principles of diffraction, as explained in Focus Box 5.10. Focus Box 10.2 shows how this is used to measure the radial mass distribution in spiral galaxies. Precision measurements of the velocity as a function of distance from the centre of a galaxy were pioneered in the spectroscopic survey of Andromeda by Vera Rubin and Kent Ford in the early 1970s [405]. In January 2020, it was announced that the Large Synoptic Survey Telescope (LSST), which is currently under construction in Chile, will be named the Vera C. Rubin Observatory.

---

[2]Stellar parallax is the apparent shift of the image of a near star with respect to a distant constellation as the Earth moves around the Sun. It can be used to triangulate the distance to the star.

[3]For a lively account of the debate around Zwicky's way of dealing with colleagues, and his daughter's defence of her father, see the article by Richard Panek in Discover Magazine [575].

The virial theorem states a relation between the mean values of kinetic and potential energies for a system in equilibrium [363]. The mean over time of the kinetic energy T for a system of N particles is:

$$2\langle T\rangle = -\sum_{i=1}^{N}\langle \vec{r}_i\vec{F}_i\rangle = \sum_{i=1}^{N}\langle \vec{r}_i\vec{\nabla}U\rangle$$

where $\vec{F}$ is the force at position $\vec{r}$. The second equation is valid for conservative forces with the potential U. If this potential is radially symmetric and follows a power law $U\propto r^k$, like the gravitational or Coulomb potentials, we get the simplified form:

$$2\langle T\rangle = k\langle U\rangle$$

On the other hand, the kinetic energy of a single particle of mass m and velocity v is $T = \frac{1}{2}mv^2$. If we assume an equal partition of the velocities over the three spatial directions, then $\langle v_x^2 + v_z^2 + v_z^2\rangle = 3\langle v^2\rangle$, where v is the mean velocity along any line of sight. For the total kinetic energy we get $2\langle T\rangle = 3M\langle v^2\rangle$ with the total mass $M = \sum m$. The total potential energy is $U = -\alpha GM^2/R$ where R is the total radius and G is the gravitational constant. The constant α depends on the radial mass distribution. For an unrealistic radial equipartition one would e.g. get a coefficient of 3/5, more realistic coefficients are also of order one. Using the standard deviation $\sigma^2 = \langle v^2\rangle - \langle v\rangle^2$ as a measure for the width of the velocity distribution and $\langle v\rangle = 0$ for an equilibrium system, we get the mass estimator which Fritz Zwicky used for the Coma cluster:

$$\frac{\alpha GM}{R} = 3\sigma^2 \quad ; \quad M = \frac{3}{\alpha}\frac{R}{G}\sigma^2$$

A wider velocity distribution of the galaxies in the cluster thus leads to a higher virial mass estimator. Zwicky estimated the total mass of the Coma cluster as 800 galaxies of a billion solar masses each, thus $M \simeq 1.6 \times 10^{42}$kg. Its radius is about a million light years, i.e. $R \simeq 10^{22}$m. The width of the velocity distribution predicted by the virial theorem is about $\sigma \simeq 100$km/s, while the observed one is of the order of 1000km/s. There is thus much more invisible mass present in the cluster than visible one.

Focus Box 10.1: Virial theorem and Zwicky's dark matter estimate

The Doppler effect causes the wavelength of light from a moving source to be shifted. When the source moves relative to the observer with velocity v, its wavelength λ is shifted from its value at rest by $\Delta\lambda$:

$$\frac{\Delta\lambda}{\lambda} = \sqrt{\frac{1 + v/c}{1 - v/c}} - 1 \equiv z$$

When the source is receding, $z > 0$, corresponding to a red shift for visible light. When it is approaching, $z < 0$ and visible light is blue shifted. Frequency standards like the bright ($n = 3 \rightarrow 2$) line of hydrogen (H-α, $\lambda = 656$nm) or the nitrogen line (NII, $\lambda = 658$nm) are typically emitted by hot gas around young stars. Their wavelength shift is often used to measure source velocities.

An example on how this is done is shown on the right. The photo shows a image of the spiral galaxy NGC3198 seen practically edge-on. Luminous matter right of the galactic centre is receding, approaching on the left. The velocity component along the line of sight is measured. Using the orientation of the rotation axis, indicated by the line, it is converted to rotational velocities.

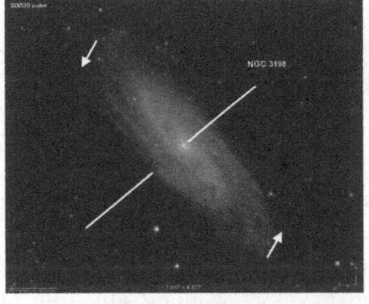

Credit: Sloan Digital Sky Survey

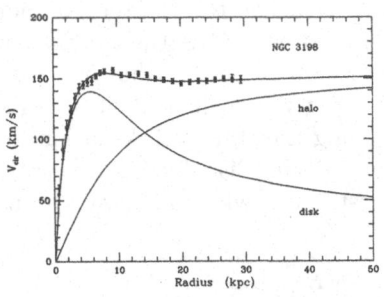

Credit: T.S. van Albada *et al.* [456]

The result is shown on the left [456], the rotational velocity v_{cir} as a function of distance to the galactic centre. Matter is moving with a velocity almost independent of distance from the galactic centre, even at points far outside the luminous region. This is not what one would expect if only luminous matter were present.

Stars and other bright objects in a spiral galaxy are moving around its axis with a roughly circular motion. The velocity as a function of distance to the axis, R, is given by Kepler's third law: $v_{cir}^2(R) = \frac{GM(R)}{R}$. Here $M(R)$ is the total mass included in the path and G the gravitational constant. Up to the galaxy's outer limit, $M(R)$ grows with radius and so does the rotational velocity. Outside the visible limit of a few kpc, one should have $M = $ const, and a decrease of the rotational velocity, $v_{cir} \sim 1/\sqrt{R}$. An example of a modern measurement of the rotational curve is shown above [456]. One observes $v_{cir}(R) \simeq$ const at large R. This means that there is an extended halo of invisible mass reaching far beyond the optical limit. The calculated velocity distribution caused by the galactic disk is shown by the curve, together with the deduced contribution by the halo of dark matter.

Focus Box 10.2: Doppler effect and rotation curves of spiral galaxies

Gravitational lensing, suspected to exist by Newton[4] and by Einstein prior to the formulation of general relativity [496], is a formidable tool to measure the mass of large astronomical structures, even when this mass does not emit light. The principle is based on the fact that light rays follow straight lines in space-time distorted by the gravity of objects. In this manner the gravity of a heavy object in the foreground causes multiple deformed images of an object in the background. Measuring the deformation of the image of a galaxy behind a cluster, for example, one can calculate the mass of the cluster in the foreground. This technique is used in the compound image of Figure 10.1 showing the cluster of galaxies 1E 0657-56, better known as the bullet cluster. What is shown in white are bright objects from optical observation, thus normal matter. Overlaid in light grey is an image of hot gas taken with X-rays, showing normal matter again. And a gravitational lensing image in dark grey, showing the distribution of all matter.

The image shows the situation after two clusters of galaxies collided. The light grey part corresponds to the luminous hot gas, and shows the deformation, deceleration by friction, and the coalescence which is expected after such a collision for ordinary matter. The collision leaves a trail of heated gas behind like a bullet passing through matter. The dark grey cloud shows that the bulk of the matter in both clusters is not luminous and extends farther than the luminous one. The smaller cluster now on the right has traversed the larger on the left with little disturbance for its dark component. Its bulk is in advance with respect to the much less abundant normal matter.

One concludes that dark matter accounts for about 85% of the mass of galaxies and their clusters, but this percentage can vary a lot. A recently discovered galaxy cluster, named Dragonfly 44, is even suspected to contain almost 100% of dark matter [631].

Since it can be observed by different techniques, it is thus clear that dark matter exists. It contributes to the gravitational confinement of galaxies and their clusters, and is probably also involved in the formation of large cosmic structures. Its gravitational interaction is the same as for normal matter, we are thus entitled to call it matter even though it is unconventional since it does not appear to interact otherwise. If it consists of particles, and there is no reason to doubt that it does, they must have the following properties:

- They must be electrically neutral, otherwise they would shine light.

- They are moving at non-relativistic speeds, thus they are probably heavy. Dark matter of this kind is called cold dark matter.

- They are probably fermions, otherwise dark matter would collapse.

- They interact very weakly with each other and with normal matter, otherwise reaction products would be abundant.

---

[4] "Query 1. Do not Bodies act upon Light at a distance, and by their action bend its Rays; and is not this action *(caeteris paribus)* strongest at the least distance?" [2, 3rd book, part I, p. 313]

Figure 10.1 A combined image, covering 7.5×5.4 arcmin, of the galaxy cluster 1E 0657-56, also known as the bullet cluster. The optical image from Magellan and HST shows galaxies as bright white spots. Hot gas in the cluster, which contains the bulk of the normal matter, is shown by the Chandra X-ray Observatory image in light grey. Most of the mass in the cluster is shown in dark grey, as measured by gravitational lensing, the distortion of background images by mass in the cluster. (Credit: X-ray: NASA/CXC/CfA/M.Markevitch et al.; Optical: NASA/STScI; Magellan/U.Arizona/D.Clowe et al.; Lensing Map: NASA/STScI; ESO WFI; Magellan/U.Arizona/D.Clowe et al.)

The search for hypothetical dark matter constituents, which we generically denote with χ, is thus well targeted. They are attacked from three sides simultaneously, as schematically depicted in Figure 10.2:

- One can try to produce dark matter pair-wise at high energy colliders like the LHC. Since they do not interact much with ordinary matter, their presence would be signalled by missing energy and momentum in the final state [654]. So far, the energy of the colliders and/or the sensitivity of experiments has been insufficient for an observation.

- One also searches for the rare interactions of dark matter particles with ordinary matter, in a scattering process. This is sometimes called direct detection. It requires large mass cryogenic detectors sensitive to the tiny

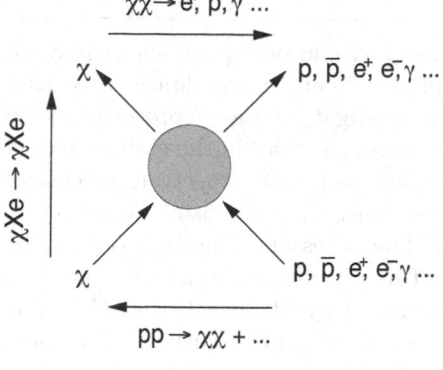

Figure 10.2 Schematic diagram of the three ways to hunt for dark matter particles χ. Top left to right: annihilation of dark matter particles into ordinary matter-antimatter pairs. Left bottom to top: interaction of dark matter with heavy nuclei like Xe. Bottom right to left: Production of dark matter particle pairs in proton-proton collisions.

recoil that a χ-nucleus scattering would cause [685]. Despite the ever increasing sensitivity of experiments, there has been no detection yet.

- Since dark matter particles are their own antiparticles, they can annihilate into pairs of ordinary particles. These can then be observed as an extraordinary contribution to cosmic rays, neutrinos or photons [606]. Since dark matter is everywhere, observatories covering a large proportion of the sky are the most promising options. The Fermi satellite observatory and ground based observatories for high energy gamma rays like H.E.S.S., Veritas and MAGIC are examples for photons. The AMS observatory for charged cosmic rays on the International Space Station is an example for charged cosmic rays (see Section 10.2). The neutrino observatory IceCube at the South Pole may detect neutrinos from dark matter annihilation [640].

Dark matter particles are thus fenced in from all sides and it seems that their discovery is only a matter of time and energy. One cannot ignore, however, that they may escape detection forever. A popular model for undetectable dark matter are the so-called sterile neutrinos [553, 555], right-handed companions of ordinary neutrinos with a modest mass of order 10keV. Because of their wrong spin orientation, their interactions with matter are heavily suppressed, such that they may hide forever. But then again the existence of sterile neutrinos may be detectable by other means [576].

Dark matter may in fact not be the only explanation for the observed phenomena. As has been shown by Mordehai Milgrom [448, 449, 450], one can also modify Newtons law of universal gravitation such that it stays valid at and below solar system dimensions, but is modified at larger scales. Such efforts have also been made to explain away dark energy, which we will discuss in Section 10.3.

10.2 COSMIC RAY REVIVAL

We have seen in the previous section that astrophysical phenomena play an increasingly important role in particle physics. They are no longer used as a source of energetic particles, but serve to investigate energetic processes in the cosmos, or the particle content of the Universe. Energies in the cosmos exceed man-made energies by many orders of magnitude. The spectrum of cosmic rays extends all the way to 10^{20}eV, there must thus be cosmic accelerators which can reach such fantastic energies. The necessary condition is that the size and magnetic field of the cosmic structure must be large enough to contain the cosmic rays. This has first been formulated by Michael Hillas [454]. The product of the magnetic field and the radius of the source must be larger than the ratio of momentum p to charge Ze of the accelerated particle. The latter, often called magnetic rigidity $R = pc/(Ze)$, is thus the relevant kinematic quantity for charged cosmic rays.

These considerations lead to the so-called Hillas diagram shown in Figure 10.3 classifying potential sources for very high energy cosmic rays. In addition, the source class must be able to provide the necessary power and it must be sufficiently abundant to sustain the observed flux of particles. In general, the flux of cosmic ray particles, i.e. the number passing near Earth per square meter and second, decreases as a power law of rigidity, with a slope of roughly three orders of magnitude per decade. This means that the flux drops from about 1 particle per m^{-2} per second at a few times 10GeV to one particle per km^2 per century at EeV energies, 10^{18}eV. Thus the size of the detector determines the observable energy range, not because very high energies cannot be measured, but because they occur so rarely. Above about 5×10^{19}eV, the range of cosmic rays travelling through interstellar matter is severely limited by the large resonant cross section of protons interacting with photons of the cosmic microwave background (p$\gamma \to \Delta^+$, see Sections 9.1 and 10.3). Due to this process, the flux cuts off rather sharply at such energies [617, 642]. The cut-off is called the Greisen-Zatsepin-Kuzmin limit [390, 393]; it is indeed observed [666, 687].

Observables for cosmic rays are their composition, charge and flux as a function of rigidity or energy, at or near Earth. The observed abundances are dominated by hydrogen nuclei, i.e. protons (87%), and helium (9%). A few percent are heavier nuclei and electrons. Antimatter, on the contrary, is surprisingly rare, at the level of per mil for positrons and 10^{-5} for antiprotons. Heavier antinuclei have not been detected with the required certainty to claim an observation. The chemical composition of cosmic ray nuclei [653] is similar to the composition of matter in the solar system [545], with notable exceptions. Due to their larger binding energy, elements with even charge are more abundant in both than those with odd charge. However, the depletion of odd atomic numbers is less pronounced in cosmic rays, since heavier nuclei

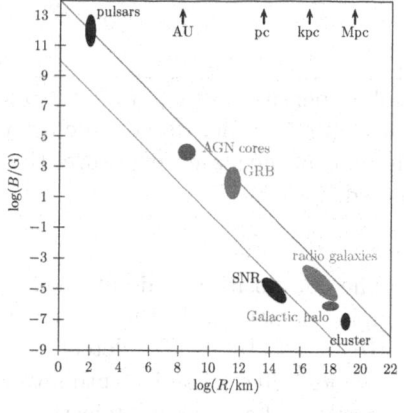

Figure 10.3 The so-called Hillas diagram [574] classifying potential sources for high-energy cosmic rays according to their size R and magnetic field B. Sources which can accelerate protons to $E > 10^{21}$eV should lie above the upper line, sources above the lower line can accelerate iron nuclei up to 10^{20}eV. (Credit: M. Kachelriess [574])

can be split by spallation reactions[5] and provide disfavoured species. This is especially visible in the case of lithium, beryllium and boron [656], intermediate products of stellar nucleosynthesis of the C-N-O group [644, 657]. They are therefore very rare in the solar system matter, much more abundant in cosmic rays due to spallation. All in all, the chemical composition supports that cosmic rays consist of stellar matter, ejected e.g. in supernova explosions. They come to us suffering important alterations of composition and spectra by interactions with interstellar matter as well as diffusion by magnetic fields at all scales, from solar to galactic and intergalactic.

At modest energies, from MeV to TeV, spectrometers and calorimeters can be deployed in space to have direct access to cosmic ray data. Detection of the rare antimatter contents, which is especially sensitive to potential contributions from dark matter, is of course restricted to magnetic spectrometers measuring the sign of the particle charge. Examples of modern space experiments are given in Focus Boxes 10.3 and 10.4. The results concerning ordinary matter and dominant contributions to the cosmic rays flux generally agree well with each other [618, 622, 648, 683]. Precision has vastly improved with respect to previous rather qualitative measurements in this energy range. The spectra at GeV to TeV energies contain important information about the sources, acceleration mechanisms and transport of cosmic rays confined in the Milky Way. For many nuclei, they reveal a new scale of the magnetic rigidity $R = E/Z$ around a few 100 GV, where the slope of the spectrum changes. Spectra of nuclei are fed into numerical models of cosmic ray propagation like GALPROP [510] and interpreted this way in terms of cosmic accelerators and particle transport through the galaxy.

In the energy range from 10^{15} to 10^{18}eV one suspects the transition from galactic to extragalactic origin [579] of cosmic rays. Spectral features like the so-called knee and ankle in the all-particle spectra mark this energy range.

---

[5]Spallation is the split of a heavy nucleus into lighter ones by inelastic scattering. Spallation products fly off with basically the same speed as the original nucleus.

The flux of cosmic ray particles (the number per unit time, surface, solid angle and energy) falls rapidly, roughly by a factor 1000 per decade in energy. The product of detector surface, angular acceptance and exposure time determines what energy range can be observed.

Credit: NASA

The composition is dominated by light nuclei, such that $e^+/p \simeq \bar{p}/e^- \simeq 10^{-4}$. To detect rare species, like cosmic antimatter, excellent background rejection is key. The photo shows the AMS detector installed on the International Space Station since May 2011.

The heart of AMS, as shown in the right graph with a high energy electron traversing, is a magnetic spectrometer, consisting of a permanent magnet with field strength 0.15T and nine layers of high resolution position detectors (Tracker). It determines the magnetic rigidity $pc/(Ze)$ of longitudinally traversing particles in the range MV to TV, as well as the sign of their charge Ze. A series of detector elements identifies the particle species multiple times. The Transition Radiation Detector (TRD) distinguishes light and heavy particles.

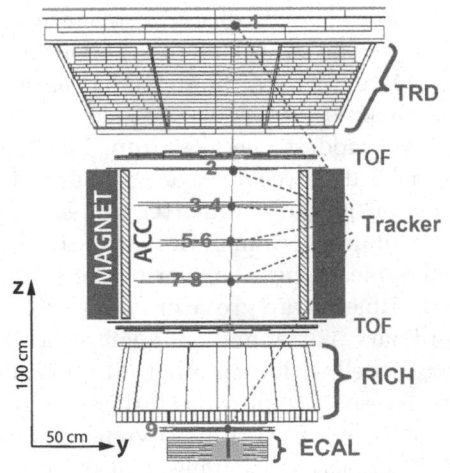

The Time-of-Flight system (TOF) determines the direction of flight, measures the velocity and the charge $|Ze|$. The Ring Imaging Cherenkov counter (RICH) also determines velocity and charge. The particle energy is finally measured in a calorimeter (ECAL) at the very bottom. AMS registered more than 100 billion cosmic rays up to May 2017, the largest sample ever recorded. After a replacement of its cooling system in early 2020, it continues to take data, probably until the end of the ISS lifetime.

Focus Box 10.3: The space borne cosmic ray spectrometer AMS

Credit: JAXA

If one gives up on charge determination, one can no longer distinguish matter from antimatter, but a calorimeter suffices. This is the principle of the CALET detector [688], accommodated on the Japanese Kibo module of the ISS, as shown on the photograph. It measures the energy of $e^{\pm}$ and nuclei in the MeV to TeV range and distinguishes the two by the shape of the shower.

The calorimeter consists of a fine grain imaging part on top and a total absorption calorimeter on the bottom, using scintillating materials as sensitive elements. The event display on the right shows a high energy electron or positron.

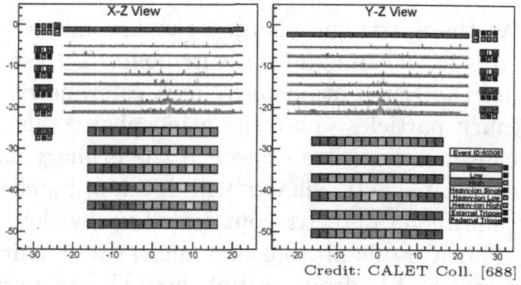

Credit: CALET Coll. [688]

Installed in October 2015, CALET registered 630 million cosmic rays up to May 2018. It will continue data taking for at least five more years.

Focus Box 10.4: Space borne cosmic rays calorimeter CALET

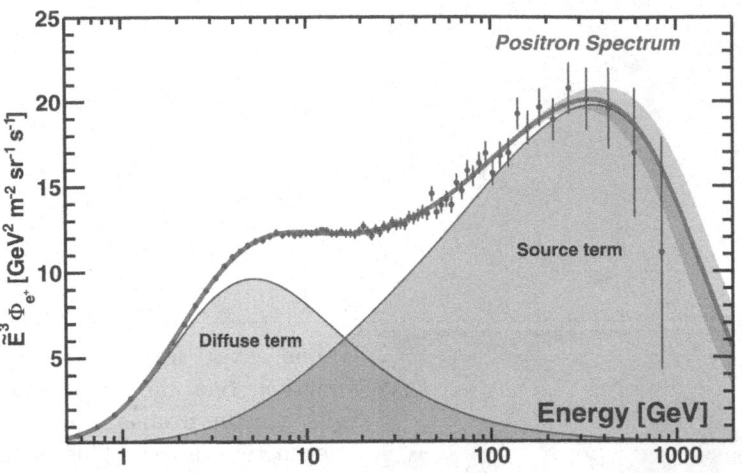

Figure 10.4 The flux $\Phi(E)$ of cosmic ray positrons, multiplied by E^3 for better visibility of the high energy part. The data from the AMS experiment (dots with error bars) [681] are compared to a generic two-component model (thick curve and error band), describing conventional sources (labelled diffuse term) and an additional unconventional source (source term). (Credit: AMS Coll.)

At these extreme energies the low fluxes require square kilometre detector sizes for meaningful measurements. Terrestrial cosmic ray observatories thus use the Earth atmosphere as a calorimetric detector. When high energy primary particles enter the atmosphere, they interact with air molecules and cause an extensive shower. If the primary particle is a photon or electron, an electromagnetic shower will develop; its characteristics are the same as in a calorimeter detector, but scaled up by the low density and light composition of air. Cosmic protons and nuclei cause hadronic showers consisting of a large number of hadrons. All of these shower particles excite air molecules, which then emit fluorescent light in the UV range. It can be detected and located by appropriate cameras. The different shapes of the showers permit to distinguish electromagnetic and hadronic ones. When the shower hits the Earth surface it has a diameter of order kilometre.

One can observe and measure air showers by two methods: fluorescence light and Cherenkov light caused by the relativistic components of the shower; and shower particles arriving on the Earth surface. Both measurement methods are briefly explained in Focus Box 10.5. While lower energy cosmic rays are scrambled by magnetic fields and lose memory of their original direction, very high rigidity particles ought to be stiff enough to roughly point back to their origin, or at least show some anisotropy. Prime suspects for sources are Active Galactic Nuclei, which have to be within $100\mathrm{Mpc} \simeq 3 \times 10^{21}\mathrm{m}$,

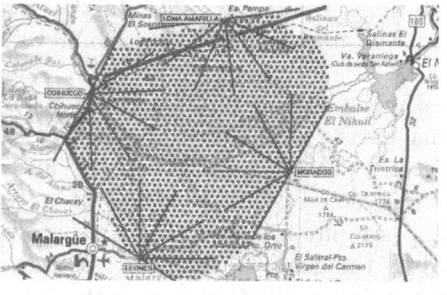

Credit: Pierre Auger Coll. [615], reprinted by permission from Elsevier

The largest air shower detector in operation is the Pierre Auger Cosmic Ray Observatory in Argentina [615]. The lefthand graph above shows the geographic area covered by the detector, about 30 times the size of Paris. The straight lines show the line of sight of 27 florescence detectors installed in four places. The photo on the right shows one overseeing the site. The dots indicate the locations of 1660 surface detector tanks filled with water. One is seen in the lower right corner of the photograph. When relativistic particles from the tail of the air shower pass the water, they emit Cherenkov light, which is detected by photomultipliers. The complementary information from the fluorescence cameras and the surface array is used to determine energy and direction of the primary cosmic ray and the starting altitude of the shower. The latter gives an idea on the identity of the primary.

CTAO/M.-A. Besel/IAC (G.P. Diaz)/ESO   Credit: Akira Okumura

Primary photons or electrons cause an electromagnetic shower. The shower electrons are relativistic and emit Cherenkov light when traversing the atmosphere. The UV light follows the original direction of the primary particle, its intensity is proportional to energy. Cherenkov light is reflected by conventional mirrors. Currently operating telescope arrays to detect these showers are H.E.S.S. in Namibia [530], MAGIC on the Canary islands [554] and VERITAS in Arizona [552]. The upper righthand photograph shows a prototype of the large telescopes for the Cherenkov Telescope Array, CTA [650]. This array, currently under construction, will finally comprise more than 100 telescopes of three sizes in the northern and southern hemisphere, so that the whole sky can be observed. An artist's impression of the array is shown in the lefthand picture.

Focus Box 10.5: Terrestrial cosmic ray and photon obsservatories

because of the Greisen-Zatsepin-Kuzmin limit. Efforts to associate the few extreme energy cosmic rays observed so far with such sources have, however, had limited success [676, 679].

As mentioned in Section 10.1, the presence of light antimatter, positrons or antiprotons, in cosmic rays may be a signal for the annihilation of dark matter particles, e.g. via $\chi\chi \to e^+e^-$, or their decay, $\chi \to e^+/\bar{p}+\dots$. Matter and antimatter would of course be balanced in such processes, but the contribution to matter fluxes will be swamped by ordinary astrophysical processes. Astrophysical antimatter production is suppressed by the general asymmetry between matter and antimatter in the Universe (see Section 10.3), so a non-standard contribution would stick out more. And indeed, both antiprotons [636, 673] and positrons [680, 681] show an excess with respect to what one expects from secondaries coming from the interaction of ordinary cosmic rays. The excess in positrons observed by the AMS cosmic ray observatory, taking data on the International Space Station since 2011, is shown in Figure 10.4. Two contributions are visible: one from conventional sources, ordinary cosmic rays diffusing through the galaxy and producing positrons by interaction, and an extra contribution taking over at some 20GeV energy and cutting off at close to 1TeV. Both contributions are described by an ad-hoc analytical formula, inspired by propagation models at low energies and bringing out the main features of the unconventional contribution.

The origin of the unconventional contribution is not clear, dark matter is a valid candidate but not the only one [664]. Galactic pulsars [600] or other astrophysical sources [623] have been proposed. Recent data from the HAWC observatory [639] indicate that pulsars are not a likely source of the observed excess. More data may allow to find out the real origin of this fascinating new source of cosmic antimatter.

10.3 DARK ENERGY

We have seen in the preceding section that astronomical observation can point towards missing elements in the Standard Model. Now we call cosmology to the witness stand. Cosmology is the theory of the Universe as a whole and at first glance it has little to do with the physics of the smallest structures of matter and forces. However, particle physics has the outrageous ambition to describe everything, from beginning to end, from microcosm to the cosmos itself. So at some point there ought to be synergy when the extreme scales meet. And like all other physics, astrophysics and cosmology should become a branch of applied particle physics.

But as long as there are giant gaps in the Standard Model the roles are rather reversed: cosmology serves to point out missing mechanisms and ingredients in our view of particles and their interactions. These are to be searched for by means of astrophysical observations, accelerator based and non-accelerator particle physics experiments. Complementarity between the two disciplines comes from the fact that large scale phenomena are dominated by gravity, usually negligible at very small scales. Let us start with a quick fly-over of Big Bang cosmology.

Observations indicate that our Universe has been created some ten billion years ago in a Big Bang, out of a point-like singularity. Since then, it has expanded under the influence of the total energy it was born with, plus known and unknown forces. There are common misconceptions about the expansion of the Universe, partially created by sloppy language used by scientists. The expansion bears no resemblance to objects flying apart after an explosion. Instead it is space itself which is expanding, such that even objects at rest move apart. A two-dimensional model like the one shown in Figure 10.5 may help your intuition. Imagine a rubber sheet of size $R \times R$ which is pulled apart at a constant rate, such that both dimensions grow by a constant percentage $\Delta R/R$ per laps of time Δt. The upper figures show what happens to a regular grid of points on the sheet. Now image that you sit at a point on the sheet, like position 1 or 2 in the lower part of Figure 10.5, and measure the recession speed of objects surrounding you. You notice that all other points move away from you, none towards you, independent of where you sit. And you notice that points further away recede at a higher speed than nearer ones. This is due to the fact that $\dot{R} = \Delta R/\Delta t \propto R$. In contrast to objects flying apart after an explosion, there is thus no privileged point from which the expansion starts, no origin. And the speed of recession is proportional to distance, regardless of where the observer is.

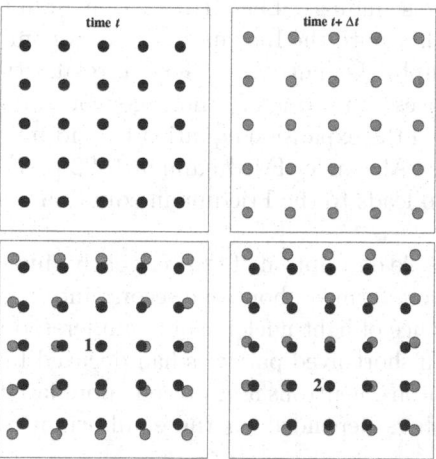

Figure 10.5 A two-dimensional model of space expansion demonstrated with a regular grid of points. Between time t (top left) and $t + \Delta t$ (top right), distances in both dimensions have grown preserving the regularity. We select two different reference points 1 and 2 and keep them fixed (bottom row). For both, all other points move away, the faster the more distant they are. There is thus no privileged reference point when space itself expands. All other objects recede from any observer, with a velocity proportional to distance.

This is what is happening to the Universe. Since the Big Bang, all lengths have constantly grown. Indeed *all* lengths, including e.g. the wavelengths of light. Thus light emitted at a certain time in the history of the Universe has changed to a longer wavelength since then. The temperature of the Universe has thus decreased over time; and conversely, it has been a very hot place in the beginning. Expansion has since cooled it down to its current rather frisky temperature of 2.7 K.

This view of our Universe is supported by a few basic observations:

- the red-shift of line spectra from far away galaxies, which leads to the Hubble-Lemaître law and deviations from it;

- the relative abundances of light elements synthesised during the first few moments of the Universe;

- the cosmic microwave radiation, a black body radiation indicating temperature, and its relation to the geometry of space-time and large scale structures.

We will discuss these pillars of modern cosmology one by one.

The first observation has a complex history about which there is no consensus among historians [603]. In any case, towards the end of the 1920s, Georges Lemaître [220] and Edwin Hubble [231], mostly based on observations of others, noticed a linear relation between the red-shift of spectral lines and distance to the source, indicating that our Universe is expanding. The observation and its relation to space expansion is explained in Focus Box 10.6. Lemaître, a Belgian priest teaching at the Louvain Catholic University, published his findings in an obscure journal; when the English translation appeared in 1931 [241], he deleted the concluding paragraphs about the implications for an expanding Universe, but published them as a separate paper [242]. Thus readers could not realise the precedence for Lemaître's findings, Lemaître himself didn't care. Hubble's name stuck to the linear law until the International Astronomical Union voted in 2018 to rename it Hubble-Lemaître law. General relativity allows to understand the expansion process, in terms of a homogeneous and isotropic Universe of constant density, via the expansion equation first formulated by the Russian and Soviet physicist Alexander Friedmann in 1922 [187]. A simple Newtonian reasoning [546] also leads to the Friedmann equation, as Focus Box 10.7 shows.

Steven Weinberg has given an admirable description of the very early times after the Big Bang [475]. Light nuclei were formed about one second into the life of the Universe. The relative abundance of light nuclei, can be understood with a simple energy argument [522]. All short-lived particles had decayed by that time, only electrons, neutrinos, photons, neutrons and protons populated the very hot soup. Their relative numbers depended on the equilibrium of weak interactions processes:

$$\nu_e \, n \leftrightarrow e^- \, p \;\; ; \;\; \bar{\nu}_e \, p \leftrightarrow e^+ \, n \;\; ; \;\; n \to p \, e^- \, \bar{\nu}_e$$

Due to the expansion, the thermal energy became smaller and smaller. At some point, it fell below the nucleon mass and nucleons became non-relativistic. At sufficiently small temperatures, the reaction rates became too small to maintain an equilibrium between the reactions converting neutrons into protons

The spectral lines from galaxies at cosmological distances, further away than about 15 Mpc $\simeq 4.6 \times 10^{23}$m, are shifted towards longer wavelengths. This is due to the Doppler effect caused by their recession with respect to us, or any other observer in the Universe. The wavelength λ' of a source moving with velocity v/c with respect to the observer is shifted with respect to the emitted wavelength λ in the source rest system:

$$\lambda' = \lambda \sqrt{\frac{1 + v/c}{1 - v/c}} \equiv \lambda(1 + z)$$

The second equation uses the red shift parameter $z = \Delta\lambda/\lambda$ often found in this context. One uses the red shift to measure the velocity of recession. The optical magnitude of a cosmic object can serve as an estimator of distance. The measured flux F of photons diminishes as the square of the distance such that a so-called luminosity distance, $D_L = \sqrt{L/(4\pi F)}$, can be deduced if the total luminosity L is known. For certain classes of objects, like Cepheids or type-Ia supernovae, it can be reliably estimated. What one finds is a linear dependence of the red shift on distance r, called the Hubble-Lemaître law:

$$v = Hr$$

The digram on the right shows the radial velocity determined from red shift versus the luminosity distance for type-Ia supernovae [544]. The small square region in the lower left corner marks the span of Hubble's original diagram from 1929.

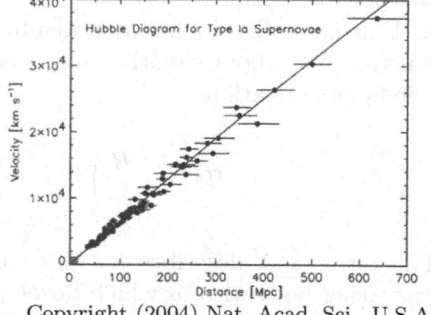

Recession velocities are thus proportional to distance, with a coefficient called the Hubble constant H. This law corresponds to a uniform expansion of all distances, including wavelengths, by a common factor $R(t)$, as demonstrated in Focus Box 10.7. Thus a length r_0 measured today compares to a length measured at another time t like $r(t) = r_0 R(t)$. Without loss of generality one can normalise all lengths to r_0; R thus becomes a generic scale factor applying to all lengths.

Hubble's constant is given by $H = \dot{R}/R$. Contrary to its name, it is not a constant, because of the decelerating effect of gravity and the acceleration by dark energy (see Focus Box 10.7). The Hubble Space Telescope key project [527] finds a current value of $H_0 = 74.0 \pm 1.4$(km/s)/Mpc from relatively near-by sources [667, 668]. Results derived from cosmic microwave background – thus the early Universe – appear to give a lower value [684]. For an up-to-date review of cosmological parameters, see:
http://pdg.lbl.gov/2019/reviews/rpp2019-rev-cosmological-parameters.pdf.

Focus Box 10.6: The Hubble-Lemaître law

The Friedmann equation for the large scale evolution of the Universe assumes a homogeneous and isotropic mass distribution characterised by the mass density ρ. A point mass m at a large distance R from Earth will feel the attraction of a mass $M = \frac{4}{3}\pi R^3 \rho$. Its Newtonian equation of motion will thus be $m\ddot{R} = -G_N M m/R^2$ with Newtons gravitational constant G_N. Integration gives:

$$\frac{1}{2}m\dot{R}^2 - G_N \frac{mM}{R} = \text{const} = -\frac{1}{2}Kmc^2$$

The integration constant K is introduced such that it is dimensionless. The two terms on the left are the kinetic and potential energy of our point mass, so the term on the right is the total energy. For $K = -1$, it is positive and expansion continues forever with the speed of light as its asymptotic velocity. $K = +1$ corresponds to a deficit of total energy, expansion will stop and reverse at some point. The very special case $K = 0$ corresponds to a perfect balance between kinetic and potential energy, with the expansion velocity tending towards zero when the kinetic energy is used up. The Universe will end up in a motionless nirvana. This defines a closed Universe.

As Einstein realised, one can add a further term, with a so-called cosmological constant Λ, without violating general relativity. One thus obtains the complete Friedmann equation:

$$H^2 = \left(\frac{\dot{R}}{R}\right)^2 = \frac{8\pi G_N \rho}{3} - \frac{Kc^2}{R^2} + \frac{\Lambda}{3}$$

The term $\propto \Lambda$ describes a constant curvature of space that the Universe was either born with or which developed dynamically during its history. Integrating the Friedmann equation for the simplest case $K = \Lambda = 0$ in the non-relativistic regime, one obtains $R(t) = (\frac{9}{2}G_N M)^{\frac{1}{3}}t^{\frac{2}{3}}$. The Hubble constant H_0 found today is thus related to the age of the Universe by $H_0^{-1} = R/\dot{R} = \frac{3}{2}t_0$. With this, one finds a rather short age of $t_0 = (9.4 \pm 1.3)$ Gy, much below the age of the oldest measured cosmic objects, estimated to be about 14 Gy. Either $K < 0$ or a non-zero value of the cosmological constant Λ would increase the predicted age.

Again for $K = \Lambda = 0$, we can integrate the Friedmann equation a second time to obtain a critical density ρ_c for which the Universe is just closed:

$$\rho_c = \frac{3H_0^2}{8\pi G_N} \simeq 10^{-26} \frac{\text{kg}}{\text{m}^3}$$

One can measure all observed densities relative to this critical one and sum them up, to see if the closure condition is indeed satisfied or even exceeded.

Focus Box 10.7: The Friedmann equation

(with a gain in energy) and protons into neutrons (requiring kinetic energy). The two baryons thus fell out of equilibrium, and neutrons started to decay freely increasing the number of protons. At the same time, deuteron bound states started to be formed. For a while, formation and dissociation stayed in equilibrium via the reaction $\gamma\,^2\mathrm{H} \leftrightarrow \mathrm{pn}$. When the temperature fell below the threshold, deuterons became stable and helium synthesis started. The last light nucleus formed in this chain process, called primordial nucleosynthesis, is lithium $^7\mathrm{Li}$. Heavier nuclei are synthesised in stars by fusion. We give a quantitative glimpse of primordial nucleosynthesis in Focus Box 10.8. The result is that the number of helium nuclei relative to hydrogen is predicted to be about 8%, as indeed observed in the Milky Way and close to what is found in cosmic rays (see Section 10.2). The hot beginning of the Universe is thus confirmed.

The total abundance of nuclei corresponds to a total baryonic mass density of $\rho_\mathrm{B} = (3.0\pm1.5)\times10^{-28}\mathrm{kg/m}^3$ or a number density of $N_\mathrm{B} = (0.18\pm0.09)/\mathrm{m}^3$. Comparing to the number density N_γ of photons found in the Cosmic Microwave Background, discussed below, on finds:

$$\frac{N_\mathrm{B}}{N_\gamma} \simeq (4 \pm 2) \times 10^{-10}$$

The numbers were of course comparable while the thermal equilibrium between radiation and matter was still intact. The vast majority of baryons has thus disappeared, leaving baryons in a minority of a few in ten billion photons.

In the equilibrium period, the reaction $\mathrm{p\bar{p}} \leftrightarrow \gamma\gamma$ also guaranteed an equilibrium between matter and antimatter. One would thus expect an almost total annihilation between matter and antimatter, occurring after the temperature fell below the threshold required to maintain this equilibrium. Today, only an even tinier but equal fraction of matter and antimatter, $N_\mathrm{B}/N_\gamma = N_{\bar{\mathrm{B}}}/N_\gamma \simeq 10^{-18}$, ought to have survived.

This is evidently not the case, otherwise we would not exist. One observes a matter fraction eight orders of magnitude larger and almost no antimatter. Man-made objects have visited almost all objects in the solar system without exploding in matter-antimatter annihilation, there are thus no close-by anti-planets. In fact also the Milky Way and its supercluster are made of matter. In cosmic rays, the abundance of antiprotons is compatible with secondary production by matter cosmic rays, up to a small excess [636, 673]. Heavier antimatter has not yet been convincingly observed, with current limits on the ratio $N_{\overline{\mathrm{He}}}/N_\mathrm{He} \simeq 10^{-7}$[588]. The only other antimatter which has been observed are positrons, which we discussed in Section 10.1.

There are two ways to reconcile the apparent absence of antimatter and the principles of a hot Big Bang, symmetric between matter and antimatter. There could be an as yet unknown mechanism which has segregated matter and antimatter into separate portions of the Universe, and we happen to live in a matter dominated region. In that case, at some point an antihelium nucleus

When the thermal energy of nuclei, kT with Boltzmann constant k and temperature T, becomes smaller than their mass, $M_\mathcal{N}$, they become non-relativistic. Their relative number is given by the factor $N_\mathrm{n}/N_\mathrm{p} = e^{-\frac{Q}{kT}}$, with $Q = M_\mathrm{n} - M_\mathrm{p} = 1.3$ MeV. At sufficiently low temperatures, $kT \simeq 0.87$ MeV, the rate of inverse reactions re-creating protons out of neutrons, which maintains the equilibrium between the two species, becomes too low and the two populations decouple from each other. Thermal thresholds are of course not very sharp because of the long tail of the Maxwell-Boltzmann distribution towards high energies.

The relative number of nucleons at that point was $N_\mathrm{n}(0)/N_\mathrm{p}(0) = e^{-\frac{Q}{kT_0}} = 0.23$, and neutrons started to decay. At time t there still were $N_\mathrm{n}(t) = N_\mathrm{n}(0)e^{-t/\tau_\mathrm{n}}$, while the number of protons increased to $N_\mathrm{p}(t) = N_\mathrm{p}(0) + N_\mathrm{n}(0)\left(1 - e^{-t/\tau_\mathrm{n}}\right)$, with the lifetime of free neutrons, $\tau_\mathrm{n} = (880 \pm 0.9)$ s. The ratio of neutrons to protons thus evolved like $N_\mathrm{n}(t)/N_\mathrm{p}(t) = (0.23e^{-t/\tau_\mathrm{n}})/(1.23 - 0.23e^{-t/\tau_\mathrm{n}})$. If nothing else had happened, neutrons would have exponentially disappeared from the Universe and only protons would be left.

But protons and neutrons form stable nuclei, and inside those the neutron lifetime is no longer limited. Nucleosynthesis starts with deuteron formation, n p $\leftrightarrow \gamma$ ^{2}H. The binding energy is very large, $E_\gamma > 2.2$ MeV, so is the cross section for this process, $\sigma \simeq 0.1$mb. The formation process thus stayed in equilibrium with the inverse photo-desintegration, down to a temperature of about $kT \simeq 0.05$ MeV. This threshold crossed, ^{2}H became stable and helium synthesis started:

$$^2\text{H n} \rightarrow \, ^3\text{H } \gamma \quad ; \quad ^2\text{H p} \rightarrow \, ^3\text{He } \gamma$$
$$^3\text{H p} \rightarrow \, ^4\text{He } \gamma \quad ; \quad ^3\text{He n} \rightarrow \, ^4\text{He } \gamma$$

At this moment, $t \simeq 400$s, the ratio of neutrons to protons was $r = N_\mathrm{n}(t)/N_\mathrm{p}(t) = 0.14$. Since one needs two protons and two neutrons to synthesise a ^{4}He, the ratio between the helium and hydrogen masses turned out to be (with $M_\mathrm{He} \simeq 4M_\mathrm{p}$, $N_\mathrm{p} = N_\mathrm{H} + 2N_\mathrm{He}$, $N_\mathrm{n} = 2N_\mathrm{He}$):

$$\frac{M_\mathrm{He}}{M_\mathrm{p}} = \frac{4N_\mathrm{He}}{4N_\mathrm{He} + N_\mathrm{p}} = \frac{2r}{1 + r} = 0.25$$

Experimentally one finds indeed a mass ratio of 0.24 ± 0.1 at the level of our Milky Way [509], and $N_\mathrm{He}/N_\mathrm{p}$ about 8%. This agreement between prediction and observation is one of the big triumphs of the Big Bang model. In the same vein, the abundances of ^{2}H, ^{3}He and ^{7}Li are predicted correctly. They exceed significantly their abundances inside stars, and indeed come mostly from primordial nucleosynthesis.

Focus Box 10.8: Primordial nucleosynthesis

might diffuse into the solar system and be observed by a cosmic ray spectrometer like AMS. Even more exciting, if one would find a heavier antinucleus, e.g. anticarbon or antioxygen, one would prove the existence of antistars somewhere in our galaxy, since nuclei heavier than lithium are synthesised in stars. As unlikely as this might be, the importance of such a possibility justifies the experimental effort.

The second possibility is that all primordial antimatter has disappeared during the history of the Universe. In 1967, Andrei Sakharov, one of the fathers of the Soviet nuclear bomb program (see Section 6.5) and laureate of the Nobel peace prize in 1975, formulated the necessary conditions for making antimatter vanish dynamically [395]:

- There must be an interaction which violates the conservation of baryon number.

- In order to tell this interaction the difference between matter and antimatter, the symmetry between them must be violated.

- The disappearance must occur during a period of non-equilibrium, such that antimatter is not recreated by an inverse reaction.

Among these three conditions, only the second has so far been experimentally established by observing the violation of CP symmetry (see Section 9.2) at a very low level. If cosmic ray experiments continue to tighten the limits for surviving antimatter, the fascinating mechanism of baryon number non-conservation must exist. Theoretically conceivable but without experimental support, this clearly challenges experimental particle physics.

The third supporting evidence for the Big Bang origin of the Universe comes from the so-called cosmic microwave background (CMB). This radiation was discovered by Arno Penzias and Robert Wilson in 1965 [387]. They worked at Bell Labs, a private research establishment with no less than nine Nobel laureates among its collaborators. The original intention was to detect radio waves reflected off Echo balloon satellites for communication. They constructed a 6m horn antenna, with a receiver cooled down to liquid helium temperatures, 4K, so that the faint signal from the balloon would be recognisable. Despite all efforts to reduce noise, an irreducible random signal remained. It is a constant radiation from the cosmos in the microwave band, with a peak wavelength of 7.35cm. Its wavelength distribution corresponds to that of a perfect black body (see Section 5.1) with a temperature of about 2.725K.

The origin of this radiation is the time when the Universe passed from an opaque to a transparent state [397, 400]. Before this time, photons and atoms were in equilibrium via the reaction $\gamma H \leftrightarrow e^- p$ and photons could not penetrate very far. When the wavelength of photons fell below the ionisation energy of hydrogen, some 390,000 years after the Big Bang, hydrogen atoms became stable and the photon radiation decoupled from matter. What we see today as cosmic microwave background is the relic of this radiation, cooled down to today's temperature by the expansion of the Universe. Photons from the CMB are the oldest electromagnetic radiation which we can observe.

Since the CMB is the ideal black body radiation, it maps the temperature distribution of the early Universe. The satellites COBE and WMAP of NASA, as well as the Planck satellite of ESA have done so with ever increasing angular and temperature resolution. The two leading scientists of the pioneering COBE mission, George Smoot and John Mather, received the Nobel prize in physics in 2006 for their work on the project. The temperature turns out to be surprisingly homogeneous, globally 2.725K with fluctuations of only 1 part in 100,000. This is hard to understand, since there are regions in the sky far enough apart so that they were never in thermal contact for an expansion with less than the speed of light. There must thus have been a period of more rapid expansion, the so-called inflation period. During this period the Universe expanded with more than the speed of light. There is no contradiction to special relativity, since it is space itself that expands.

The angular structure of the tiny temperature fluctuations can measure the curvature of space-time, since it acts just as a gravitational lens. Roughly speaking, the curvature is measured by the size of cold or hot spots in microwave background temperature maps. The larger the curvature the smaller the physical scale of the spots. The most important result is probably that the observable Universe has no overall curvature, its geometry is flat with percent precision [602]. There is of course local curvature transmitting gravity, but here we are talking about global features, at distances much larger than a galaxy cluster, characterising the observable Universe as a hole. Flat space means that the global geometry is Euclidean, the three angles in any triangle[6] sum up to 180°. The clumps in the distribution of large visible structures in the Universe corroborate this finding [566].

This means that the overall matter and energy density must be exceedingly close, if not identical, to the critical density defined in Focus Box 10.7, of the order of 10^{-26}kg/m^3 or roughly 10 hydrogen atoms per cubic metre. Since the matter density, luminous and dark taken together, is no more than about 32% of the critical density [602], a dominating component must be missing. It has been identified in the late 1990s as dark energy.

Dark energy is again detected using the relation between recession speed and distance. It turns out that for very distant sources, the recession velocity is less than for small distances [502, 516]. Since photons from far away sources were emitted earlier, his means that the Universe is now expanding faster than in the past. For this discovery, Saul Perlmutter, Adam Riess and Brian Schmidt received the 2011 Nobel prize in physics. A few more details on the evidence are given in Focus Box 10.9. For a more in-depth discussion, I rec-

---

[6]If the curvature were positive, the total density would be larger than the critical one, and angles in a triangle would sum up to more than 180° like in spherical geometry. If it were negative, the total density would fall short of the critical one and angles would sum up to less than 180° in a hyperbolic geometry.

ommend Ruth Durrer's article [583]. A comprehensive review of observational methods is given in [596].

Since the gravitational force is attractive, one would assume that the expansion of the Universe slows down as a function of time, but the opposite seems to be the case. There ought be a counteracting form of energy that supplies a repulsive force; it is called dark energy because again it is not subject to other forces. The cosmological constant Λ in the Friedmann equation would represent such an energy. It would mean that space-time was born with the necessary curvature. Cosmology with this constant as well as cold dark matter, abbreviated ΛCDM, is often considered the Standard Model of cosmology.

Alternatively, new scalar fields have been proposed to generate such an effect dynamically. So far, measurements of recession speed as a function of distance have stayed the only indication for the existence of dark energy and it is difficult to find more telling data [570].

While the flatness of space-time requires an additional energy density of non-standard origin, it is not unanimously accepted that an acceleration of space-time expansion exists or is due to an evenly distributed dark energy. The astronomers and theorists around Subir Sarkar of Oxford University, for example, have re-examined the evidence for an accelerated expansion with a large sample of type-Ia supernovae [634]. They argued that the significance for an observed effect is much less that previously thought. More recently, Sarkar and others also observed evidence for an anisotropy in the acceleration [675], roughly aligned with the dipolar anisotropy in the cosmic microwave background. This could indicate a peculiar local flow of our reference system. Thus some additional energy density ought to exist to make space-time flat, but it is unclear if cosmic acceleration has to do with it. What stays a matter of fact is that most of the Universe is dark.

Like in the case of dark matter, one can alternatively invoke modifications of the laws of gravity to explain a possible acceleration of the Universe's expansion without dark energy. The ways to do this and the experimental methods which may distinguish between dark energy and modified gravity are beyond the scope of this book. An excellent review has been recently presented by the Theory Working Group for the Euclid space mission [655], ESA's major future project to look into the expanding Universe.

Gravity thus plays a major role wherever the Standard Model of particle physics is known to be incomplete. Finally understanding this interaction, which men and women have known the longest, is a clear priority.

10.4 QUANTUM GRAVITY

Do we need a quantum theory of gravity? Can we not just live with its coverage by general relativity, as outlined in Section 3.5, summarised by Wheeler's slogan [409]: "Mass tells space-time how to curve, and space-time tells mass how to move?" In general relativity, no particles are exchanged, the gravitational

The luminosity distance, D_L, is related to the astronomical magnitude m, a conventional measure of brightness, by the relation $m(z) = 5\log_{10} D_L + 25$. At non-relativistic velocities, the red shift is $z \simeq v/c$. For a Hubble constant which is independent of time, $z \propto D_L$, one would thus expect a relation between magnitude and red shift of the form $m(z) = a + b\log_{10} z$. If one samples a standard light source, with constant emitted luminosity, one can plot magnitudes against red shifts and test this hypothesis. Such a standard sample is supposedly given by type-Ia supernovae, thermonuclear explosions of white dwarf stars in binary systems. Their explosion releases a well defined amount of energy, such that its luminosity can be used to derive an astronomical standard candle with appropriate corrections.

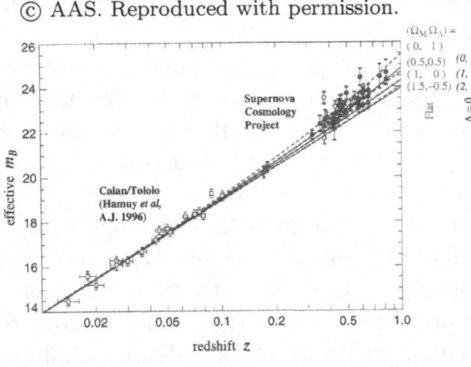

The figure on the left [516] shows the magnitude as a function of red shift for 42 such supernovae with high red shift (full dots). The data accumulate above the straight line characterising the Hubble-Lemaître law in these coordinates. This means that the red shift of distant sources is less than expected, they recede less fast than they should.

At the same time, the light from these sources took longer to arrive than that from closer sources (empty dots) which lie on the Hubble-Lemaître line. Thus recession speed is faster today than in the past. The data are compatible with domination by dark energy (uppermost dashed curve).

According to these data, the global energy density of the Universe is composed of about 5% ordinary matter, 27% dark matter and 68% dark energy. The total density is very close to the critical one. Dark energy is very much diluted, with a density of the order of $(1.67 \times 10^{-27}$ kg/m$^3)$, such that the solar system up to Pluto contains only about 6 tons. But its distribution is supposedly uniform, such that it dominates evolution at cosmic scales.

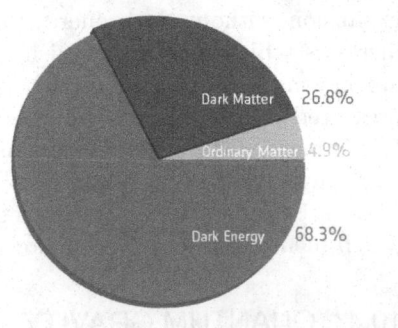

Credit: ESA Science and Technology

Focus Box 10.9: Dark energy

field is the metric of space-time itself. Macroscopic measurements confirm the validity of general relativity [556], including tools that you use every day like the Global Positioning System [541]. After all, if quantum effects for this exceedingly weak interaction exist, they are only relevant when we approach the extreme distance of the Planck scale[7], $l_P = \sqrt{\hbar G/c^3} \simeq 1.6 \times 10^{-35}$m, where all interactions are of comparable strength. We can obviously safely ignore the practical consequences of gravity at any energy reachable by even the most powerful accelerators. But why should gravity as the only force not be quantised? And why should we refrain from attempting to unify its description with that of the other elementary forces?

There are two complementary approaches to quantum gravity, which relate to these two questions differently. The first one, very much in line with the arguments we have presented in this book so far, treats gravity as yet another force. It tries to construct a quantum "theory of everything" on curved space-time. Gravity would then be transmitted by fluctuations of the metric over a flat four-dimensional Minkowski background metric, the one we used to describe the other Standard Model forces. The direct implementation of this idea suffers from the fact that the gravitational constant G is not dimensionless (see Section 2.1). It has the dimension of an inverse squared energy, leading to infinities. The theory is not perturbatively renormalisable in the sense we superficially discussed in Section 9.2, it makes no predictions when using the usual tools. A successful implementation requires to eliminate infinities in a different way. If particles are not point-like objects, but extended one-dimensional structures called strings, non-renormalisable infinities disappear. String theory [677] sees particles as excitations of these strings, including a tensor-like particle as the potential quantum of gravity, the graviton. String theory requires extra dimensions beyond the four of Minkowski space; we would then live on a four-dimensional surface of this space. Continuity of the probability density requires the extra dimensions to be compact: a detour of the fields through an extra dimension would come back to the same amplitude and phase. Otherwise particles could disappear somewhere and reappear in a different space-time point.

Gravitons, the quanta of such a field theory of gravity, are exceedingly difficult if not impossible to detect. One should not confuse the recently detected gravitational waves with the action of gravitons. These waves relate to gravitons like an intense laser beam to a single photon[8]. And they exist. In 1982, Joseph Taylor Jr. and Joel M. Weisberg found that the orbit of a pulsar gyrating around a neutron star was slowly shrinking due to energy loss by gravitational waves [440]. The waves themselves were first detected in 2015 by the twin Laser Interferometer Gravitational-wave Observatory (LIGO) detectors, located in Livingston, Louisiana, and Hanford, Washington, USA as well

---

[7]A simple argument to derive this scale is found in Sabine Hossenfelder's review of the experimental search for quantum gravity [585].

[8]The number of photons emitted by a 1mW laser pointer is of the order of 10^{15} per second.

as the Virgo interferometer near Pisa, Italy [630]. Given the incredible smallness of the effect, this is an admirable experimental break-though all by itself. However, if gravitons exist, gravitational waves correspond to immense numbers of them, $10^{38}/\nu^3$ according to one estimate [565], with ν the frequency of the wave. Single gravitons, to be detected in analogy to the photoelectric effect or Compton scattering for photons, have an incredibly small cross section. Even a detector of the size of Jupiter would not be able to detect a single graviton from conceivable astrophysical sources [565]. This led Freeman Dyson to argue that a graviton may be undetectable [597] in principle. He writes [559, p. 221][9]:

> According to my hypothesis, the gravitational field described by Einstein's theory is a purely classical field without any quantum behavior. Gravitational waves exist, and can be detected, but they are classical waves and not collections of gravitons. If this hypothesis is true, we have two separate worlds, the classical world of gravitation and the quantum world of atoms, described by separate theories. The two theories are mathematically different and cannot be applied simultaneously. But no inconsistency can arise from using such theories, because any differences between their predictions are physically undetectable.

This would mean to give up on the reductionist picture of Figure 1, reluctantly giving up the dream of a final unified theory. But all may not be lost for experimentalists and effects of quantum gravity other than the detection of single gravitons may be observable, especially in the vicinity of black holes [585].

A complementary theoretical approach does not try to unify forces but takes general relativity more seriously. Its proponents argue that space-time may itself be quantised. Gravity thus stays a property of space-time and is not just another force. This approach is represented by loop quantum gravity [508]. Here quantisation means that space and time themselves are quantised, that there is a minimum distance. At dimensions comparable to the Planck length, space and time are then a network with an extremely fine mesh made of so-called loops[10]. As a consequence, surface and volume are also quantised and divergences coming from infinite densities disappear; the minimum distance provides a natural cut-off for infrared divergences. There is thus no need for a perturbative approach, neither to gravity nor its matter couplings. Space-time is a dynamical object of the theory, with no need of a fixed background metric. I admit that the mathematics of this fascinating approach is not within my reach and refrain from venturing further.

What needs attention, though, is the different role of time in the two approaches sketched above. In the field theoretical approach, time keeps its role as the principle parameter of system evolution. Matter and interactions

---

[9] © F. Dyson, with permission of New York Review Books.
[10] Please do not confuse these loops of space-time with the particle loops in vacuum polarisation, shown in Figures 9.12 and 9.13.

evolve in space as a function of time. In the loop approach to quantum gravity, time loses that special role. In fact Carlo Rovelli argues that it disappears from consideration [660]. It has no unity, since it is different for every observer; no present which can be unambiguously defined; and no direction since past and future are equivalent, CP violation not withstanding. It is not a parameter, but an aspect of a dynamical field. And it is intrinsically ours, a human concept.

It is thus entirely possible that the reductionist diagram, Figure 1 in Chapter 1, which served as a guiding line throughout much of this book, has a basic flaw. If gravity is not a force like the others, it will not join the Standard Model of particle physics but stay a property of space-time, probably together with dark energy. Maybe the whole reductionist approach is facing its limits and I will at some point have to reluctantly join Freeman Dyson in his harsh judgement [559, p. 13][11]:

> A reductionist philosophy, arbitrarily proclaiming that the growth of understanding must go only in one direction, makes no scientific sense. Indeed, dogmatic philosophical beliefs of any kind have no place in science.

FURTHER READING

Lars Bergström and Ariel Goobar, *Cosmology and Particle Astrophysics*, Wiley, 1999.

Hans Reichenbach, *The Direction of Time*, Dover Publications, 1999.

Donald H. Perkins, *Particle Astrophysics*, Oxford University Press, 2003.

Steven Weinberg, *Cosmology*, Oxford University Press, 2008.

Michele Maggiore, *Gravitational Waves*, Oxford University Press, 2008 (Vol. 1) and 2018 (Vol. 2).

Maurizio Spurio, *Particles and Astrophysics*, Springer, 2015.

Thomas K. Gaisser, Ralph Engel and Elisa Resconi, *Cosmic Rays and Particle Physics*, Cambridge University Press, 2016.

Sabine Hossenfelder (Edt.), *Experimental Search for Quantum Gravity*, Springer, 2018.

Carlo Rovelli, *The Order of Time*, Allan Lane, 2018.

Elias Kiritsis, *String Theory in a Nutshell*, Princeton University Press, 2019.

---

[11] © F. Dyson, with permission of New York Review Books.

Bibliography

[1] I. Newton. *Philosophiæ Naturalis Principia Mathematica*. London, 1687.

[2] I. Newton. *Opticks: or, A Treatise of the Reflexions, Refractions, Inflexions and Colours of Light*. London, 1704.

[3] D. Bernoulli. *Hydrodynamica*. Basel, 1738.

[4] A. Lavoisier. *Méthode de nomenclature chimique*. Paris, 1787.

[5] J.-L. Lagrange. *Mécanique analytique*. Paris, 1788.

[6] A. Lavoisier. *Traité élémentaire de la chimie*. Paris, 1789.

[7] T. Young. The Bakerian Lecture: On the Theory of Light and Colours. *Phil. Trans. Royal Soc. London*, 92:12–48, 1801.

[8] J. Dalton. On the Absorption of Gases by Water and other Liquids. *Memoirs of the Lit. and Phil. Soc. of Manchester*, 6:271–287, 1805.

[9] J. Dalton. *A New System of Chemical Philosophy*. Manchester, 1808.

[10] J.L. Gay-Lussac. Mémoire sur la combinaison des substances gazeuses, les unes avec les autres. *Mémoires de la Société d'Arcueil*, 2:207–234, 1809.

[11] A. Avogadro. Essai d'une manière de déterminer les masses relatives des molecules élémentaires des corps, et les proportions selon lesquelles elles entrent dans ces combinaisons. *Journal de Physique*, 73:58–76, 1811.

[12] H. Kater. An account of experiments for determining the length of the pendulum vibrating seconds in the latitude of London. *Phil. Trans. R. Soc. London*. 2:83–85, 1818.

[13] R. Brown. A brief account of microscopical observations made in the months of June, July and August 1827, on the particles contained in the pollen of plants; and on the general existence of active molecules in organic and inorganic bodies. *Phil. Mag.* 4:161–173, 1828.

[14] R. Brown. Additional Remarks on Active Molecules. *Phil. Mag.* 6:161–166, 1829.

[15] J.R. von Mayer. Bemerkungen über die Kräfte der unbelebten Natur. *Ann. der Chem. und Pharm.* 42:233–240, 1842.

[16] J.P. Joule. On the calorific effects of magneto-electricity, and on the mechanical value of heat. *Phil. Mag.* 23:435–443, 1843.

[17] R. Clausius. Über die bewegende Kraft der Wärme und die Gesetze, welche sich daraus für die Wärmelehre selbst ableiten lassen. *Ann. Phys.* 155:368–397, 1850.

[18] J.P. Joule. On the Mechanical Equivalent of Heat. *Phil. Mag.* 140:61–82, 1850.

[19] H. Fizeau. Sur les hypothèses relatives à l'éther lumineux. *Compt. Ren. Hebd. Seances Acad. Sci.* 33:349–365, 1851.

[20] W.J.M. Rankine. XVIII. On the general law of the transformation of energy. *Phil. Mag. Ser. 6*, 5:106–117, 1853.

[21] A. de la Rive. *Treatise on Electricity in Theory and Practice*. Longman, Brown, Green and Longmans, 1853.

[22] A.J. Ångström. Optical researches. *Phil. Mag. Ser. 4*, 9:327–342, 1855.

[23] J.C. Maxwell. *On the Stability of the Motion of Saturn's Rings*. Macmillan and Co., Cambridge, 1859.

[24] G. Kirchhoff. Ueber das Verhältniss zwischen dem Emissionsvermögen und dem Absorptionsvermögen der Körper für Wärme und Licht. *Ann. Phys.* 19:275–301, 1860.

[25] G. Kirchhoff and R. Bunsen. Chemical analysis by spectrum-observations. *Phil. Mag. Ser. 4*, 20:89–109, 1860.

[26] J.C. Maxwell. Illustrations of the dynamical theory of gases. Part I. On the motions and collisions of perfectly elastic spheres. *Phil. Mag.* 19:19–32, 1860.

[27] J.C. Maxwell. Illustrations of the dynamical theory of gases. Part II. On the process of diffusion of two or more kinds of moving particles among one another. *Phil. Mag.* 20:21–37, 1860.

[28] J.A.R. Newlands. On Relations among the Equivalents. *Chemical News*, 7:70–72, 1863.

[29] C. Wiener. Erklärung des atomistischen Wesens des tropfbar-flüssigen Körperzustandes, und Bestätigung desselben durch die sogenannten Molecularbewegungen. *Ann. Phys.* 194:79–94, 1863.

[30] J.A.R. Newlands. On Relations among the Equivalents. *Chemical News*, 10:94–95, 1864.

[31] J.A.R. Newlands. Relations among the Equivalents. *Chemical News*, 10:59–60, 1864.

[32] J.W. Hittorf and J. Plücker. On the Spectra of Ignited Gasses and Vapours with Especial Regard to the Same Elementary Gaseous Substance. *Phil. Trans. Royal Soc.* 155:1, 1865.

[33] J.A.R. Newlands. On the Law of Octaves. *Chemical News*, 12:83, 1865.

[34] Anonymous. Minutes of the Meeting on March 1, 1866. *Proc. of Soc., Chem. Soc.* 13:113, 1866.

[35] D. Mendeleev. Ueber die Beziehungen der Eigenschaften zu den Atomgewichten der Elemente. *Z. f. Chem.* 12:405–406, 1869.

[36] J.L. Meyer. Die Natur der chemischen Elemente als Funktion ihrer Atomgewichte. *Ann. Chem. Suppl.* 7:354–364, 1870.

[37] L. Boltzmann. Weitere Studien über das Wärmegleichgewicht unter Gasmolekülen. *Sitzungsber. der Kaiserli. Akad. der Wiss. in Wien, math.-natur. Classe*, 66:275–370, 1872.

[38] J. C. Maxwell. *A Treatise on Electricity and Magnetism.* Macmillan and Co., publishers to the University of Oxford, 1873.

[39] Anonymous Book Review. Clerk-Maxwell's Electricity and Magnetism. *Nature* 7:478, 1873.

[40] A. de la Rive Eulogy. *Proceedings of the American Academy of Arts and Sciences, 669th Meeting (May 12, 1874).* 1874.

[41] E. Goldstein. Vorläufige Mittheilungen über elektrische Entladungen in verdünnten Gasen. In *Monatsberichte der Königl. Preuss. Akad. der Wiss.* 279. 1876.

[42] L. Boltzmann. Über die Beziehung zwischen dem zweiten Hauptsatz der mechanischen Wärmetheorie und der Wahrscheinlichkeitsrechnung respektive den Sätzen über das Wärmegleichgewicht. *Sitzungsber. der Kaiserli. Akad. der Wiss. in Wien, math.-natur. Classe*, 76:373–435, 1877.

[43] A.M. Mayer. On the Morphological Laws of the Configurations Formed by Magnets Floating Vertically and Subjected to the Attraction of a Superposed Magnet; with Notes on Some of the Phenomena in Molecular Structure Which These Experiments May Serve to Explain and Illustrate. *Am. J. of Science*, 16:247–256, 1878.

[44] M. Planck. *Über den zweiten Hauptsatz der mechanischen Wärmetheorie.* Ackermann, München, 1879.

[45] A.A. Michelson. The Relative Motion of the Earth and the Luminiferous Ether. *Am. J. of Science*, 22:120–129, 1881.

[46] J.J. Balmer. Notiz über die Spectrallinien des Wasserstoffs. *Ann. Phys.* 25:80–87, 1883.

[47] H. Hertz. Versuche über die Glimmentladung. *Ann. Phys.* 19:782–816, 1883.

[48] A.A. Michelson and E.W. Morley. Influence of Motion of the Medium on the Velocity of Light. *Am. J. of Sci.* 31:377–386, 1886.

[49] H. Hertz. Über einen Einfluss des ultravioletten Lichtes auf die electrische Entladung. *Ann. Phys.* 267:983–1000, 1887.

[50] H. Hertz. Über sehr schnelle elektrische Schwingungen. *Ann. Phys.* 267:421–448, 1887.

[51] A.A. Michelson and E.W. Morley. On the Relative Motion of the Earth and the Luminiferous Ether. *Am. J. of Science*, 34:333–345, 1887.

[52] W. Hallwachs. Über den Einflug des Lichtes auf elektrostatisch geladene Körper. *Ann. Phys.* 33:301, 1888.

[53] J.R. Rydberg. On the structure of the line-spectra of the chemical elements. *Phil. Mag.* 29:331–337, 1890.

[54] R. Eötvös. Über die Anziehung der Erde auf verschiedene Substanzen. *Math. und Naturwiss. Berichte aus Ungarn*, 8:65–68, 1891.

[55] O. Lummer and F. Kurlbaum. Bolometrische Untersuchungen. *Ann. Phys.* 46:204–224, 1892.

[56] W. Weber. *Wilhelm Weber's Werke. Bd. 3 Galvanismus und Elektro-dynamik.* Herausgegeben von der Königlichen Gesellschaft der Wissenschaften zu Göttingen, Julius Springer, Berlin, Germany, 1892/94

[57] W. Wien. Eine neue Beziehung der Strahlung schwarzer Körper zum zweiten Hauptsatz der Wärmetheorie. *Sitzungsber. der Königl. Pr. Akad. der Wiss. zu Berlin*:55, 1893.

[58] G.J. Stoney. Of the "Electron" or Atom of Electricity. *Phil. Mag. 38*:418–420, 1894.

[59] J.J. Thomson. On the Velocity of Cathode Rays. *Phil. Mag. 38*:358–385, 1894.

[60] H. A. Lorentz. *Versuch einer Theorie der electrischen und optischen Erscheinungen.* E.J. Brill, Leiden, 1895.

[61] J.B. Perrin. Nouvelles propriétés des Rayons Cathodiques. *Compt. Ren. Hebd. Seances Acad. Sci.* 121:1130, 1895.

[62] W.C. Röntgen. Über eine neue Art von Strahlen, I. Mittheilung. *Sitzungsberichte der phys.-med. Ges. zu Würzburg*:132, 1895.

[63] L. Boltzmann. *Vorlesungen über Gastheorie, Teil I und II.* J.A. Barth, Leipzig, 1896 and 1898

[64] H. Becquerel. Emission des radiations nouvelles par l'Uranium métallique. *Compt. Rend. Hebd. Seances Acad. Sci.* 122:1086, 1896.

[65] H. Becquerel. Sur diverses propriétés des rayons uraniques. *Compt. Rend. Hebd. Seances Acad. Sci.* 123:855, 1896.

[66] H. Becquerel. Sur les propriétés différentes des radiations invisibles émises par les sels d'Uranium, et du rayonnement de la paroi anti-cathodique d'un tube de Crookes. *Compt. Rend. Hebd. Seances Acad. Sci.* 122:689, 1896.

[67] H. Becquerel. Sur les radiations invisibles émises par les corps phosphorescence. *Compt. Rend. Hebd. Seances Acad. Sci.* 122:501, 1896.

[68] H. Becquerel. Sur les radiations émises par phosphorescence. *Compt. Rend. Hebd. Seances Acad. Sci.* 122:420–421, 1896.

[69] H. Becquerel. Sur les radiations émises par phosphorescence. *Compt. Rend. Hebd. Seances Acad. Sci.* 122:762, 1896.

[70] H. Becquerel. Sur quelques propriétés des radiations invisibles émises par divers corps. *Compt. Rend. Hebd. Seances Acad. Sci.* 122:559, 1896.

[71] W.C. Röntgen. Über eine neue Art von Strahlen, II. Mittheilung. *Sitzungsberichte der phys.-med. Ges. zu Würzburg*:110, 1896.

[72] W. Kaufmann. Die magnetische Ablenkbarkeit der Kathodenstrahlen und ihre Abhängigkeit vom Entladungspotential. *Ann. d. Phys. u. Chem.* 61:544–552, 1897.

[73] W. Kaufmann and A. Aschkinass. Über die Deflexion der Kathodenstrahlen (und Nachtrag). *Ann. d. Phys. u. Chem.* 62:588–598, 1897.

[74] J.J. Thomson. Cathode Rays. *Proceedings of the Royal Institution*, 15:419, 1897.

[75] J.J. Thomson. Cathode Rays. *Phil. Mag.* 44:295, 1897.

[76] E. Wiechert. Experimentelles über die Kathodenstrahlen). *Schriften der Phys.-ökon. Ges. zu Königsberg in Pr.* 3:3–18, 1897.

[77] P. Zeeman. Doublets and triplets in the spectrum produced by external magnetic forces. *Phil. Mag.* 44:55–60, 1897.

[78] P. Zeeman. On the influence of magnetism on the nature of the light emitted by a substance. *Phil. Mag.* 43:226–239, 1897.

[79] P. Zeeman. The effect of magnetisation on the nature of light emitted by a substance. *Nature*, 55:347, 1897.

[80] G.C. Schmidt. Über die von den Thorverbindungen und einigen anderen Substanzen ausgehende Strahlung. *Ann. Phys. Chem.* 65:141, 1898.

[81] M. Sklodowska Curie. Rayons émis par les composés de l'uranium et du thorium. *Compt. Rend. Hebd. Seances Acad. Sci.* 126:1101, 1898.

[82] P. Curie, M. Curie, M.G. Bémont. Sur une nouvelle substance fortement radio-active, contenue dans la pechblende. *Compt. Rend. Hebd. Seances Acad. Sci.* 127:1215–1217, 1898.

[83] P. Curie, M. Curie, M.G. Bémont. Sur une substance nouvelle radio-active contenue dans la pechblende. *Compt. Rend. Hebd. Seances Acad. Sci.* 127:175, 1898.

[84] J.J. Thomson. On the charge of electricity carried by the ions produced by Röntgen rays. *Phil. Mag.* 46:528, 1898.

[85] J.J. Thomson. *The Discharge of Electricity through Gases.* Scribner, 1898.

[86] E. Rutherford. Uranium radiation and the electrical conduction produced by it. *Phil. Mag. 47*:109, 1899.

[87] O. Lummer and E. Pringsheim. Die Verteilung der Energie im Spektrum des schwarzen Körpers. *Verh. der DPG*, 1:23–41, 1899.

[88] O. Lummer and E. Pringsheim. Die Verteilung der Energie im Spektrum des schwarzen Körpers und des blanken Platins. *Verh. der DPG*, 1:215–235, 1899.

[89] F. Paschen. Über die Vertheilung der Energie im Spectrum des schwarzen Körpers bei niederen Temperaturen. *Sitzungsber. der Königl. Pr. Akad. der Wiss. zu Berlin*:405–420, 1899.

[90] F. Paschen and H. Wanner. Eine photometrische Methode zur Bestimmung der Exponentialkonstanten der Emissionsfunktion. *Sitzungsber. der Königl. Pr. Akad. der Wiss. zu Berlin*:5–11, 1899.

[91] J.J. Thomson. On the masses of the ions in gases at low pressures. *Phil. Mag. 48*:547, 1899.

[92] H. Becquerel. Sur la dispersion du rayonnement du radium dans un champ magnétique. *Compt. Rend. Hebd. Seances Acad. Sci.* 130:372–276, 1900.

[93] H. Becquerel. Sur la transparence de l'aluminium pour le rayonnement du radium. *Compt. Rend. Hebd. Seances Acad. Sci.* 130:1154, 1900.

[94] O. Lummer and E. Pringsheim. Über die Strahlung des schwarzen Körpers für lange Wellen. *Verh. der DPG*, 2:163–180, 1900.

[95] P. Villard. Sur la réflexion et la réfraction des rayons cathodiques déviables du radium. *Compt. Rend. Hebd. Seances Acad. Sci.* 130:1010, 1900.

[96] P. Villard. Sur le rayonnement du radium. *Compt. Rend. Hebd. Seances Acad. Sci.* 130:1178, 1900.

[97] M. Planck. Zur Theorie des Gesetzes der Energieverteilung im Normalspektrum. *Verh. der DPG*, 2:202, 1900.

[98] Lord Rayleigh. Remarks upon the law of complete radiation. *Phil. Mag.* 49:539–540, 1900.

[99] J.J. Thomson. Some Speculations as to the Part Played by Corpuscles in Physical Phenomena. *Nature 62*:31–32, 1900.

[100] H. Wanner. Photometrische Messung der schwarzen Strahlung. *Ann. Phys.* 2:141, 1900.

[101] H. Rubens and F. Kurlbaum. On the Heat Radiation of Long Wave-Length Emitted by Black Bodies at Different Temperatures. *Astrophys. J.* 14:335–348, 1901.

[102] W. Kaufmann. Die Entwicklung des Elektronenbegriffs. *Phys. Z.* 3:9–15, 1901.

[103] Right. Hon. Lord Kelvin. Nineteenth Century Clouds over the Dynamical Theory of Heat and Light. *Phil. Mag.* 2:1–40, 1901.

[104] M. Planck. Ueber das Gesetz der Energieverteilung im Normalspectrum. *Ann. Phys.* 4:553–563, 1901.

[105] W. Voigt. Ueber das Elektrische Analogon des Zeemaneffectes. *Ann. Phys.* 309:197–208, 1901.

[106] W. Kaufmann. Über die lichtelektrische Wirkung. *Ann. Phys.* 8:149–198, 1902.

[107] H. Becquerel. Recherches sur une Propriété Nouvelle de la Matière. Activité Radiante Spontanée ou Radioactivité de la Matière. *Mémoires de l'Académie des Sciences,* 46:1–360, 1903.

[108] E. Rutherford. The Magnetic and Electric Deviation of the Easily Absorbed Rays from Radium. *Phil. Mag.* 21:669, 1903.

[109] E. Rutherford and F. Soddy. (Radioactive Change). *Phi. Mag.* 5:576–591, 1903.

[110] J.J. Thomson. On the Charge of Electricity carried by a Gaseous Ion. *Phil. Mag.* 5:346–355, 1903.

[111] F.T. Trouton and R.H. Noble. The Forces Acting on a Charged Condenser moving through Space. *Proc. Roy. Soc.* 37:132–133, 1903.

[112] F.T. Trouton and R.H. Noble. The mechanical forces acting on a charged electric condenser moving through space. *Phil. Trans. Royal Soc.* A 202:165–181, 1903.

[113] H.A. Wilson. A determination of the charge on the ions produced in air by Röntgen rays. *Phil. Mag.* 5:429, 1903.

[114] H.A. Lorentz. Electromagnetic phenomena in a system moving with any velocity smaller than that of light. *Proc. Roy. Neth. Acad. od Arts and Sci. (KNAW),* 6:809–831, 1904.

[115] H. Nagaoka. Kinetics of a system of particles illustrating the line and the band spectrum and the phenomena of radioactivity. *Phil. Mag.* 7:445–455, 1904.

[116] J.J. Thomson. On the structure of the atom: an investigation of the stability and periods of oscillation of a number of corpuscles arranged at equal intervals around the circumference of a circle; with applications of the result to the theory of atomic structure. *Phil. Mag.* 7:237–265, 1904.

[117] A. Einstein. Ist die Trägheit eines Körpers von seinem Energieinhalt abhängig? *Ann. Phys.* 18:639–641, 1905.

[118] A. Einstein. Zur Elektrodynamik bewegter Körper. *Ann. Phys.* 17:891–921, 1905.

[119] A. Einstein. Über die von der molekularkinetischen Theorie der Wärme geforderte Bewegung von in ruhenden Flüssigkeiten suspendierten Teilchen. *Ann. Phys.* 322:549–560, 1905.

[120] A. Einstein. Über einen die Erzeugung und Verwandlung des Lichtes betreffenden heuristischen Gesichtspunkt. *Ann. Phys.* 17:132–148, 1905.

[121] H. Poincaré. Sur la dynamique de l'électron. *Compt. Rend.* 140:1504–1508, 1905.

[122] A. Einstein. Zur Theorie der Lichterzeugung und Lichtabsorption. *Ann. Phys.* 20:199–, 1906.

[123] Ph. Lenard. *Über Kathodenstrahlen.* Wentworth Press (reprint 2018), 1906.

[124] M. Smoluchowski. Zur kinetischen Theorie der Brownschen Molekularbewegung und der Suspensionen. *Ann. Phys.* 21:756–780, 1906.

[125] P. Langevin. Sur la théorie du mouvement Brownien. *Compt. Rend.* 146:530–533, 1908.

[126] E. Rutherford and H. Geiger. An electrical method of counting the number of α particles from radioactive substances. *Proc. Royal Soc. A,* 81:141–161, 1908.

[127] H. Geiger and E. Marsden. On a Diffuse Reflection of the α-Particles. *Proc. Roy. Soc.* 19:495, 1909.

[128] H. Minkowski. *Raum und Zeit.* B.G. Teubner, Berlin und Leipzig, 1909.

[129] E. Rutherford and T. Royds. The nature of the α particle from radioactive substances. *Phil. Mag.* 17:281–286, 1909.

[130] Th. Wulf. Über die in der Atmosphäre vorhandene Strahlung von hoher Durchdringungsfähigkeit. *Phys. Z.* 10:152–157, 1909.

[131] F. Ehrenhaft. Über die Messung von Elektrizitätsmengen, die kleiner zu sein scheinen als die Ladung des einwertigen Wasserstoffions oder Elektrons und von dessen Vielfachen abweichen. *Kais. Akad. Wiss. Wien, Sitzber. math.-nat. Kl.* 119:815–867, 1910.

[132] A. Gockel. Luftelektrische Beobachtungen bei einer Ballonfahrt. *Phys. Z.* 11:280, 1910.

[133] A. Haas. Der Zusammenhang des Planckschen elementaren Wirkungsquantums mit den Grundgrossen der Electronentheorie. *Jahrbuch der Radioaktivitaet und Elektronik,* 7:261–268, 1910.

[134] R.A. Millikan. A New Modification of the Cloud Method of determining the Elementary Electrical Charge and the Most Probable Value of the Charge. *Phil. Mag.* 19:209–228, 1910.

[135] J. Perrin. *Brownian Movement and Molecular Reality.* Taylor and Francis, London, 1910.

[136] Th. Wulf. Beobachtungen über Strahlung hoher Durchdringungsfähigkeit auf dem Eiffelturm. *Phys. Z.* 11:811–813, 1910.

[137] L. Meitner O. v. Baeyer and O. Hahn. Über die Beta-Strahlen des aktiven Niederschlags des Thorium. *Phys. Z.* 12:273–279, 1911.

[138] C.G. Barkla. The spectra of the fluorescent Röntgen radiations. *Phil. Mag.* 22:396–412, 1911.

[139] R.A. Millikan. The Isolation of an Ion, a Precision Measurement of its Charge, and the Correction of Stokes's Law. *Phys. Rev.* 32:350–397, 1911.

[140] E. Rutherford. The Scattering of α and β Particles by Matter and the Structure of the Atom. *Phi. Mag.* 21:669–688, 1911.

[141] C.T.R. Wilson. On a method of making visible the paths of ionising particles through a gas. *Proc. Roy. Soc.* A85:285–288, 1911.

[142] V.F. Hess. Über Beobachtungen der durchdringenden Strahlung bei sieben Freiballonfahrten. *Phys. Z.* 13:1084–1091, 1912.

[143] J.W. Nicholson. The Constitution of the Solar Corona II. *Monthly Notices of the Royal Astr. Soc.* 72:677–692, 1912.

[144] D. Pacini. Penetrating Radiation at the Surface of and in Water. *Nuovo Cim.* 6:93, 1912.

[145] C.T.R. Wilson. On an Expansion Apparatus for Making Visible the Tracks of Ionising Particles in Gases and Some Results Obtained by Its Use. *Proc. Roy. Soc. A*, 87:277–292, 1912.

[146] N. Bohr. On the constitution of atoms and molecules. *Phil. Mag.* 26:857–875, 1913.

[147] N. Bohr. On the constitution of atoms and molecules I. *Phil. Mag.* 26:1–25, 1913.

[148] N. Bohr. On the constitution of atoms and molecules II. *Phil. Mag.* 26:476–502, 1913.

[149] N. Bohr. On the theory of the decrease of velocity of moving electrified particles on passing through matter. *Phil. Mag.* 25:10–31, 1913.

[150] H. Geiger and E. Marsden. The laws of deflexion of a particles through large angles. *Phil. Mag.* 25:604–623, 1913.

[151] W. Kolhörster. Über eine Neukonstruktion des Apparates zur Messung der durchdringenden Strahlung nach Wulf und die damit bisher gewonnenen Ergebnisse. *Phys. Z.* 14:1066–1069, 1913.

[152] R.A. Millikan. On the Elementary Electrical Charge and the Avogadro Constant. *Phys. Rev.* 2:109–143, 1913.

[153] J. Chadwick. Intensitätsverteilung im magnetischen Spektrum der β-Strahlen von Radium B + C. *Verh. der DPG*, 16:383, 1914.

[154]   E. Rutherford and E. Andrade. The Spectrum of Penetrating γ Rays from Radium B and Radium C. *Phil. Mag. Ser. 6 28*:263, 1914.

[155]   J. Franck and G. Hertz. Über die Erregung der Quecksilberresonanzlinie 253,6 nm durch Elektronenstöße. *Verh. der DPG*, 16:512–517, 1914.

[156]   J. Franck and G. Hertz. Über Zusammenstöße zwischen Elektronen und Molekülen des Quecksilberdampfes und die Ionisierungsspannung desselben. *Verh. der DPG*, 16:457–467, 1914.

[157]   R.A. Millikan. A Direct Determination of h. *Phys. Rev.* 4:73–75, 1914.

[158]   H.G.J. Moseley. The high-frequency spectra of the elements. Part II. *Phil. Mag.* 27:703–713, 1914.

[159]   J. Stark. Beobachtungen über den Effekt des elektrischen Feldes auf Spektrallinien I. Quereffekt. *Ann. Phys.* 43:965–983, 1914.

[160]   Various. *German mobs' vengeance on Jews*. The Telegraph, November 11. 1914.

[161]   Various. *Reply to the German professors – Reasoned Statement by British Scholars*. The Times, October 21. 1914.

[162]   P. Zeeman. Fresnel's coefficient for light of different colours. (First part). *Proc. Kon. Acad. Van Weten.* 17:445–451, 1914.

[163]   J.L. Heilbron. "My courage is ablaze so wildly", Niels Bohr *en route* to his quantum atom. In *One Hundred Years of the Bohr Atom*, 27–50. The Royal Danish Academy of Sciences and Letters, 1915.

[164]   H. Kellermann. *Der Krieg der Geister: Eine Auslese deutscher und ausländischer Stimmen zum Weltkriege 1914*. Vereinigung Heimat und Welt, Weimar, 1915.

[165]   H. Minkowski. Das Relativitätsprinzip (Vortrag vor der Göttinger Mathematischen Gesellschaft, 5.11.1907). *Ann. Phys.* 352:927–938, 1915.

[166]   Various. *The New York Times Current History of the European War, Vol. 1, January 9, 1915*. Vol. 1. The Project Gutenberg eBook #16702, 1915.

[167]   P. Zeeman. Fresnel's coefficient for light of different colours. (Second part). *Proc. Kon. Acad. Van Weten.* 18:398–408, 1915.

[168]   A. Einstein. *Meine Meinung über den Krieg*. Quoted from: Die Zeit No. 10/2014. 1916.

[169]   A. Einstein. Näherungsweise Integration der Feldgleichungen der Gravitation. *Sitzungsber. der Königl. Pr. Akad. der Wiss.* 1:688–696, 1916.

[170]   P.S. Epstein. Zur Theorie des Starkeffektes. *Ann. Phys.* 50:489–520, 1916.

[171] R.A. Millikan. A Direct Photoelectric Determination of Planck's h. *Phys. Rev.* 7:355–388, 1916.

[172] K. Schwarzschild. Zur Quantenhypothese. *Sitzungsber. der Königl. Pr. Akad. der Wiss.* 1:548–568, 1916.

[173] A. Sommerfeld. Zur Quantentheorie der Spektrallinien. *Ann. Phys.* 356:1–94, 1916.

[174] R.A. Millikan. *The Electron.* Univ. of Chicago Press, 1917.

[175] A. Einstein. Über Gravitationswellen. *Sitzungsber. der Königl. Pr. Akad. der Wiss.* 1:154–167, 1918.

[176] E. Noether. Invariante Varianzprobleme. *Nachr. d. König. Gesellsch. d. Wiss. zu Göttingen, Math-Phys. Klasse*:235, 1918.

[177] F.W. Aston. A positive ray spectrograph. *Phil. Mag.* 38:707–714, 1919.

[178] I. Langmuir. The Arrangement of Electrons in Atoms and Molecules. *J. of the Am. Chem. Soc.* 41:868–934, 1919.

[179] E. Rutherford. Collision of α particles with light atoms. An anomalous effect in nitrogen. *Phil. Mag.* 37:581–587, 1919.

[180] H. Schäfer. Das Abhorchen von Ferngesprächen und die Erdtelegraphie im Felde. *Polytechnisches Journal*, 334:93–97, 1919.

[181] A. Sommerfeld. *Atombau und Spektrallinien.* Braunschweig, Vieweg und Sohn, 1919.

[182] F.W. Aston. Isotopes and Atomic Weights. *Nature*, 105:617–619, 1920.

[183] N. Bohr. Über die Serienspektra der Elemente. *Z. Phys.* 2:423–469, 1920.

[184] A.S. Eddington F.W. Dyson and C. Davidson. A Determination of the Deflection of Light by the Sun's Gravitational Field, from Observations Made at the Total Eclipse of May 29, 1919. *Phil. Trans. Royal Soc. A*, 220:291–333, 1920.

[185] E. Rutherford. Bakerian Lecture: Nuclear Constitution of Atoms. *Proc. Roy. Soc. A*, 97:374–400, 1920.

[186] A. Landé. Über den anomalen Zeemaneffekt. *Z. Phys.* 5:231–241, 1921.

[187] A. Friedmann. Über die Krümmung des Raumes. *Z. Phys.* 10:377–386, 1922.

[188] G. Hettner. Die Bedeutung von Rubens' Arbeiten für die Plancksche Strahlungsformel. *Die Naturwissenschaften*, 10:1033–1038, 1922.

[189] H. Rubens and G. Michel. Prüfung der Planck'schen Strahlungsformel. *Phys. Z.* 22:569–577, 1922.

[190] P. Stark. *Die gegenwärtige Krise in der deutschen Physik.* J.A. Barth, Leipzig, 1922.

[191] O. Stern. Ein Weg zur experimentellen Prüfung der Richtungsquantelung im Magnetfeld. *Z. Phys.* 7:249–253, 1922.

[192] A.H. Compton. A Quantum Theory of the Scattering of X-Rays by Light Elements. *Phys. Rev.* 21:483–502, 1923.

[193] M. von Laue. Johannes Stark: Die gegenwärtige Krise in der deutschen Physik. *Naturwissenschaften,* 11:29–30, 1923.

[194] A. Sommerfeld. *Atomic structure and spectral lines.* E.P. Dalton and Company, New York, 1923.

[195] H.A. Kramers. The law of dispersion and Bohr's theory of spectra. *Nature,* 133:673, 1924.

[196] H.A. Kramers. The quantum theory of dispersion. *Nature,* 114:310, 1924.

[197] P. Lenard and J. Stark. *Hitlergeist und Wissenschaft.* Grossdeutsche Zeitung, May 8. 1924.

[198] M. Born and P. Jordan. Zur Quantenmechanik. *Z. f. Phys.* 34:858–888, 1925.

[199] L. de Broglie. Recherches sur la théorie des quanta. *Ann. de Phys.* 3:22, 1925.

[200] W. Heisenberg. Über die quantentheoretische Umdeutung kinematischer und mechanischer Beziehungen. *Z. f. Phys.* 33:879–893, 1925.

[201] H.A. Kramers and W. Heisenberg. Über die Streuung von Strahlen durch Atome. *Z. Phys,* 31:681, 1925.

[202] W. Heisenberg M. Born and P. Jordan. Zur Quantenmechanik II. *Z. f. Phys.* 35:557–615, 1925.

[203] W. Pauli. Über den Zusammenhang des Abschlusses der Elektronengruppen im Atom mit der Komplexstruktur der Spektren. *Z. Phys.* 31:765–783, 1925.

[204] G.E. Uhlenbeck and S. Goudsmit. Ersetzung der Hypothese vom unmechanischen Zwang durch eine Forderung bezüglich des Verhaltens jedes einzelnen Elektrons. *Naturwissenschaften,* 47:953, 1925.

[205] G.E. Uhlenbeck and S. Goudsmit. Spinning Electrons and the Structure of Spectra. *Nature,* 117:264, 1925.

[206] M. Born. Quantenmechanik der Stoßvorgänge. *Z. f. Phys.* 38:803–827, 1926.

[207] M. Born. Zur Quantenmechanik der Stoßvorgänge. *Z. f. Phys.* 37:863–867, 1926.

[208] P.A.M. Dirac. On the Theory of Quantum Mechanics. *Proc. Roy. Soc. A,* 112:661–677, 1926.

[209] W. Gordon. Der Comptoneffekt nach der Schrödingerschen Theorie. *Z. f. Phys.* 39:117–133, 1926.

[210] O. Klein. Quantentheorie und fünfdimensionale Relativitätstheorie. *Z. f. Phys.* 37:895, 1926.

[211] W. Pauli. Über das Wasserstoffspektrum vom Standpunkt der neuen Quantenmechanik. *Z. f. Phys.* 36:336–363, 1926.

[212] E. Schrödinger. An Undulatory Theory of the Mechanics of Atoms and Molecules. *Phys. Rev.* 28:1049–1070, 1926.

[213] E. Schrödinger. Quantisierung als Eigenwertproblem. *Ann. Phys.* 79:361, 489, 734, 1926.

[214] E. Schrödinger. Quantisierung als Eigenwertproblem. *Ann. Phys.* 81:109, 1926.

[215] L.H. Thomas. The Motion of the Spinning Electron. *Nature*, 107:514, 1926.

[216] F.W. Aston. Bakerian Lecture: A new mass-spectrograph and the whole number rule. *Proc. Roy. Soc. A*, 115:487–514, 1927.

[217] M. Born. Physical Aspects of Quantum Mechanics. *Nature*, 119:354–357, 1927.

[218] L. de Broglie. La mécanique ondulatoire et la structure atomique de la matière et du rayonnement. *J. Phys. Radium*, 8:225–241, 1927.

[219] C. Davisson and L.H. Germer. Diffraction of Electrons by a Crystal of Nickel. *Phys. Rev.* 30:705, 1927.

[220] G.Lemaître. Un Univers homogène de masse constante et de rayon croissant rendant compte de la vitesse radiale des nébuleuses extragalactiques. *Ann. Soc. Sci. Brux.* 61:49–59, 1927.

[221] W. Heisenberg. Über den anschaulichen Inhalt der quantentheoretischen Kinematik und Mechanik. *Z. Phys.* 43:172–198, 1927.

[222] E.H. Kennard. Zur Quantenmechanik einfacher Bewegungstypen. *Z. Phys.* 44:326–352, 1927.

[223] E. Rutherford. Structure of the radioactive atom and origin of the α rays. *Phil. Mag.* 4:580–605, 1927.

[224] C. Davisson and L.H. Germer. Reflection of Electrons by a Crystal of Nickel. *Proc. Nat. Acad. Sci. USA*, 14:317–322, 1928.

[225] G. Gamow. Zur Quantentheorie des Atomkernes. *Z. Phys.* 51:204–121, 1928.

[226] H. Geiger and W. Müller. Elektronenzählrohr zur Messung schwächster Aktivitäten. *Die Naturwissenschaften*, 6:617–618, 1928.

[227] R.W. Gurney and E.U. Condon. Wave Mechanics and Radioactive Disintegration. *Nature*, 122:439, 1928.

[228] D. Pekár. Die Entwicklung der Eötvösschen Originaldrehwagen. *Die Naturwissenschaften*, 16:1079–1088, 1928.

[229] H. Weyl. *Gruppentheorie und Quantenmechanik*. Hirzel, Leipzig, 1928.

[230] W. Bothe and W. Kolhörster. Das Wesen der Höhenstrahlung. *Z. f. Phys.* 56:751–777, 1929.

[231] E.P. Hubble. A Relation between Distance and Radial Velocity among Extra-Galactic Nebulae. *Proc. Natl. Acad. Sci. USA*, 15:168–173, 1929.

[232] N.F. Mott. The scattering of fast electrons by atomic nuclei. *Proc. Roy. Soc. A*, 124:425–442, 1929.

[233] H. Bethe. Zur Theorie des Durchgangs schneller Korpuskularstrahlen durch Materie. *Ann. Phys.* 397:325–400, 1930.

[234] W. Bothe. Zur Vereinfachung von Koinzidenzzählungen. *Z. f. Phys.* 59:1–5, 1930.

[235] W. Bothe and H. Becker. Künstliche Erregung von Kern-γ-Strahlen. *Z. f. Phys.* 60:289–306, 1930.

[236] G. Gamow and E. Rutherford. Mass defect curve and nuclear constitution. *Proc. Roy. Soc. A*, 126:632–644, 1930.

[237] W. Heisenberg. *The physical principles of the quantum theory*. Dover Publications, 1930.

[238] G. Joos. Die Jenaer Wiederholung des Michelson Versuchs. *Ann. Phys.* 7:385–407, 1930.

[239] A. Einstein. Maxwell's Influence on the Evolution of the Idea of Physical Reality. In *James Clerk Maxwell: A Commemoration Volume*. Cambridge University Press, 1931.

[240] E. Hubble and M.L. Humason. The Velocity-Distance Relation among Extra-Galactic Nebulae. *Astrophys. J.* 74:43–80, 1931.

[241] G. Lemaître. Expansion of the universe, A homogeneous universe of constant mass and increasing radius accounting for the radial velocity of extra-galactic nebulae. *Mon. Not. R. Astron. Soc.* 91:483–490, 1931.

[242] G. Lemaître. The Expanding Universe. *Mon. Not. R. Astron. Soc.* 91:490–501, 1931.

[243] C.D. Anderson. The Apparent Existence of Easily Deflectable Positives. *Science*, 76:238–239, 1932.

[244] J. Chadwick. The Existence of a Neutron. *Proc. Roy. Soc. A*, 136:692–708, 1932.

[245] J. D. Cockcroft and E. T. S. Walton. Experiments with High Velocity Positive Ions. (I) Further Developments in the Method of Obtaining High Velocity Positive Ions. *Proc. Roy. Soc. A*, 136:610–630, 1932.

[246] J. D. Cockcroft and E. T. S. Walton. Experiments with High Velocity Positive Ions. (II) The Disintegration of Elements by High Velocity Protons. *Proc. Roy. Soc. A*, 137:229–242, 1932.

[247] I. Curie and F. Joliot. Émission de protons de grande vitesse par les substances hydrogénées sous l'influence des rayons γ très pénétrants. *Comptes rendus*, 194:273–275, 1932.

[248] E. Gapon and D. Iwanenko. Zur Bestimmung der Isotopenzahl. *Naturwissenschaften*, 20:792–793, 1932.

[249] W. Heisenberg. Über den Bau der Atomkerne I. *Z. Phys.* 77:1–11, 1932.

[250] W. Heisenberg. Über den Bau der Atomkerne II. *Z. Phys.* 78:156–164, 1932.

[251] N.F. Mott. The polarisation of electrons by double scattering. *Proc. Roy. Soc. A*, 135:429–458, 1932.

[252] C.D. Anderson. The Positive Electron. *Phys. Rev.* 43:491–494, 1933.

[253] P.M.S. Blackett and G.P.S. Occhialini. Some Photographs of the Tracks of Penetrating Radiation. *Proc. Roy. Soc. A*, 139:699, 1933.

[254] K.T. Compton R.J. Van de Graaf and L.C. Van Atta. The Electrostatic Production of High Voltage for Nuclear Investigations. *Phys. Rev.* 43:149–157, 1933.

[255] W. Heisenberg. Über den Bau der Atomkerne III. *Z. Phys.* 80:587–596, 1933.

[256] F. Zwicky. Die Rotverschiebung von extragalaktischen Nebeln. *Helv. Phys. Acta*, 6:110–127, 1933.

[257] H. Bethe and R. Peierls. The "Neutrino". *Nature*, 133:532, 1934.

[258] J. Chadwick and M.A. Goldhaber. A Nuclear Photo-effect: Disintegration of the Diplon by γ-Rays. *Nature*, 133:532, 1934.

[259] E. Fermi *et al.* Artificial Radioactivity produced by Neutron Bombardment. *Proc. Roy. Soc. A*, 146:483–500, 1934.

[260] E. Fermi. Versuch einer Theorie der β-Strahlen. I. *Z. f. Phys.* 88:161, 1934.

[261] E. Hubble. The Distribution of Extra-Galactic Nebulae. *Astrophys. J.* 79:8–76, 1934.

[262] W. Pauli and V.F. Weisskopf. On Quantization of the Scalar Relativistic Wave Equation. *Helv. Phys. Acta*, 7:709–731, 1934.

[263] B. Podolsky A. Einstein and N. Rosen. Can Quantum-Mechanical Description of Physical Reality be Considered Complete? *Phys. Rev.* 47:777–780, 1935.

[264] G. Breit and E. Feenberg. The Possibility of the Same Form of Specific Interaction for All Nuclear Particles. *Phys. Rev.* 50:850–856, 1935.

[265] N.P. Heydenburg M.A. Tuve and L.R. Hafstad. The Scattering of Protons by Protons. *Phys. Rev.* 50:806–825, 1935.

[266] C.F. von Weizsäcker. Zur Theorie der Kernmassen. *Z. Phys.* 96:431–458, 1935.

[267] H. Weyl. *Memorial Address for Emmy Noether, delivered in Goodhart Hall, Bryn Mawr College, on April 26, 1935.* 1935. URL: https://celebratio.org/Noether_E/article/111/

[268] H. Yukawa. On the Interaction of Elementary Particles I. *Proc. Phys. Math. Soc. Japan*, 17:48–57, 1935.

[269] C.D. Anderson and S.H. Neddermeyer. Cloud Chamber Observations of Cosmic Rays at 4300 Meters Elevation and Near Sea-Level. *Phys. Rev.* 50:263–271, 1936.

[270] H.A. Bethe and R.F. Bacher. Nuclear Physics A: Stationary States of Nuclei. *Rev. Mod. Phys.* 8:82–229, 1936.

[271] P. Lenard. *Deutsche Physik.* Vol. 1–4. J.F. Lehmann-Verlag, München, 1936.

[272] S. Smith. The Mass of the Virgo Cluster. *Astrophys. J.* 83:23–30, 1936.

[273] C.D. Anderson and S.H. Neddermeyer. Note on the Nature of Cosmic Ray Particles. *Phys. Rev.* 51:884–886, 1937.

[274] H.A. Bethe. Nuclear Physics B: Nuclear Dynamics, Theoretical. *Rev. Mod. Phys.* 9:69–245, 1937.

[275] I. Curie and P. Savitch. Sur les radioéléments formés dans l'uranium irradié par les neutrons. *J. Phys. Radium*, 8:385–387, 1937.

[276] M.S. Livingston and H.A. Bethe. Nuclear Physics C: Nuclear Dynamics, Experimental. *Rev. Mod. Phys.* 9:245–398, 1937.

[277] J.C. Street and E.C. Stevenson. New Evidence for the Existence of a Particle of Mass Intermediate Between the Proton and Electron. *Phys. Rev.* 52:1003–1004, 1937.

[278] J.A. Wheeler. On the Mathematical Description of Light Nuclei by the Method of Resonating Group Structure. *Phys. Rev.* 52:52, 1937.

[279] F. Zwicky. On the Masses of Nebulae and of Clusters of Nebulae. *Astrophys. J.* 86:217–246, 1937.

[280] A. Einstein and L. Infeld. *The Evolution of Physics.* Cambridge Library of Modern Science, 1938.

[281] O. Hahn and F. Strassmann. Über die Entstehung von Radiumisotopen aus Uran durch Bestrahlen mit schnellen und verlangsamten Neutronen. *Naturwissenschaften*, 26:755–756, 1938.

[282] N. Bohr and J.A. Wheeler. The Mechanism of Nuclear Fission. *Phys. Rev.* 56:426–450, 1939.

[283] S. Flügge. Kann der Energieinhalt der Atomkerne technisch nutzbar gemacht werden? *Naturwiss.* 27:402–410, 1939.

[284] O.R. Frisch. Further evidence for the Division of Heavy Nuclei under Neutron Bombardment. *Nature*, 143:276, 1939.

[285] O. Hahn and F. Strassmann. Nachweis der Entstehung aktiver Bariumisotope aus Uran und Thorium durch Neutronenbestrahlung; Nachweis weiterer aktiver Bruchstücke bei der Uranspaltung. *Naturwissenschaften*, 27:89–95, 1939.

[286] O. Hahn and F. Strassmann. Über den Nachweis und das Verhalten der bei der Bestrahlung des Urans mittels Neutronen entstehenden Erdalkalimetalle. *Naturwissenschaften*, 27:11–15, 1939.

[287] W. Heisenberg. *Die Möglichkeit der technischen Energiegewinnung aus der Uranspaltung*. 1939. URL: http://www.deutsches-museum.de/archiv/archiv-online/Geheimdokumente/forschungszentren/leipzig/energie-aus-uran/

[288] L. Meitner and O.R. Frisch. Disintegration of Uranium by Neutrons: a New Type of Nuclear Reaction. *Nature*, 143:239–240, 1939.

[289] A. Einstein. Considerations Concerning the Fundaments of Theoretical Physics. *Science*, 91:487, 1940.

[290] A.O. Nier em et al. Further Experiments on Fission of Separated Uranium Isotopes. *Phys. Rev.* 57:748, 1940.

[291] A.O. Nier em et al. Nuclear Fission of Separated Uranium Isotopes. *Phys. Rev.* 57:546, 1940.

[292] C.F. von Weizsäcker. Eine Möglichkeit der Energiegewinnung aus Uran 238. *KPFB*, G-59 1940.

[293] W. Bothe and P. Jensen. Die Absorption thermischer Neutronen in Elektrographit. *KPFB*, G-71 1941.

[294] F. Rasetti. Disintegration of Slow Mesotrons. *Phys. Rev.* 60:198, 1941.

[295] B. Rossi and D.H. Hall. Variation of the Rate of Decay of Mesotrons with Momentum. *Phys. Rev.* 59:223–228, 1941.

[296] C. Döpel R. Döpel and W. Heisenberg. *Der experimentelle Nachweis der effektiven Neutronenvermehrung in einem Kugel-Schichten-System aus D2O und Uranmetall, secret report G-136 (Ref. No. 47)*. 1942. URL: http://www.deutsches-museum.de/archiv/archiv-online/Geheimdokumente/forschungszentren/leipzig/neutronenvermehrung

[297] E.C.G. Stückelberg. La mécanique du point matériel en théorie de relativité et en théorie des quanta. *Helv. Phys. Acta*, 15:23–36, 1942.

[298] M. Planck. Zur Geschichte der Auffindung des physikalischen Wirkungsquantums. *Die Naturwissenschaften*, 31:153–159, 1943.

[299] R. Serber. *The Los Alamos Primer*. Lightning Source UK Ltd. 1943.

[300] L. Landau. On the energy loss of fast particles by ionization. *J. Phys. USSR*, 8:201, 1944.

[301] V.I. Veksler. A new method of accelerating relativistic particles. *Comptes Rendus de l'Académie Sciences de l'URSS*, 43:329–331, 1944.

[302] W. Bothe and P. Jensen. Die Absorption thermischer Neutronen in Kohlenstoff. *Z. Phys.* 122:749–755, 1945.

[303] O. Piccioni M. Conversi E. Pancini. On the Decay Process of Positive and Negative Mesons. *Phys. Rev.* 68:232, 1945.

[304] E.M. McMillan. The synchrotron – a proposed high energy accelerator. *Phys. Rev.* 68:143, 1945.

[305] H. DeWolf Smyth. *Atomic Energy for Military Purposes: The Official Report on the Development of the Atomic Bomb under the Auspices of the United States Government, 1940-1945*. Princeton University Press, 1945.

[306] D. Acheson and D. Lilienthal. *A Report on the International Control of Atomic Energy. Prepared for the Secretary of State's Committee on Atomic Energy*. U. S. Government Printing Office, Washington, D. C., March 16, 1946. Department of State. Publication 2498. 1946.

[307] W. Heisenberg. Über die Arbeiten zur technischen Ausnutzung der Atomkernenergie in Deutschland. *Naturwiss.* 33:325–329, 1946.

[308] W. Heisenberg. Research in Germany on the Technical Application of Atomic Energy. *Nature*, 160:211–215, 1947.

[309] O. Piccioni M. Conversi E. Pancini. On the Disintegration of Negative Mesons. *Phys. Rev.* 71:209–210, 1947.

[310] N. Metropolis and S. Ulam. The Monte Carlo Method. *J. Am. Stat. Assoc.* 44:335–341, 1947.

[311] R.P. Feynman. Space-time approach to nonrelativistic quantum mechanics. *Rev. Mod. Phys.* 20:367–387, 1948.

[312] M. Goldhaber and G. Scharff-Goldhaber. Identification of Beta-Rays with Atomic Electrons. *Phy. Rev.* 73:1472, 1948.

[313] W. Ehrenberg and R. E. Siday. The Refractive Index in Electron Optics and the Principles of Dynamics. *Proceedings of the Physical Society. Section B*, 62:8–21, 1949.

[314] M. Goeppert-Mayer. On Closed Shells in Nuclei. II. *Phys. Rev.* 75:1969–1970, 1949.

[315] J.H.D. Jensen O. Haxel and H.E. Suess. On the "Magic Numbers" in Nuclear Structure. *Phys. Rev.* 75:1766, 1949.

[316] M.N. Rosenbluth. High Energy Elastic Scattering of Electrons on Protons. *Phys.Rev.* 79:615–619, 1949.

[317] M. Goeppert-Mayer. Nuclear configurations in the spin-orbit coupling model. I. Empirical Evidence. *Phys. Rev.* 78:16–21, 1950.

[318] M. Goeppert-Mayer. Nuclear Configurations in the Spin-Orbit Coupling Model. II. Theoretical Considerations. *Phys. Rev.* 78:22–23, 1950.

[319] D. Bohm. A Suggested Interpretation of the Quantum Theory in Terms of 'Hidden Variables' I. *Phys. Rev.* 85:166–179, 1952.

[320] D.A. Glaser. Some Effects of Ionizing Radiation on the Formation of Bubbles in Liquids. *Phys. Rev. B*, 87:665, 1952.

[321] P. F. A. Klinkenberg. Tables of Nuclear Shell Structure. *Rev. Mod. Phys.* 24:63–73, 1952.

[322] H. Bethe and J. Ashkin. "Penetration of Beta-Rays through Matter". In *Experimental Nuclear Physics*. Ed. by E. Segré. New York: J. Wiley, 1953. Pp. 252–303

[323] F.J. Dyson. *On the relation between Scattering Matrix Elements and Cross-Section. Les Houches: Ecole d'été de physique théorique.* 1954.

[324] A.A. Abrikosov L.D. Landau and I.M. Khalatnikov. An Asymptotic Expression for the Electron Green Function in Quantum Electrodynamics. *Dokl. Akad. Nauk SSSR*, 95:773, 1954.

[325] A.A. Abrikosov L.D. Landau and I.M. Khalatnikov. An Asymptotic Expression for the Photon Green Function in Quantum Electrodynamics. *Dokl. Akad. Nauk SSSR*, 95:1177, 1954.

[326] A.A. Abrikosov L.D. Landau and I.M. Khalatnikov. The Removal of Infinities in Quantum Electrodynamics. *Dokl. Akad. Nauk SSSR*, 95:497, 1954.

[327] R.D. Woods and D.S. Saxon. Diffuse Surface Optical Model for Nucleon-Nuclei Scattering. *Phys. Rev.* 95:577–578, 1954.

[328] C.N. Yang and R.L. Mills. Conservation of Isotopic Spin and Isotopic Gauge Invariance. *Phys. Rev.* 96:191–195, 1954.

[329] M. Goeppert-Mayer and J.H.D. Jensen. *Elementary Theory of Nuclear Shell Structure.* John Wiley & Sons, 1955.

[330] W. Heisenberg. "The Development of the Interpretation of the Quantum Theory". In *Niels Bohr and the Development of Physics*. Ed. by W. Pauli. London: Pergamon, 1955. Pp. 12–29

[331] R. Hofstadter and R.W. McAllister. Electron Scattering From the Proton. *Phys. Rev.* 98:217–218, 1955.

[332] C.L. Cowan *et al.* Detection of the Free Neutrino: a Confirmation. *Science*, 124:103–104, 1956.

[333] R. Jungk. *Heller als tausend Sonnen.* Scherz, Bern, 1956.

[334] T. D. Lee and C. N. Yang. Question of Parity Conservation in Weak Interactions. *Phys. Rev.* 104:254–258, 1956.

[335] E.J. Lofgren. *Experiences with the Bevatron.* 1956. URL: https://digital.library.unt.edu/ark:/67531/metadc880580/m1/1/

[336] R.W. McAllister and R. Hofstadter. Elastic Scattering of 188 MeV Electrons From the Proton and the α Particle. *Phys. Rev.* 102:851–856, 1956.

[337] G.K. O'Neill. Storage-Ring Synchrotron: Device for High-Energy Physics Research. *Phys. Rev.* 102:1418–1419, 1956.

[338] C.S. Wu *et al.* Experimental Test of Parity Conservation in Beta Decay. *Phys. Rev.* 105:1413, 1957.

[339] M. Gell-Mann and A. H. Rosenfeld. Hyperons and Heavy Mesons (Systematics and Decay). *Ann. Rev. Nucl. Sci.* 7:407–478, 1957.

[340] B. Pontecorvo. Inverse beta processes and nonconservation of lepton charge. *J. Exptl. Theoret. Phys. (USSR)*, 34:247, 1957.

[341] F. Villars. The Collective Model of Nuclei. *Ann. Rev. Nucl. Phys.* 7:185–230, 1957.

[342] Y. Aharonov and D. Bohm. Significance of electromagnetic potentials in the quantum theory. *Phys. Rev.* 115:485, 1959.

[343] S. Fukui and S. Miyamoto. A new type of particle detector: the "discharge chamber". *Nuovo Cim.* 11:113–115, 1959.

[344] B. Pontecorvo. Electron and Muon Neutrinos. *J. Exptl. Theoret. Phys. (USSR)*, 37:1751, 1959.

[345] R.V. Pound and G.A. Rebka. Gravitational Red-Shift in Nuclear Resonance. *Phys. Rev. Lett.* 3:439–411, 1959.

[346] R.G. Chambers. Shift of an Electron Interference Pattern by Enclosed Magnetic Flux. *Phys. Rev. Lett.* 5:3, 1960.

[347] R.V. Pound and G.A. Rebka. The apparent weight of photons. *Phys. Rev. Lett.* 4:337–341, 1960.

[348] M. Schwartz. Feasibility of using high energy neutrino to study weak interactions. *Phys. Rev. Lett.* 4:306, 1960.

[349] Y. Aharonov and D. Bohm. Further Considerations on Electromagnetic Potentials in the Quantum Theory. *Phys. Rev.* 123:1511–1524, 1961.

[350] S. Fukui. The Discharge Chamber and Its Characteristics. *J. Phys. Soc. Jpn.* 16:2574–2585, 1961.

[351] Murray Gell-Mann. The Eightfold Way: A Theory of strong interaction symmetry. *SLAC Report* 1961.

[352] S. Glashow. Partial Symmetries of Weak Interactions. *Nucl. Phys.* 22:579, 1961.

[353] R. Hofstadter. The electron-scattering method and its application to the structure of nuclei and nucleons. *Nobel Lecture, December 11, 1961*:560–581, 1961.

[354] Y. Ne'eman. Gauges, Groups And An Invariant Theory Of The Strong Interactions. *Nucl. Phys.* 26:222, 1961.

[355] P. W. Anderson. Plasmons, Gauge Invariance, and Mass. *Phys. Rev.* 130:439–442, 1962.

[356] N. Cabibbo. Unitary Symmetry and Leptonic Decays. *Phys. Rev. Lett.* 10:531–533, 1962.

[357] G. Danby *et al.* Observation of High-Energy Neutrino Reactions and the Existence of Two Kinds of Neutrinos. *Phys. Rev. Lett.* 9:36–44, 1962.

[358] M. Gell-Mann. Symmetries of baryons and mesons. *Phys. Rev.* 125:1067, 1962.

[359] W. Heisenberg. *Physics and Philosophy: The Resolution in Modern Science.* Harper & Row, New York, 1962.

[360] R.G. Hewlett and O.E. Anderson. *The New World, Vol. 1: A History of the United States Atomic Energy Commission.* Penn. State Univ. Press, 1962.

[361] A. Salam J.Goldstone and S. Weinberg. Broken Symmetries. *Phys. Rev,* 127:965–970, 1962.

[362] Th. S. Kuhn. *The Structure of Scientific Revolutions.* University of Chicago Press, 1962.

[363] L.D. Landau and E.M. Lifshitz. *The Classical Theory of Fields.* Addison-Wesley, 1962.

[364] G. Möllenstedt and W. Bayh. Messung der kontinuierlichen Phasen-schiebung von Elektronenwellen im kraftfeldfreien Raum durch das magnetische Vektorpotential einer Luftspule. *Die Naturwissenschaften,* 49:81–82, 1962.

[365] S. Okubo. Note on Unitary Symmetry in Strong Interaction. *Progr. Theor. Phys.* 27:949–966, 1962.

[366] S. Okubo. Note on Unitary Symmetry in Strong Interaction. II Excited States of Baryons. *Progr. Theor. Phys.* 28:24–32, 1962.

[367] S. Sakata Z. Maki M. Nakagawa. Remarks on the Unified Model of Elementary Particles. *Prog. Theo. Phys.* 28:870–880, 1962.

[368] M. Roos. Tables of Elementary Particles and Resonant States. *Rev. Mod. Phys.* 35:314, 1963.

[369] J.D. Bjorken and S.L. Glashow. Elementary Particles and SU(4). *Phys. Lett.* 11:255–257, 1964.

[370] F. Englert and R. Brout. Broken Symmetry and the Mass of Gauge Vector Mesons. *Phys. Rev. Lett.* 13:321–323, 1964.

[371] A. H. Rosenfeld *et al.* Data on Elementary Particles and Resonant States. *Ann. Rev. Nucl. Sci.* 36:977–1004, 1964.

[372] V.E. Barnes *et al.* Observation of a Hyperon with Strangeness Minus Three. *Phys. Rev. Lett.* 12:204, 1964.

[373] M. Gell-Mann. A Schematic Model of Baryons and Mesons. *Phys. Lett.* 8:214, 1964.

[374] M. Gell-Mann. The Symmetry group of vector and axial vector currents. *Physics Physique Fizika*, 1:63, 1964.

[375] C.R. Hagen G.S. Guralnik and T.W.B. Kibble. Global Conservation Laws and Massless Particles. *Phys. Rev. Lett.* 13:585–587, 1964.

[376] Schopper H. Ein neues Elementarteilchen – das Ω^-. *Phys. Blätter*, 20:227–228, 1964.

[377] P. Higgs. Broken Symmetries and the Masses of Gauge Bosons. *Phys. Rev. Lett.* 13:508–509, 1964.

[378] P. Higgs. Broken symmetries, massless particles and gauge fields. *Phys. Lett.* 12:132–133, 1964.

[379] S.D. Drell J.D. Bjorken. *Relativistic quantum mechanics.* McGraw-Hill, 1964.

[380] V.L. Fitch J.H. Christenson J.W. Cronin and R. Turlay. Evidence for the 2π Decay of the K^0. *Phys. Rev. Lett.* 13:138–140, 1964.

[381] R.V. Pound and J.L. Snider. Effect of Gravity on Nuclear Resonance. *Phys. Rev. Lett.* 13:539–540, 1964.

[382] M. Roos. Data on elementary particles and resonant states, November 1963. *Nucl. Phys.* 52:1–24, 1964.

[383] S. Weinberg. Photons and Gravitons in S Matrix Theory: Derivation of Charge Conservation and Equality of Gravitational and Inertial Mass. *Phys. Rev.* 135:B1049–B1056, 1964.

[384] G. Zweig. *An SU(3) model for strong interaction symmetry and its breaking.* CERN Report 8182/TH401 and 8419/TH412. 1964.

[385] W.C. Barber *et al.* Wide Angle Electron-Electron Scattering On The Princeton-Stanford Storage Rings. In *5th International Conference On High-Energy Accelerators (HEACC 65)*, 266. 1965.

[386] L.D. Landau. "Paper Number 78, 79, 80". In *Collected Papers of L.D. Landau.* Ed. by D. ter Haar. London: Pergamon, 1965. Pp. 607, 6011, 6016

[387] A.A. Penzias. and R.W. Wilson. A Measurement of Excess Antenna Temperature at 4080 Mc/s. *Astrophys. J.* 142:418–421, 1965.

[388] L. Pearce Williams. *Michael Faraday, A Biography.* Da Capo Press, New York, 1965.

[389] G. F. Chew. *The Analytic S Matrix: A Basis for Nuclear Democracy.* W. Benjamin, New York, 1966.

[390] K. Greisen. End to the Cosmic-Ray Spectrum? *Phys. Rev. Lett.* 16:748–750, 1966.

[391] F. Low. Dynamics of Strong Interactions. In *Proceedings of the 13th International Conference on High Energy Physics (ICHEP66)*, 241. 1966.

[392] B. Rossi. *Cosmic Rays*. George Allen and Unwin Ltd., London, 1966.

[393] G.T. Zatsepin and V.A. Kuzmin. Upper Limit of the Spectrum of Cosmic Rays. *J. Exp. Theo. Phys. Lett.* 4:78–80, 1966.

[394] O.R. Frisch and J.A. Wheeler. The discovery of fission. *Physics Today*, 20:43–52, 1967.

[395] A.D. Sakharov. Violation of CP invariance, C asymmetry, and baryon asymmetry of the universe. *Journal of Experimental and Theoretical Physics Letters*, 5:24–27, 1967.

[396] S. Weinberg. A Model of Leptons. *Phys. Rev. Lett.* 19:1547, 1967.

[397] P.J.E. Peebles. Recombination of the Primeval Plasma. *Astrophys. J.* 153:1–11, 1968.

[398] A. Salam. Weak and Electromagnetic Interactions. In *Nobel Symposium 8, Elementary Particle Theory: Relativistic Groups and Analyticity*, 367. Almqvist & Wiksell, Stockholm, 1968.

[399] B.L. van der Waarden. *Sources of Quantum Mechanics*. Dover Publications, New York, 1968.

[400] V.G. Kurt Y.B. Zeldovich and R.A. Syunyaev. Recombination of Hydrogen in the Hot Model of the Universe. *Zh. Eksp. Teor. Fiz.* 55:278–286, 1968.

[401] J.D. Bjorken and E.A. Paschos. Inelastic Electron Proton and Gamma Proton Scattering, and the Structure of the Nucleon. *Phys. Rev.* 185:1975–1982, 1969.

[402] W. Heisenberg. *Der Teil und das Ganze*. R. Piper & Co, München, 1969.

[403] J.W. van Spronsen. *The Periodic System of Chemical Elements: a History of the First Hundred Years*. Elsevier, Amsterdam, 1969.

[404] J. Habermas. "Technology and Science as 'Ideology'". In *Toward a Rational Society*. Ed. by J. Shapiro. Boston: Beacon Press, 1970.

[405] V.C. Rubin and Jr. W. K. Ford. Rotation of the Andromeda Nebula from a Spectroscopic Survey of Emission Regions. *Astrophys. J.* 158:379–403, 1970.

[406] J. Iliopoulos S.L. Glashow and L. Maiani. Weak Interactions with Lepton-Hadron Symmetry. *Phys. Rev.* D2:1285–1292, 1970.

[407] M. A. Tavel (transl.) E. Noether. Invariant variation problems. *Transport Theory and Statistical Physics*, 1:186–207, 1971.

[408] W.C. Barber *et al.* Test Of Quantum Electrodynamics By Electron Electron Scattering. *Phys. Rev.* D3:2796–2800, 1971.

[409] K.S. Thorne C.W. Misner and J.A. Wheeler. *Gravitation.* H.W. Freeman, 1973.

[410] F.J. Hasert *et al.* (Gargamelle Coll.) Observation of Neutrino Like Interactions Without Muon Or Electron in the Gargamelle Neutrino Experiment. *Phys. Lett.* 46B:138–140, 1973.

[411] F.J. Hasert *et al.* (Gargamelle Coll.) Search for Elastic ν_μ Electron Scattering. *Phys. Lett.* 46B:121–124, 1973.

[412] M. Gell-Mann H. Fritzsch and H. Leutwyler. Advantages of the Color Octet Gluon Picture. *Phys. Lett.* 47B:365–368, 1973.

[413] M. Kobayashi and T. Maskawa. CP Violation in the Renormalizable Theory of Weak Interaction. *Progr. Theor. Phys.* 49:652–657, 1973.

[414] F.E. Mills. Isabelle design study. *IEEE Transactions on Nuclear Science*, 20:1036–1038, 1973.

[415] A. Benvenuti *et al.* (HPWF Coll.) Further Observation of Muonless Neutrino Induced Inelastic Interactions. *Phys. Rev. Lett.* 32:1454–1457, 1974.

[416] A. Benvenuti *et al.* (HPWF Coll.) Measurement of Rates for Muonless Deep Inelastic Neutrino and anti-neutrino Interactions. *Phys. Rev. Lett.* 32:1457, 1974.

[417] A. Benvenuti *et al.* (HPWF Coll.) Observation of Muonless Neutrino Induced Inelastic Interactions. *Phys. Rev. Lett.* 32:800–803, 1974.

[418] J.-E. Augustin *et al.* Discovery of a Narrow Resonance in $e^+ e^-$ Annihilation. *Phys. Rev. Lett.* 33:1406–1408, 1974.

[419] J.J. Aubert *et al.* Experimental Observation of a Heavy Particle J. *Phys. Rev. Lett.* 33:1404–1406, 1974.

[420] E. Fermi. *Nuclear Physics, Course given at University of Chicago in 1949.* University of Chicago Press, 1974.

[421] F.J. Hasert *et al.* (Gargamelle Coll.) Observation of Neutrino Like Interactions without Muon or Electron in the Gargamelle Neutrino Experiment. *Phys. Lett.* B73:1–22, 1974.

[422] R. Newbury. Fresnel Drag and the Principle of Relativity. *Isis*, 65:379–386, 1974.

[423] I. Estermann and S.N. Foster. History of molecular beam research: Personal reminiscences of the important evolutionary period 1919-1933. *Am. J. of Phys.* 4:661–671, 1975.

[424] M.L. Perl *et al.* Evidence for Anomalous Lepton Production in e^+ - e^- Annihilation. *Phys. Rev. Lett.* 35:1489–1492, 1975.

[425] H. Harari. A New Quark Model for Hadrons. *Phys. Lett.* B57:265, 1975.

[426] G.B. Lubkin. Bohr, Mottelson and Rainwater win Nobel physics prize. *Physics Today*, 28:69–72, 1975.

[427] W. Heisenberg. Was ist ein Elementarteilchen? *Naturwissenschaften*, 63:1–7, 1976.

[428] J. Blietschau *et al.* (Gargamelle Coll.) Evidence for the Leptonic Neutral Current Reaction $\bar{\nu}_\mu e^- \to \bar{\nu}_\mu e^-$. *Nucl. Phys.* B114:189–198, 1976.

[429] S.R. Weart. Scientists with a secret. *Physics Today*, 29:23–30, 1976.

[430] J.L. Heilbron. J.J. Thomson and the Bohr atom. *Physics Today*, 30:23–30, 1977.

[431] S.W. Herb *et al.* (Fermilab E288 Coll.) Observation of a Dimuon Resonance at 9.5 GeV in 400 GeV Proton-Nucleus Collisions. *Phys. Rev. Lett.* 39:252–255, 1977.

[432] B.R. Wheaton. Philipp Lenard and the Photoelectric Effect, 1889-1911. *Historical Studies in the Physical Sciences*:299–322, 1977.

[433] R.D. Field and R.P. Feynman. A Parametrization of the Properties of Quark Jets. *Nucl. Phys.* D18:1–76, 1978.

[434] R.D. Field R.P. Feynman and G.C. Fox. A Quantum Chromodynamic Approach for the Large Transverse Momentum Production of Particles and Jets. *Phys. Rev.* D18:3320–3343, 1978.

[435] S.M. Guralnik. The Context of Faraday's Electrochemical Laws. *ISIS*, 70:59–75, 1979.

[436] F. Scully. An experimenter's history of neutral currents. *Progr. Part. Nucl. Phys.* 2:41–87, 1979.

[437] L. Thorndahl D. Mohl G. Petrucci and S. van der Meer. Physics and Technique of Stochastic Cooling. *Phys. Rept.* 58:73–119, 1980.

[438] R.P. Feynman. "Los Alamos from Below". In *Reminiscences of Los Alamos, 1943-1945.* Ed. by J.O. Hirschfelder L. Badash and H.P. Broida. Dordrecht: D. Reidel, 1980. Pp. 105–132

[439] M. Foucault. *Power/Knowledge: Selected Interviews and Other Writings 1972-1977.* Pantheon Books, New York, 1980.

[440] J.H. Taylor and J.M. Weisberg. A New Test of General Relativity – Gravitational Radiation and the Binary Pulsar PSR 1913+16. *Astrophys. J.* 253:908–920, 1982.

[441] V.F. Weisskopf. The places where quantum mechanics was born. *J. de Phys.* 43 C8:325–328, 1982.

[442] R. Widerøe. Some Memories and Dreams from the Childhood of Particle Accelerators. *Europhysics News*, 15:9–11, 1982.

[443] R.K. DeKosky. William Crookes and the quest for absolute vacuum in the 1870s. *Ann. Sci.* 40:1–18, 1983.

[444] G. Arnison *et al.* (UA1 Coll.) Experimental Observation of Isolated Large Transverse Energy Electrons with Associated Missing Energy at $\sqrt{s} = 540$ GeV. *Phys. Lett.* 122B:103–116, 1983.

[445] G. Arnison *et al.* (UA1 Coll.) Experimental Observation of Lepton Pairs of Invariant Mass Around 95 GeV/c^2 at the CERN SPS Collider. *Phys. Lett.* 126B:398–410, 1983.

[446] P. Galison. The Discovery of the Muon and the Failed Revolution against Quantum Electrodynamics. *Centaurus*, 26:262, 1983.

[447] M. Banner *et al.* (UA2 Coll.) Evidence for the Existence of New Unstable Elementary ParticlesObservation of Single Isolated Electrons of High Transverse Momentum in Events with Missing Transverse Energy at the CERN $\bar{p}p$ Collider. *Phys. Lett.* 122B:476–485, 1983.

[448] M. Milgrom. A modification of the Newtonian dynamics - Implications for galaxies. *Astrophys. J.* 270:371–383, 1983.

[449] M. Milgrom. A modification of the Newtonian dynamics - Implications for galaxy systems. *Astrophys. J.* 270:384–389, 1983.

[450] M. Milgrom. A modification of the Newtonian dynamics as a possible alternative to the hidden mass hypothesis. *Astrophys. J.* 270:365–370, 1983.

[451] P. Bagnaia *et al.* (UA2 Coll.) Evidence for $Z^0 \to e^+e^-$ at the CERN $\bar{p}p$ Collider. *Phys. Lett.* 129B:130–140, 1983.

[452] E. Amaldi. From the discovery of the neutron to the discovery of nuclear fission. *Phys. Rep.* 111:1–331, 1984.

[453] W. Heisenberg. "Considérations théoriques générales sur la Structure du Noyau". In *Scientific Review Papers, Talks, and Books*. Ed. by H.P. Dürr W. Blum and H. Rechenberg. Heidelberg: Springer, 1984. Pp. 179–225

[454] A.M. Hillas. The Origin of Ultra-High-Energy Cosmic Rays. *Ann. Rev. Astron. Astrophys.* 22:425–444, 1984.

[455] H. Kragh. Equation with many Fathers: The Klein-Gordon Equation in 1926. *Am. J. of Phys.* 52:1024–1033, 1984.

[456] T.S. van Albada *et al.* Distribution of dark matter in the spiral galaxy NGC 3198. *Astrophys. J.* 295:305–313, 1985.

[457] R.P. Feynman. *QED: The Strange Theory of Light and Matter*. Princeton Univ. Press, 1985.

[458] S.A. Goudsmit. *Alsos*. American Institute of Physics, 1985.

[459] G. Binnig and H. Rohrer. Scanning tunneling microscopy. *IBM Journal of Research and Development*, 30:355–369, 1986.

[460] A. Tonomura *et al.* Evidence for Aharonov-Bohm Effect with Magnetic Field Completely Shielded from Electron wave. *Phys. Rev. Lett.* 56:792–795, 1986.

[461] R. Eckhardt. Stan Ulam, John von Neumann, and the Monte Carlo Method. *Los Alamos Science*, 15:131–141, 1987.

[462] I. Falconer. Corpuscles, Electrons and Cathode Rays: J.J. Thomson and the 'Discovery of the Electron. *BJHS*, 20:241–276, 1987.

[463] J.D. Jackson. The Superconducting Super Collider. In *High-energy physics. Proceedings, International Europhysics Conference, Uppsala, Sweden, June 25 - July 1, 1987. Vols. 1, 2*. 1987.

[464] N. Metropolis. The Beginning of the Monte Carlo Method. *Los Alamos Science*, 15:125–130, 1987.

[465] J. Ziegler. Texas wins a bright start for physics. *New Scientist*, 120:18, 1988.

[466] A. Pais. George Uhlenbeck and the Discovery of Electron Spin. *Physics Today*, 42:34–40, 1989.

[467] E.G. Adelberger *et al.* Eötvös experiments, lunar ranging and the strong equivalence principle. *Nature*, 347:261–263, 1990.

[468] L.W. Whitlow *et al.* Precise measurements of the proton and deuteron structure functions from a global analysis of the SLAC deep inelastic electron scattering cross-sections. *Phys.Lett.* B282:475–482, 1992.

[469] M. Riordan. The Discovery of quarks. *Science*, 256:1287, 1992.

[470] M. Walker. Physics and Propaganda: Werner Heisenberg's Foreign Lectures under National Socialism. *Historical Studies in the Physical and Biological Sciences*, 22:339–389, 1992.

[471] D.B. Cline. 30 Years of Weak Neutral Currents. In *Proceedings Int. Symp. On 30 Years of Neutral Currents, Santa Monica CA*, 643–673. 1993.

[472] Project on Government Oversight. *The Superconducting Super Collider's Super Excesses*. 1993. URL: http://www.pogoarchives.org/m/co/super-collider-1993.pdf

[473] D. Haidt. Discovery of Weak Neutral Currents in Gargamelle. In *Proceedings Int. Symp. On 30 Years of Neutral Currents, Santa Monica CA*, 187–206. 1993.

[474] S. Weinberg. *Dreams of a Final Theory*. Vintage Books, 1993.

[475] S. Weinberg. *The First Three Minutes: A Modern View Of The Origin Of The Universe*. Basic Books, 1993.

[476] H.P. Bonzel and Ch. Kleint. On the History of Photoemission. *Prof. Surf. Sci.* 49:107–153, 1995.

[477] F. Abe *et al.* (CDF Coll.) Observation of top quark production in $p\bar{p}$ collisions. *Phys. Rev. Lett.* 74:2626–2631, 1995.

[478] S. Abachi *et al.* (D0 Coll.) Observation of the top quark. *Phys. Rev. Lett.* 74:2632–2637, 1995.

[479] S. Weinberg. *The Quantum Theory of Fields*. Vol. 1. Cambridge Univ. Press, 1995.

[480] V.V. Ezhela et al. *Particle Physics: One Hundred Years of Discoveries: An Annotated Chronological Bibliography*. Springer, 1996.

[481] J.G. Branson. Gluon Jets. In *History of Original Ideas and Basic Discoveries in Particle Physics*. Ed. by H.B. Newman and Th. Ypsilantis, 101–121. New York: Plenum, 1996.

[482] N. Byers. "The Life and Times of Emmy Noether". In *History of Original Ideas and Basic Discoveries in Particle Physics*. Ed. by H.B. Newman and Th. Ypsilantis. New York: Plenum, 1996. Pp. 945–964

[483] D. Derbes. Feynman's derivation of the Schrödinger equation. *Am. J. of Phys.* 64:881–884, 1996.

[484] J. Krige (Edt.) *History of CERN*. Vol. 3. Elsevier, Amsterdam, 1996.

[485] S. Flügge. "Exploiting Atomic Energy: From Laboratory Experiment to the Uranium Machine – Research Results in Dahlem". In *Physics and National Socialism*. Ed. by K. and M.A. Hentschel. Basel: Birkhäuser, 1996. Pp. 197–206

[486] D.J. Gross. Asymptotic Freedom, Confinement and QCD. In *History of Original Ideas and Basic Discoveries in Particle Physics*. Ed. by H.B. Newman and Th. Ypsilantis, 75–99. New York: Plenum, 1996.

[487] K. Hentschel and A.M. Hentschel (Edts.) *Physics and National Socialism: An Anthology of Primary Sources*. Birkhäuser Verlag Basel, 1996.

[488] J. and W. von Ungern-Sternberg. *Der Aufruf 'An die Kulturwelt!': Das Manifest der 93 und die Anfänge der Kriegspropaganda im Ersten Weltkrieg*. Steiner, Stuttgart, 1996.

[489] G. Metzler. "Welch ein deutscher Sieg!" *Vierteljahreshefte für Zeitgeschichte*, 44:173–200, 1996.

[490] O. Piccioni. "The Discovery of the Muon". In *History of Original Ideas and Basic Discoveries in Particle Physics*. Ed. by H.B. Newman and Th. Ypsilantis. New York: Plenum, 1996. Pp. 143–159

[491] F. Wilczek. *Asymptotic freedom: Lecture on receipt of the Dirac medal for 1994*. Princeton Institute for Advanced Studies Report Report IASSNS-HEP-96-95, arXiv hep-th/9609099. 1996.

[492] Anonymus. *The Reines-Cowan Experiments, Detecting the Poltergeist*. Los Alamos Science 25. 1997.

[493] D.B. Cline. *Weak Neutral Currents: The Discovery of the Electroweak Force*. Routledge, New York, 1997.

[494] P.O. Enquist. "Ein politisches Gedicht". In *Die Kartenzeichner*. Fischer, 1997. Pp. 211–241

[495] A. Einstein *et al.* "Manifesto to the Europeans". In *Collected Papers of Albert Einstein, Vol. 6: The Berlin Years.* New York: Princeton Univ. Press, 1997. Pp. 28–29

[496] T. Sauer J. Renn and J. Stachel. The Origin of Gravitational Lensing: A Postscript to Einstein's 1936 *Science* Paper. *Science*, 275:184–186, 1997.

[497] D. Perkins. "Gargamelle and the Discovery of Weak Neutral Currents". In *The Rise of the Standard Model.* Ed. by L. Hoddeson *et al.* Cambridge: Cambridge Univ, Press, 1997. Pp. 428–446

[498] I. Prigogine and I. Stenger. *The end of Certainty.* The Free Press, New York, 1997.

[499] R.H. Stuewer. "Gamow, Alpha Decay, and the Liquid-Drop Model of the Nucleus". In *George Gamow Symposium.* Ed. by W.C. Parke E. Harper and D. Anderson. ASP Conference Series 129, 1997. Pp. 29–43

[500] M. Veltman. "The Path to Renormalzability". In *The Rise of the Standard Model.* Ed. by L. Hoddeson *et al.* Cambridge: Cambridge Univ, Press, 1997. Pp. 145–178

[501] S.L. Wu. "Hadron Jets and the Discovery of Gluons". In *The Rise of the Standard Model.* Ed. by L. Hoddeson *et al.* Cambridge: Cambridge Univ, Press, 1997. Pp. 600–624

[502] A.G. Riess *et al.* (Supernova Search Team Coll.) Observational Evidence from Supernovae for an Accelerating Universe and a Cosmological Constant. *Astron. J.* 116:1009–1038, 1998.

[503] R. Bock and A. Vasilescu. *The Particle Detector BriefBook.* Springer, 1998.

[504] A.H. Boozer. What is a Stellarator? *Physics of Plasmas,* 5:1647–1665, 1998.

[505] F. Dyson. *Imagined Worlds.* Harvard Univ. Press, 1998.

[506] M. Pohl. *Teilchen, Kräfte und das Vakuum.* vdf Zürich, 1998.

[507] P.L. Rose. *Heisenberg and the Nazi Atomic Bomb Project: A Study in German Culture.* Univ. of California Press, 1998.

[508] C. Rovelli. Loop Quantum Gravity. *Living Rev. Rel.* 1:1, 1998.

[509] D.H. Schramm and M.S. Turner. Big-bang nucleosynthesis enters the precision era. *Rev. Mod. Phys.* 70:303–318, 1998.

[510] A.W. Strong and I.V. Moskalenko. Propagation of Cosmic-Ray Nucleons in the Galaxy. *Astrophys. J.* 509:212–228, 1998.

[511] P. Day (Edt.) *The philosopher's tree: a selection of Michael Faraday's writings.* CRC Press, 1999.

[512] Richard P. Feynman. *The Meaning of It All: Thoughts of a Citizen Scientist.* Perseus Books, Reading, 1999.

[513] J. Chapront, M. Chapront-Touzé and G. Francou. Determination of the lunar orbital and rotational parameters and of the ecliptic reference system orientation from LLR measurements and IERS data. *Astron. and Astrophys.* 343:624–633, 1999.

[514] H. Kragh. *Quantum Generations.* Princeton Univ. Press, 1999.

[515] M.J. Nye. *Before big science: the pursuit of modern chemistry and physics, 1800-1940.* Harvard University Press, 1999.

[516] S. Perlmutter *et al.* (Supernova Cosmology Project Coll.) Measurements of Ω and Λ from 42 High-Redshift Supernovae. *Astrophys. J.* 565:517–586, 1999.

[517] Anonymous. 100 Jahre Quantenphysik. *Phys. Blätter*, 56:36, 2000.

[518] M. Frayn. *Copenhagen.* Anchor Books, 2000.

[519] J.L. Heilbron. *The Dilemmas of an Upright Man: Max Planck and the Fortunes of German Science.* Harvard Univ. Press, 2000.

[520] D. Hoffmann. Schwarze Körper im Labor. *Phys. Blätter*, 56:43–47, 2000.

[521] H. Kragh. Max Planck: the reluctant revolutionary. *Physics World*, 13:31–35, 2000.

[522] D.H. Perkins. *Introduction to High Energy Physics, 4th edition.* Cambridge Univ. Press, 2000.

[523] T. Powers. *The Unanswered Question.* New York Review of Books, May 25. 2000.

[524] P.L. Rose and T. Powers. *Heisenberg in Copenhagen.* New York Review of Books, October 19. 2000.

[525] J. Bernstein. *Hitler's uranium club : the secret recordings at Farm Hall.* Copernicus Books, New York, 2001.

[526] D. Cyraniski. Imploding detectors shatter plans for Japan's neutrino experiments. *Nature*, 414:381–382, 2001.

[527] W.L. Freedman *et al.* Final Results from the Hubble Space Telescope Key Project to Measure the Hubble Constant. *Astrophys. J.* 553:47–72, 2001.

[528] M. Frayn J. Bernstein and T. Powers. *Heisenberg in Copenhagen: An Exchange.* New York Review of Books, February 8. 2001.

[529] K. Kodama *et al.* (DONUT Coll.) Observation of tau neutrino interactions. *Phys. Lett.* B504:218, 2001.

[530] W. Hoffmann (H.E.S.S. Coll.) Status of the High Energy Stereoscopic System (H.E.S.S.) Project. In *Proceedings, 27th International Cosmic Ray Conference, 07 - 15 August, 2011, Hamburg, Germany*, 2785–2788. 2001. URL: http://icrc2001.uni-wuppertal.de

[531] S.L. Wolff. Physiker im "Krieg der Geister". *Working Paper, Munich Center for the History of Science and Technology* 2001.

[532] Niels Bohr Archive. *Release of documents relating to 1941 Bohr-Heisenberg meeting.* 2002. URL: https : / / www . nbarchive . dk / collections/bohr-heisenberg/

[533] M. Frayn. *'Copenhagen' Revisited.* New York Review of Books, March 28. 2002.

[534] M. Frayn G. Holton J. Logan and T. Powers. *'Copenhagen': An Exchange.* New York Review of Books, April 11. 2002.

[535] N.P. Landsman. Getting even with Heisenberg. *Studies in History and Philosophy of Modern Physics*, 33:297–325, 2002.

[536] W. Tittel N. Gisin G. Ribordy and H. Zbinden. Quantum cryptography. *Rev. Mod. Phys.* 74:145–195, 2002.

[537] R. Polenberg. *In the Matter of J. Robert Oppenheimer: The Security Clearance Hearing.* Cornell Univ. Press, 2002.

[538] T. Powers. *What Bohr Remembered.* New York Review of Books, March 28. 2002.

[539] P.L. Rose and T. Powers. *Copenhagen, Cont'd.* New York Review of Books, May 9. 2002.

[540] D. Zimmerman. Paul Langevin and the Discovery of Active Sonar or Asdic. *The Northern Mariner/Le marin du nord*, 12:39–52, 2002.

[541] N. Ashby. Relativity in the Global Positioning System. *Living Rev. in Rel.* 6:1–42, 2003.

[542] B. Friedrich and D. Herschbach. Stern and Gerlach: How a Bad Cigar Helped Reorient Atomic Physics. *Physics Today*, 56:53–59, 2003.

[543] P. Galison. *Einstein's Clocks, Poincaré's Maps.* W.W. Norton & Co., 2003.

[544] R.P. Kirschner. Hubble's diagram and cosmic expansion. *Pro. Natl. Acad. Sci. USA*, 101:8–13, 2003.

[545] K. Lodders. Solar System Abundances and Condensation Temperatures of the Elements. *Astrophys. J.* 591:1220–1247, 2003.

[546] D.H. Perkins. *Particle Astrophysics.* Oxford Univ. Press, 2003.

[547] T. Powers. *A Letter from Copenhagen.* New York Review of Books, August 14. 2003.

[548] S. Weinberg. *The Discovery of Fundamental Particles.* Cambridge Univ. Press, 2003.

[549] J.G. Williams and J.O. Dickey. Lunar Geophysics, Geodesy, and Dynamics. In *Proceedings From the Science Session and Full Proceedings CD-ROM, NASA/CP-2003-212248, 2003.* Ed. by C. Noll R. Noomen S. Klosko and M. Pearlman. 2003.

[550] M.L. Perl *et al.* A Brief review of the search for isolatable fractional charge elementary particles. *Mod. Phys. Lett.* A19:2595, 2004.

[551] C. Rovelli. *Quantum Gravity.* Cambridge Univ. Press, 2004.

[552] F. Kennrich *et al. (VERITAS Coll.)* VERITAS: the Very Energetic Radiation Imaging Telescope Array System. *New Astronomy Reviews,* 48:345–349, 2004.

[553] T. Asaka and M. Shaposhnikov. The νMSM, Dark Matter and Baryon Asymmetry of the Universe. *Phys. Lett B,* 620:17–26, 2005.

[554] D. Ferenc (MAGIC Coll.) The MAGIC gamma-ray observatory. *Nucl. Instr. Meth. A,* 553:274–281, 2005.

[555] S. Blanchet T. Asaka and M. Shaposhnikov. The νMSM, dark matter and neutrino masses. *Phys. Lett B,* 631:151–156, 2005.

[556] C.M. Will. The Confrontation between General Relativity and Experiment. *Living Rev. Rel.* 9:3–100, 2005.

[557] G. Audi. The History of Nuclidic Masses and of their Evaluation. *Int. J. Mass Spect.* 251:85–94, 2006.

[558] P.Roy. Chowdhury and D.N. Basu. Nuclear matter properties with the re-evaluated coefficients of liquid drop model. *Acta Phys. Polon. B,* 37:1833–1846, 2006.

[559] F. Dyson. *The Scientist as Rebel.* New York Review Books, 2006.

[560] D. Novković *et al.* Testing the exponential decay law of gold ^{198}Au. *Nucl. Instr. and Meth.* A 566:477–480, 2006.

[561] LEP Electroweak Working Group *et al.* Precision Electroweak Measurements on the Z Resonance. *Phys. Rept.* 427:257, 2006.

[562] M.W. Jackson. *Harmonious Triads: Physicists, Musicians and Instrument Makers in Nineteenth-Century Germany.* MIT Press, 2006.

[563] F.E. James. *Statistical Methods in Experimental Physics.* World Scientific, 2006.

[564] J. Peidle R. Newburgh and W. Rueckner. Einstein, Perrin, and the reality of atoms: 1905 revisited. *Am. J. Phys.* 74:478–481, 2006.

[565] T. Rothman and S. Boughn. Can Gravitons Be Detected? *Found. Phys.* 36:1801–1825, 2006.

[566] C. Frenk V. Springel and S. White. The large-scale structure of the Universe. *Nature,* 440:1137–1144, 2006.

[567] C.A. Bertulani. *Nuclear Physics in a Nutshell.* Princeton Univ. Press, 2007.

[568] S. Bethke. Experimental tests of asymptotic freedom. *Progress in Particle and Nuclear Physics,* 58:351–386, 2007.

[569]   C.R. Browning. *The Origins of the Final Solution: The Evolution of Nazi Jewish Policy, September 1939 - March 1942.* Univ. of Nebraska Press, 2007.

[570]   G. Brumfiel. A constant problem. *Nature*, 448:245–248, 2007.

[571]   J. Müller and L. Biskupek. Variations of the gravitational constant from lunar laser ranging data. *Classical and Quantum Gravity*, 24:4533, 2007.

[572]   F. Wilczek. Hard-core revelations. *Nature*, 445:156–157, 2007.

[573]   S. Durr *et al.* Ab-Initio Determination of Light Hadron Masses. *Science*, 322:1224, 2008.

[574]   M. Kachelriess. Lecture notes on high energy cosmic rays. In *17th Jyvaskyla Summer School.* 2008.

[575]   R. Panek. *The Father of Dark Matter Still Gets No Respect.* 2008. URL: https://www.discovermagazine.com/the-sciences/the-father-of-dark-matter-still-gets-no-respect

[576]   M. Shaposhnikov. Sterile neutrinos in cosmology and how to find them in the lab. *Journal of Physics: Conference Series*, 136:022045, 2008.

[577]   L. Evans (Edt.) *The Large Hadron Collider: A Marvel of Technology.* EPFL Press, 2009.

[578]   J.B.R. Battat *et al.* The Apache Point Observatory Lunar Laser-ranging Operation (APOLLO): Two Years of Millimeter-Precision Measurements of the Earth-Moon Range. *Publications of the Astronomical Society of the Pacific*, 121:29–40, 2009.

[579]   R. Engel J. Blümer and J.R. Hörandel. Cosmic Rays from the knee to the highest energies. *Progr. Part. Nucl. Phys.* 63:293–338, 2009.

[580]   R. Pohl *et al.* The Size of the Proton. *Nature*, 466:213–261, 2010.

[581]   M. Shaposhnikov and C. Wetterich. Asymptotic Safety of Gravity and the Higgs Boson Mass. *Phys. Lett. B*, 683:196–200, 2010.

[582]   Ph. Bryant and K. Hübner. CERN's ISR: the world's first hadron collider. *CERN Courier Jan. 2011* 2011.

[583]   R. Durrer. What do we really know about Dark Energy? *Phil. Trans. Roy. Soc. Lond. A*:5102–5114, 2011.

[584]   G. Bressi *et al.* Testing the neutrality of matter by acoustic means in a spherical resonator. *Phys. Rev.* A83:06052101, 2011.

[585]   S. Hossenfelder. "Experimental Search for Quantum Gravity". In *Classical and Quantum Gravity: Theory, Analysis and Applications.* Ed. by V.R. Frignanni. Nova Publishers, 2011.

[586]   M White. *The Great Big Book of Horrible Things.* W.N. Norton & Co., 2011.

[587] R. Catinaud. Which physics for a new institute? Albert Gockel, Joseph Kowalski and the early years of the Fribourg Institute of Physics. *SPS Communications*, 36:24–27, 2012.

[588] K. Abe *et al.* Search for Antihelium with the BESS-Polar Spectrometer. *Phys. Rev. Lett.* 108:131301, 2012.

[589] R.G.A. Fricke and K. Schlegel. 100th anniversary of the discovery of cosmic radiation: the role of Günther and Tegetmeyer in the development of the necessary instrumentation. *Hist. Geo Space Sci.* 3:151–158, 2012.

[590] H. Fritzsch. The history of QCD. *CERN Courier Oct. 2012*:2, 2012.

[591] S. Aoki *et al. (HAL QCD Collaboration)*. Lattice QCD approach to Nuclear Physics. *Prog. Theor. Exp. Phys.*01A105, 2012.

[592] D. Hanna. Early Muon-Physics Measurements with Cosmic Rays. *Physics in Canada*, 68:7, 2012.

[593] J. Lacki. Albert Gockel: from atmospheric electricity to cosmic radiation. *SPS Communications*, 38:25–29, 2012.

[594] F. Aaserud and J.L. Heilbron. *Love, Literature, and the Quantum Atom: Niels Bohr's 1913 Trilogy Revisited.* Oxford University Press, 2013.

[595] C. Baily. Atomic Modelling in the Early 20th Century: 1904-1913. *Eur. Phys. J. H*, 38:1–38, 2013.

[596] D. Weinberg *et al.* Observational Probes of Cosmic Acceleration. *Phys. Rep.* 530:87–256, 2013.

[597] F. Dyson. Is a Graviton Detectable? *Int. J. Mod. Phys. A28*:1330041, 2013.

[598] M. Eckert. Die Geburt der modernen Atomtheorie. *Physik in Unserer Zeit*, 44:168–173, 2013.

[599] LEP Electroweak Working Group *et al.* Electroweak Measurements in Electron-Positron Collisions at W-Boson-Pair Energies at LEP. *Phys. Rept.* 532:119, 2013.

[600] P.F. Yin *et al.* Pulsar interpretation for the AMS-02 result. *Phys. Rev. D*, 88:023001, 2013.

[601] P. Massey and M.M. Hanson. "Astronomical Spectroscopy". In *Planets, Stars and Stellar Systems*. Ed. by T.D. Oswalt and H.E. Bond. Springer, 2013. Pp. 35–98

[602] P.A.R. Ade *et al.* (Planck Coll.) Planck 2013 results. XVI. Cosmological parameters. *Astron. and Astrophys.* 571:A16, 2013.

[603] M.J. Way. Dismantling Hubble's Legacy? In *Origins of the Expanding Universe*. Ed. by M.W. Way and D. Hunter. Astronomical Society of the Pacific, 2013.

[604] M. G. Alford and K. Schwenzer. Gravitational Wave Emission and Spin-Down of Young Pulsars. *The Astrophysical Journal*, 781:26–48, 2014.

[605] Alisa Bokulich. "Bohr's Correspondence Principle". In *The Stanford Encyclopedia of Philosophy*. Ed. by Edward N. Zalta. Spring 2014. Metaphysics Research Lab, Stanford University, 2014. URL: https://plato.stanford.edu/archives/spr2014/entries/bohr-correspondence/

[606] J. Conrad. Indirect Detection of WIMP Dark Matter: a compact review. In *Interplay between Particle and Astroparticle Physics (IPA 2014) (18-22 August 2014)*. 2014. URL: https://arxiv.org/pdf/1411.1925.pdf

[607] J. Ellis. The Discovery of the Gluon. *Int. J. Mod. Phys.* A29:1430072, 2014.

[608] D. Guthleben. *Les scientifiques, entre tranchées et paillasses*. 2014. URL: https://lejournal.cnrs.fr/billets/les-scientifiques-entre-tranchees-et-paillasses

[609] H. Leutwyler. On the history of the strong interaction. In *Proceedings, 50th International School of Subnuclear Physics : What we would like LHC to give us (ISSP 2012)*. Vol. A29, 1430023. 2014.

[610] Ch. Spiering M. Walter and J. Knapp (Edts.) Centenary of cosmic ray discovery. *Astroparticle Physics*, 53:1–190, 2014.

[611] G. Neuneck. Physiker im Ersten Weltkrieg. *Wissenschaft & Frieden*, 3:41–45, 2014.

[612] M. Pohl. Particle detection technology for space-borne astroparticle experiments. In *Technology and Instrumentation in Particle Physics*. Proceedings of Science, 2014. URL: https://pos.sissa.it/213/013/pdf

[613] M. Hildebrandt R.E. Robson and R.D. White. Ein Grundstein der Atomphysik. *Physik Journal*, 13:43–49, 2014.

[614] B.C. Reed. *The History and Science of the Manhattan Project*. Springer, 2014.

[615] A. Aab *et al.* (Auger Coll.) The Pierre Auger Cosmic Ray Observatory. *Nucl. Instr. Meth. A*, 798:173–213, 2015.

[616] F. Close. *Half Life: The Divided Life of Bruno Pontecorvo, Physicist or Spy*. Basic Books, 2015.

[617] D. Ivanov (Telescope Array Coll.) TA Spectrum Summary. In *Proceedings, 34th International Cosmic Ray Conference, 30 July - 6 August, 2015, The Hague, The Netherlands*, 349. Proceedings of Science, 2015.

[618] M. Aguilar *et al.* Precision Measurement of the Helium Flux in Primary Cosmic Rays of Rigidities 1.9 GV to 3 TV with the Alpha Magnetic Spectrometer on the International Space Station. *Phys. Rev. Lett.* 115:211101, 2015.

[619] J. Ferrari. *Le principe.* Actes Sud, 2015.

[620] J.-G. Hagmann. Mobilizing US physics in World War I. *Physics Today,* 70:44–50, 2015.

[621] K. Jakobs and C. Seez. The Higgs Boson discovery. *Scholarpedia,* 10(9):32413, 2015. URL: http://www.scholarpedia.org/article/The_Higgs_Boson_discovery

[622] M. Aguilar *et al.* (AMS Coll.) Precision Measurement of the Proton Flux in Primary Cosmic Rays from Rigidity 1 GV to 1.8 TV with the Alpha Magnetic Spectrometer on the International Space Station. *Phys. Rev. Lett.* 114:171103, 2015.

[623] A. Neronov M. Kachelriess and D.V. Semikoz. Signatures of a Two Million Year Old Supernova in the Spectra of Cosmic Ray Protons, Antiprotons and Positrons. *Phys. Rev. Lett.* 115:181103, 2015.

[624] L. Hoddeson M. Riordan and A. W. Kolb. *Tunnel Visions: The Rise and Fall of the Superconducting Super Collider.* University of Chicago Press, 2015.

[625] S. Mele. The Measurement of the Number of Light Neutrino Species at LEP. *Adv. Ser. Direct. High Energy Phys.* 23:89–106, 2015.

[626] C.H. Meyer and G. Schwarz. *The Theory of Nuclear Explosives that Heisenberg did not Present to the German Military.* 2015. URL: https://www.mpiwg-berlin.mpg.de/Preprints/P467.PDF

[627] S. Weinberg. *Lectures on Quantium Mechanics.* Cambridge University Press, 2015.

[628] S. Weinberg. *To Explain the World.* Penguin, 2015.

[629] F. Wilczek. *A Beautiful Question.* Penguin, 2015.

[630] B.P. Abbott *et al.* (LIGO Scientific Coll. and Virgo Coll.) Observation of Gravitational Waves from a Binary Black Hole Merger. *Phys. Rev. Lett.* 116:061102, 2016.

[631] P. Van Dokkum *et al.* A High Stellar Velocity Dispersion and ~ 100 Globular Clusters For The Ultra-Diffuse Galaxy Dragonfly 44". *Astrophys. J. Lett.* 828:L6, 2016.

[632] T. Filburn and S. Bullard. *Three Mile Island, Chenobyl and Fukushima: Curse of the Nuclear Genie.* Springer, 2016.

[633] V. Chekelian for the H1 and ZEUS collaborations. Parton distribution functions and hard QCD at HERA. In *Proceedings, 38th International Conference on High Energy Physics (ICHEP 2016): Chicago, IL, USA, August 3-10, 2016.* Vol. ICHEP2016, 617–625. 2016.

[634] J. Nielsen, A. Guffanti and S. Sarkar. Marginal evidence for cosmic acceleration from Type Ia supernovae. *Sci. Rep.* 6:35596, 2016.

[635] G. King and P. Wilson. *Lusitania: Triumph, Tragedy and the End of the Edwardian Age.* St. Martin's Griffin, 2016.

[636] M. Aguilar *et al.* (AMS Coll.) Antiproton Flux, Antiproton-to-Proton Flux Ratio, and Properties of Elementary Particle Fluxes in Primary Cosmic Rays Measured with the Alpha Magnetic Spectrometer on the International Space Station. *Phys. Rev. Lett.* 117:091103, 2016.

[637] T. Powers. *The Private Heisenberg and the Absent Bomb.* New York Review of Books, December 22. 2016.

[638] M. Riordan. A Bridge Too Far: The Demise of the Superconducting Supercollider. *Physics Today,* 69:48–54, 2016.

[639] A.U. Abeysekara *et al.* (HAWC Coll.) Extended gamma-ray sources around pulsars constrain the origin of the positron flux at Earth. *Science,* 358:911–914, 2017.

[640] M. Danninger. Review of indirect detection of dark matter with neutrinos. *J. Phys. Conf. Ser.* 888:012039, 2017.

[641] S. Weinberg *et al. Steven Weinberg and the Puzzle of Quantum Mechanics.* The New York Review of Books, April 6. 2017.

[642] F. Fenu (Pierre Auger Coll.) The cosmic ray energy spectrum measured using the Pierre Auger Observatory. In *Proceedings, 35th International Cosmic Ray Conference, 10-20 July, 2017, Bexco, Busan, Korea,* 486. Proceedings of Science, 2017.

[643] R.G.A. Fricke and K. Schlegel. Julius Elster and Hans Geitel – Dioscuri of physics and pioneer investigators in atmospheric electricity. *Hist. Geo. Space. Sci.* 8:1–7, 2017.

[644] M. Aguilar *et al.* (AMS Coll.) Observation of Identical Rigidity Dependence of He, C and O Cosmic Rays at High Rigidities by the Alpha Magnetic Spectrometer in the International Space Station. *Phys. Rev. Lett.* 119:251101, 2017.

[645] S. Weinberg. *The Trouble with Quantum Mechanics.* The New York Review of Books, Jan. 19. 2017.

[646] O.A. Westad. *The Cold War: A World History.* Hachette Book Group, 2017.

[647] O.A. Westad. *The Cold War and America's Dilusion of Victory.* The New York Times, Aug. 28. 2017.

[648] Y. S. Yoon *et al.* (CREAM-III Coll.) Proton and Helium Spectra from the CREAM-III Flight. *Astrophys. J.* 839:5, 2017.

[649] G. Bertone and D. Hooper. History of Dark Matter. *Rev. Mod. Phys.* 90:045002, 2018.

[650] CTA Consortium. *Science with the Cherenkov Telescope Array.* World Scientific, 2018.

[651] R. Davenport-Hines. *Enemies Within: Communists, the Cambridge Spies and the Making of Modern Britain.* Harper Collins, 2018.

[652] H. Fleurbaey *et al.* New Measurement of the $1S - 3S$ Transition Frequency of Hydrogen: Contribution to the Proton Charge Radius Puzzle. *Phys. Rev. Lett.* 120:183001, 2018.

[653] M. H. Israel *et al.* Elemental Composition at the Cosmic-Ray Source Derived from the ACE-CRIS Instrument. I. ^{6}C to ^{28}Ni. *Astrophys. J.* 865:69–79, 2018.

[654] M. Felcini. Searches for Dark Matter Particles at the LHC. In *53rd Rencontres de Moriond on Cosmology La Thuile, Italy, March 17-24, 2018.* 2018. eprint: `arXiv1809.06341`

[655] L. Amendola *et al.* (The Euclid Theory Working Group. Cosmology and fundamental physics with the Euclid satellite. *Living Rev. Relativ.* 21:2, 2018.

[656] M. Aguilar *et al.* (AMS Coll.) Observation of New Properties of Secondary Cosmic Rays Lithium, Beryllium and Boron by the Alpha Magnetic Spectrometer in the International Space Station. *Phys. Rev. Lett.* 120:021101, 2018.

[657] M. Aguilar *et al.* (AMS Coll.) Precision Measurement of Cosmic-Ray Nitrogen and its Primary and Secondary Components with by the Alpha Magnetic Spectrometer in the International Space Station. *Phys. Rev. Lett.* 121:051103, 2018.

[658] M. Tanabashi *et al.* (Particle Data Group). Review of Particle Properties. *Phys. Rev.* D98:03001, 2018.

[659] N. Agafonova *et al.* (OPERA Coll.) Final Results of the OPERA Experiment on ν_τ Appearance in the CNGS Neutrino Beam. *Phys. Rev. Lett.* 120:211801, 2018. Erratum: Phys. Rev. Lett. 121 (2018) 139901

[660] C. Rovelli. *The Order of Time.* Allan Lane, 2018.

[661] M. Schaaf. *Heisenberg, Hitler und die Bombe: Gespräche mit Zeitzeugen.* GNT Verlag, Diepholz, 2018.

[662] M. Walker. Bombe oder Reaktor? *Physik Journal,* 17:55–59, 2018.

[663] J. Banville. *What Made the Old Boys Turn?* The New York Review of Books, March 7. 2019.

[664] A. Cho. Space magnet homes in on clue to dark matter. *Science,* 363:572–573, 2019.

[665] K.H. Cui and B.L. Wardle. Breakdown of Native Oxide Enables Multifunctional, Free-Form Carbon Nanotube-Metal Hierarchical Architectures. *ACS Appl. Mater. Interf.* 11:35212–35220, 2019.

[666] D. Ivanov *et al.* (Telescope Array Coll.) Energy Spectrum Measured by the Telescope Array Experiment. In *Proceedings, 36th International Cosmic Ray Conference, July 24th - August 1st, 2019 Madison, WI, U.S.A.* 298. Proceedings of Science, 2019.

[667] A. G. Riess *et al.* Large Magellanic Cloud Cepheid Standards Provide a 1% Foundation for the Determination of the Hubble Constant and Stronger Evidence for Physics beyond ΛCDM. *Astrophys. J.* 876:85, 2019.

[668] K.C. Wong *et al.* H0LiCOW XIII. A 2.4% measurement of H_0 from lensed quasars: 5.3σ tension between early and late-Universe probes. arXiv:1907.04869v2. 2019.

[669] N. Bezginov *et al.* A measurement of the atomic hydrogen Lamb shift and the proton charge radius. *Science*, 365:1007–1012, 2019.

[670] P. Flowers *et al.* Chemistry 2e. OpenStax, 2019.

[671] W. Xiong *et al.* A small proton charge radius from an electron-proton scattering experiment. *Nature*, 575:147–150, 2019.

[672] B. Friedrich. Fritz Haber at one hundred fifty: Evolving views of and on a German Jewish Patriot. *Bunsen-Magazin*, 21:130–144, 2019.

[673] T. Linden I. Cholis and D. Hooper. A robust excess in the cosmic-ray antiproton spectrum: Implications for annihilating dark matter. *Phys. Rev. D*, 99:103026, 2019.

[674] Stockholm International Peace Research Institute. *SIPRI Yearbook 2019: Armament, Disarmament and International Security*. Oxford University Press, 2019.

[675] J. Colin, R. Mohayaee, M. Ramirez and S. Sarkar. Evidence for anisotropy of cosmic acceleration. *Astron. and Astrophys. Lett.* 631:L13, 2019.

[676] K. Kawata *et al.* (Telescope Array Coll.) Updated Results on the UHECR Hotspot Observed by the Telescope Array Experiment. In *Proceedings, 36th International Cosmic Ray Conference, July 24th - August 1st, 2019 Madison, WI, U.S.A.* 310. Proceedings of Science, 2019.

[677] E. Kiritsis. *String Theory in a Nutshell*. Princeton Univ. Press, 2019.

[678] H. Kubbinga. A Tribute to Lise Meitner (1887-1968). *Europhysics News*, 50:22–26, 2019.

[679] L. Caccianiga *et al.* (Pierre Auger Coll.) Anisotropies of the Highest Energy Cosmic-Ray Events Recorded by the Pierre Auger Observatory in 15 Years of Operation. In *Proceedings, 36th International Cosmic Ray Conference, July 24th - August 1st, 2019 Madison, WI, U.S.A.* 206. Proceedings of Science, 2019.

[680] M. Aguilar *et al.* (AMS Coll.) Towards Understanding the Origin of Cosmic-Ray Electrons. *Phys. Rev. Lett.* 122:101101, 2019.

[681] M. Aguilar *et al.* (AMS Coll.) Towards Understanding the Origin of Cosmic-Ray Positron. *Phys. Rev. Lett.* 122:0411102, 2019.

[682] M. Aker *et al.* (KATRIN Coll.) An improved upper limit on the neutrino mass from a direct kinematic method by KATRIN. *Phys. Rev. Lett.* 123:221802, 2019.

[683] O. Adriani *et al.* (CALET Coll.) Direct Measurement of the Cosmic-Ray Proton Spectrum from 50 GeV to 10 TeV with the Calorimetric Electron Telescope on the International Space Station. *Phys. Rev. Lett.* 122:181102, 2019.

[684] N. Aghanim *et al. (Planck Coll.) Planck 2018 results. VI. Cosmological parameters.* arXiv:1807.06209v2. 2019.

[685] M. Schumann. Direct Detection of WIMP Dark Matter: Concepts and Status. *J. Phys. G*, 46:103003, 2019.

[686] J.H.E. Smith. *Irrationality: A History of the Dark Side of Reason.* Princeton University Press, 2019.

[687] V. Verci *et al.* (Pierre Auger Coll.) Measurement of the energy spectrum of ultra-high energy cosmic rays using the Pierre Auger Observatory. In *Proceedings, 36th International Cosmic Ray Conference, July 24th - August 1st, 2019 Madison, WI, U.S.A.* 450. Proceedings of Science, 2019.

[688] Y. Asaoka *et al.* (CALET Coll.) The CALorimetric Electron Telescope (CALET) on the International Space Station: Results from the First Two Years On Orbit. *J. Phys.: Conf. Ser.* 1181:012003, 2019.

[689] M. Claessens. *ITER: The Giant Fusion Reactor.* Springer, 2020.

[690] F. Dyson. *The Power of Morphological Thinking.* New York Review of Books, January 16. 2020.

Index

Printed in the United States
by Baker & Taylor Publisher Services